Diagnostic Electron Microscopy
Volume 2

Diagnostic Electron Microscopy

Volume 2

Edited by

Benjamin F. Trump, M.D.
Professor and Chairman

Raymond T. Jones, Ph.D.
Assistant Professor

Department of Pathology
University of Maryland School of Medicine
Baltimore, Maryland

A Wiley Medical Publication
JOHN WILEY & SONS
New York · Chichester · Brisbane · Toronto

Library of Congress Cataloging in Publication Data (Revised)

Main entry under title:

Diagnostic electron microscopy.

  (A Wiley medical publication)
  Includes index.
  1. Diagnosis, Electron microscopic. I. Trump,
Benjamin F. II. Jones, Raymond T. [DNLM:
1. Microscopy, Electron. 2. Diagnosis, Laboratory
—Instrumentation. QH212.E4 D536]
RB43.5.D53   616.07′58   77-12817
ISBN 0-471-89195-9 (v. 1)
     0-471-89196-7 (v. 2)

Printed in the United States of America

10 9 8 7 6 5 4 3 2 1

This volume is dedicated to Dr. Robert E. Stowell. Dr. Stowell, presently Professor of Pathology at the University of California at Davis, was instrumental in developing the role of electron microscopy in pathology. As early as 1955, Dr. Stowell had installed an electron microscope in his Department of Pathology at the University of Kansas. Consequently that laboratory began making significant contributions to normal and abnormal cell biology. These efforts resulted in the stimulation of a large number of individuals, many of them presently in academic pathology, such as myself and my student, Dr. Jones. As a medical student, I can remember Dr. Stowell's intuition that the use of electron microscopy was the next wave of development in pathology. I followed his advice and received his support for training in this area with Dr. H. Stanley Bennet and later with Dr. E. P. Benditt at the University of Washington in Seattle and then returned with Dr. Stowell at the Armed Forces Institute of Pathology for more investigations. It is, of course, through the efforts of men like Robert E. Stowell that pathology continues to advance. His many contributions to pathology education, research, and service will be long remembered. We all stand on the shoulders of giants such as this who have paved the way for present and future developments in the field.

BENJAMIN F. TRUMP

Contributors

Henry A. Azar, M.D., Chief, Laboratory Service, Veterans Administration Hospital, and Professor of Pathology, University of South Florida, College of Medicine, Tampa, Florida

Alex Ferenczy, M.D., Associate Professor of Pathology and Obstetrics and Gynecology, McGill University; Director, Division of Gynecologic Pathology, Jewish General Hospital, Montreal, Quebec, Canada

Ramon L. Font, M.D., Assistant Chief, Department of Ophthalmic Pathology, Armed Forces Institute of Pathology, Clinical Professor of Pathology and Ophthalmology, Georgetown University Medical Center; Washington, D.C.

Julio H. Garcia, M.D., Professor and Director, Anatomic Laboratories; Head, Division of Neuropathology, Department of Pathology, University of Maryland School of Medicine, Baltimore, Maryland

Frederick A. Jakobiec, M.D., D.Sc., Instructor in Clinical Ophthalmology, College of Physicians and Surgeons, Columbia University; Director, Algernon B. Reese Laboratory of Ophthalmic Pathology, Edward S. Harkness Eye Institute, New York, New York

Peter W. Lampert, M.D., Department of Pathology, University of California, San Diego, La Jolla, California

Hernando Mena, M.D., Instructor, Department of Pathology, University of Maryland School of Medicine, Baltimore, Maryland

Ralph M. Richart, M.D., Professor of Pathology, Columbia University College of Physicians and Surgeons; Director, Division of Obstetric and Gynecologic Pathology, The Sloane Hospital for Women, New York, New York

Sydney S. Schochet, Jr. M.D., Department of Pathology, University of Texas Medical Branch, Galveston, Texas

Kyuichi Tanikawa, M.D., PhD., Professor, Department of Medicine, and Chief of Electron Microscopy Laboratory, Kurume University School of Medicine, Kurume, Japan

Myron Tannenbaum, M.D., Ph.D., Departments of Pathology and Urology, College of Physicians and Surgeons, Columbia University, New York, New York

Marjorie J. Williams, M.B., ChB., Director of Pathology and Allied Sciences, Veterans Administration Central Office, Washington, D.C.

Preface to Volume 1

The term diagnostic electron microscopy implies the utilization of electron microscopy and its associated methods in all of their ramifications for the study of human disease. At the present time, crucial diagnostic information is indeed provided in many cases by electron microscopy, and often information of confirmatory nature or of great educational value to the pathologist and clinician can be obtained. But since the increased amount of information obtainable by electron microscopy (compared with that obtainable by light microscopy) is measured in orders of magnitude, electron microscopical data cannot be judged by, or be fit into, previous concepts based on gross examination and light microscopy. It is probable, then, that we have not yet witnessed the full value of the diagnostic methods of electron microscopy simply because an appropriate data base has not as yet been generated. Therefore, it is important that these methods be used whenever possible, as it will be necessary to define new concepts and classifications of disease.

In the last few years the technology involved in electron microscopy has progressed to the point where methods have become standardized and the instrumentation routine. The quality of instrumentation is now comparable to many of the advanced types of instrumentation present in a well-equipped laboratory. It is possible for any general laboratory to maintain an instrument and associated laboratory staff, which means that the pathologist need no longer be concerned about the difficulty of the techniques and instrumentation. Furthermore, the personnel normally available in the clinical laboratory, such as the electronics technician and the senior medical technologist, are quite capable, with some additional training, of maintaining the instruments.

The field of diagnostic electron microscopy cannot be separated from the larger field of cellular pathobiology. The power of concepts derived from cellular pathobiology and applied to human disease biology cannot be overestimated. Much of the rapid, current progress in this field is due to the essential unity of cell biology in diverse animal and plant species as well as in different organ systems. Information developed on one especially suitable cell type can be rapidly applied, with a minimum of experimentation to other cell types and other organisms leading to much more rapid progression of knowledge. An important concept in this renaissance of general pathology is the correlation between structure and function at the cellular level, which has been made observable through the integration of methods in the fields of microscopy, immunology, biochemistry and physiology. The electron microscope is, of course, a fundamental tool in these investigations because it is at the level of resolution provided by this instrument that most structural correlations with function and metabolism are visible.

Implicit in the development of diagnostic electron microscopy is the study of material from human patients, which makes possible, in addition to diagnosis, studies on the basic biology of human disease. This is currently a very important trend in pathology, and one which needs world-wide support, for it is only by studying human cellular metabolism that we will be able ultimately to improve our understanding of human disease.

This book, the first in our series of treatises on diagnostic electron microscopy, outlines concepts of cellular pathology and laboratory management, and their applications to specific organ systems. Succeeding volumes will concentrate on particular organ systems. We wish to instill in the reader a sense of excitement for the growing field of diagnostic electron microscopy and to convey the urgency of expanding our knowledge of human disease from the equally important perspectives of diagnosis, education, and research.

BENJAMIN F. TRUMP, M.D.
RAYMOND T. JONES, PH.D.

Baltimore, Maryland

Preface to Volume 2

In this volume we shall continue the systematic presentation of diagnostic electron microscopy as it is currently used. The volume begins with a chapter by Dr. Marjorie J. Williams in which she describes the evolution of the diagnostic electron microscopy program in the Veterans Administration. This chapter is an important one, as it provides an excellent data base against which other programs can be compared and illustrates the evolution of the program over the last several years. The program began with three units and has developed into 42 units. The experience in this program, which is the largest organized electron microscopy program in the world, is of great interest to practitioners in the field.

The chapter on the liver by Dr. Tanikawa introduces the use of electron microscopy in liver disease. This is an area that is rapidly increasing in importance as it is becoming evident that many hepatic diseases are impossible to diagnose precisely by light microscopy. Furthermore, new diagnostic entities are being revealed through the use of electron microscopy. There is also an important place in liver disease for estimation of overall hepatic parenchymal cell damage using the electron microscope, and in the future, as morphometric techniques are applied, it should be possible to provide more accurate correlations with liver function tests done on peripheral blood.

In the chapter on hematopoietic and lymphoid systems by Dr. Azar, the many uses of electron microscopy in this field are discussed. In this area the use of the electron microscope is capable both of solving diagnostic problems that cannot be resolved by routine light microscopy and of introducing a greater appreciation of cellular detail, supplementing light microscopy and leading to a better understanding of pathophysiology. Once again, there are numerous examples of diseases of the hematopoietic and lymphoreticular systems which can only be diagnosed in this way. Electron microscopic studies can then be correlated with other studies, including immunofluorescence, and with clinical findings.

The chapter on ocular pathology by Drs. Font and Jakobiec is of interest for diagnosis of diseases of the eye and is also of general interest because of a wide variety of disease processes in conditions that affect this organ. The chapter on the bladder by Dr. Tannenbaum illustrates the application of electron microscopy to an easily accessible human tissue that can be accurately observed by the clinician and the pathologist and that offers the opportunity for careful systematic and sequential studies on the development of human cancer. It is probable that both transmission and scanning electron microscopy, together with cytology, will materially assist in the evolution of knowledge concerning this important human cancer.

In their chapter on gynecology, Drs. Ferenczy and Richart summarize specific ultrastructural features that are of diagnostic help in understanding the histogenesis of gynecological disease. There is an almost indefinite variety of neoplastic and non-neoplastic conditions involving this organ system, many are easy to diagnose, but many more are not. Electron microscopy promises to be of substantial help in elucidating the histogenesis of several important neoplasms in this system.

The chapter on the peripheral nerve by Drs. Lampert and Schochet discusses the importance of electron microscopy in morphological studies of nerve biopsies. Many conditions are readily diagnosed in this area which had previously been extremely obscure for the general pathologist. The importance of plastic embedding for even light microscopy of the nerve is brought out. Here, the electron microscope has truly replaced the special stains of the past because of the ease of production and reproducibility of the results.

In the final chapter by Drs. Garcia and Mena, the intricacies of neuropathology of the central nervous system are explored. Obviously, this is a vast subject that could itself occupy volumes; however, in this chapter we are attempting to present some of the more important current applications.

The editors sincerely hope that the users of this book will find it helpful in their studies of ultrastructure, especially as it applies to diagnostic pathology.

BENJAMIN F. TRUMP, M.D.
RAYMOND T. JONES, PH.D.

Baltimore, Maryland

Contents

1

Diagnostic Electron Microscopy in the Veterans Administration

Marjorie J. Williams, M.B., Ch.B.

Director
The Pathology Service
Veterans Administration Central Office
Washington, D.C.

Diagnostic electron microscopy (EM) in the Veterans Administration (VA) must be viewed as a part of the agency's medical care program and not as an isolated entity. Therefore, a brief description of the scope and organization of the VA's medical care program provides a necessary introduction to the more detailed discussion of the EM program. A summary of medical care in VA facilities is shown in Table 1.

Table 1. Summary of Medical Care in VA Facilities (July 1, 1975 through June 30, 1976) [a]

Facilities	Beds	Patients Treated (Episodes of Care)
171 Hospitals	93,822	1,178,894
88 Nursing home units	7,585	10,941
18 Domiciliaries	10,152	18,408
215 Outpatient clinics	0	14,223,206 visits

[a] US Senate Committee Print No. 7, 95th Congress, first session: *Veterans Administration Response to the Study of Health Care of American Veterans, September 22, 1977.* US Government Printing Office, Washington DC, 1977

Medical care is the responsibility of the Department of Medicine and Surgery (DM&S), which is one of the VA's three major departments. The other two departments are concerned with Veterans Benefits and Data Management. An outline of the DM&S organization as it relates to the EM program is shown in Figure 1. The chief medical director heads DM&S, and the line authority goes from him through the associate deputy chief medical director to the executive councils of the 28 medical districts, the hospital directors, the chiefs of staff, and the chiefs of the various professional services such as the laboratory service in the hospitals. The other organizational elements shown in the figure serve in a staff capacity to the chief medical director.

Diagnostic electron microscopy is one of the VA's some 23 designated special medical services, which also include such modalities as renal dialysis, renal transplant, and cardiopulmonary bypass surgery. These services are supported by specific appropriations and exist only in selected hospitals. The office of the assistant chief medical director for professional services has the responsibility for the nurture, planning, site selection, management, and evaluation of all the special services, and this duty is delegated to the appropriate professional service. For example, diagnostic electron microscopy is delegated to the pathology service in the VA's central office.

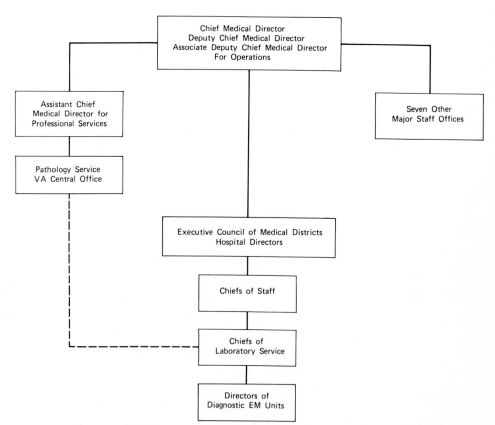

Figure 1. Veterans Administration Department of Medicine and Surgery organization in relation to the diagnostic electron microscopy program.

In the hospitals, the diagnostic EM units are organizationally part of the laboratory services. In some instances, the chief of the laboratory service may also be the director of the local EM program, but more frequently the responsibility is assigned to another pathologist with particular interest and skill in this field.

Another topic that requires brief description is the VA's extensive affiliations with medical schools (there are also affiliations with schools for dentistry and many allied health professions and occupations). There are now affiliate relations between 133 VA hospitals and 103 of the nation's 120 medical schools. Twenty-six percent of the graduate medical education in the United States occurs in VA facilities. Most of the pathology residency training in VA hospitals is conducted through integrated programs with the affiliated medical schools. The diagnostic EM units are all located in hospitals that are affiliated and provide graduate medical education in pathology.

The VA has statutory authority to enter into agreements with affiliated medical schools and other hospitals to share specialized and scarce patient care services. These agreements are established most frequently for the sharing of the special medical services. Sharing agreements for diagnostic electron microscopy are discussed later in detail.

THE VA DIAGNOSTIC EM PROGRAM: ORGANIZATION AND GOALS

The VA program for diagnostic electron microscopy was established in 1966 with the advice and encouragement of the national pathology consultants. It began with the purchase of three instruments and since then has grown to 42 diagnostic EM units dispersed throughout the continental United States and in Puerto Rico.*

The EM program is characterized by centralized planning, site selection, funding, and evaluation, with retention of considerable autonomy by the local unit directors for the daily operations. Details will be presented in this chapter on the management and operational experiences with diagnostic electron microscopy. When the program was begun in 1966, the goals were as follows:

- Provision of a relatively new and specialized modality for the enhancement of histopathological diagnosis.
- Resolution of certain diagnostic problems that could not be solved by light microscopy.

---

*Locations of VA diagnostic EM units: Albany, NY; Allen Park, MI; Ann Arbor, MI; Baltimore, MD; Birmingham, AL; Boston, MA; Bronx, NY; Charleston, SC; Chicago, IL; Cleveland, OH; Columbia, MO; Dallas, TX; Decatur (Atlanta), GA; Denver, CO; Durham, NC; East Orange, NJ; Gainesville, FL; Hines, IL; Houston, TX; Kansas City, MO; Lexington, KY; Little Rock, AR; Long Beach, CA; Los Angeles, CA; Miami, FL; Minneapolis, MN; New Orleans, LA; New York, NY; Northport, NY; Philadelphia, PA; Pittsburgh, PA; Richmond, VA; Salt Lake City, UT; San Diego, CA; San Francisco, CA; San Juan, PR; Seattle, WA; Shreveport, LA; Tampa, FL; Washington, DC; West Haven, CT; West Roxbury, MA. An additional unit at Madison, WI, will become operational in 1978.

- Affording opportunity for training and experience in ultrastructural pathology.
- Rapid development and application of research in ultrastructure to diagnostic pathology.
- Strengthening pathology graduate medical education programs.

Some 10 years later these goals are still valid. They have been expanded, however, in line with the thrust for use of specialized resources on a regional basis, to include EM support for other VA hospitals in the same medical district.

There are also in certain VA hospitals EM facilities that are part of the research and not the pathology program. The research facilities are funded separately from the diagnostic units, and their selection and evaluation are handled through different channels. A limited number of diagnostic EM studies are carried out on research equipment in some hospitals that do not have diagnostic units, but information about such examinations is not collected by the pathology service in central office. In addition, some specimens are referred elsewhere, frequently to an affiliated medical school, from VA hospitals lacking EM resources. Therefore, the present 42 diagnostic EM units do not meet all the VA's needs for ultrastructural study in support of diagnosis. The number of specimens referred within the VA from one hospital to another is not yet large but should increase as regionalization gains momentum.

VA Electron Microscopy Advisory Group

The VA electron microscopy advisory group was established in 1970 and usually has six members, all of whom are pathologists with experience and interest in electron microscopy. Currently, three members are VA pathologists and three are full-time university faculty members. The present chairman of the group is Dr. Benjamin H. Spargo, professor of pathology at the University of Chicago, and his predecessor was Dr. John R. Carter, professor and director of the Institute of Pathology at the Case Western Reserve University.

The group meets at least once each year with the pathology service at the VA central office in Washington and assists in the review of applications for new EM programs and in the evaluation of existing programs. When a site visit is considered necessary to obtain additional information, one or more of the group are members of the visiting team. At the regular meetings, there are thoughtful, analytical discussions of the VA diagnostic EM program as a whole that may lead to suggestions for changes in direction or emphasis. The fresh insights into the diagnostic electron microscopy program that this group is able to provide as a result of the breadth of experience and knowledge of its members constitute a major contribution.

Funding of VA EM Programs

The EM programs receive their basic funding from the appropriation that supports the special medical services. The support for each approved program is specifically identified and provides for both the initial and the recurring costs. Initial costs include such items as the purchase of the electron microscope and other necessary equipment and the construction or remodeling of space to accommodate the unit. Recurring costs consist of the salaries for the staff, the supplies, and the service contract for maintenance of the electron microscope. The indirect costs are supported from the general operating budget of the hospital.

The selection of the electron microscope and other equipment is made locally, although the decision must be reviewed and approved in the central office. In general, only high resolution transmission electron microscopes are acquired, but a few instruments that may be used also for scanning have been purchased. The salary allocation is sufficient to support on a full-time basis one pathologist and two technologists. This level of staffing is considered satisfactory for a diagnostic EM unit examining at least 250 specimens annually.

Replacement of existing equipment, acquisition of new capital equipment, and salaries for additional personnel, including secretarial support, must be provided by the hospital through regular budgetary procedures.

Application Procedures
to Establish an EM Program

The first diagnostic EM units, particularly in 1966 and 1967, were established in VA hospitals where, in the judgment of the pathology service in the central office, they would be used effectively. In the following two years such judgment was still the prevailing basis of selection but was modified to promote appropriate geographical distribution in the VA. By 1970, there were 22 diagnostic EM units established in laboratory services, and the size of the program was such that more formal selection procedures were considered necessary. The EM advisory group was therefore formed.

A formal application procedure has been developed to collect information in a relatively uniform manner. The application must include specific detailed plans for use of EM in diagnostic pathology and training both at the parent hospital and in the medical district; the anticipated contributions to patient care; the name and curriculum vitae of the proposed program director; and a description of the desired equipment. Statements of endorsement from the chief of the laboratory service, the hospital director, the chairman of the pathology department at the affiliated medical school, and the executive council of the medical district must be appended to the application.

The applications are reviewed by the EM advisory group, and it makes recommendations to the director of the pathology service in the central office. Final decisions on the selection depend on several factors, including a favorable review by the consultants, the suitability of the location of the proposed program in the context of broader VA policies and needs, and the availability of appropriated funds.

VA PERFORMANCE CRITERIA
FOR DIAGNOSTIC EM UNITS

In the current era of rising medical costs, considerable attention is being directed toward evaluation of the effectiveness of the newer and more complex modalities. However, useful evaluation requires criteria against which performance can be measured. Because the VA diagnostic EM units share many common features, such as receiving the same basic funding and serving the same type of patient population, the pathology service in the central office has concluded that a single set of performance criteria can be used to evaluate all the programs.

After considerable deliberation, the following basic performance criterion was established in 1973: a minimum of 250 specimens *accessioned* annually, with occasional downward deviation to not less than 225 accessions when justified by special circumstances. Such circumstances might be a change in the program director or significant problems with equipment. On October 1, 1977, the basic performance criterion was changed to a minimum of 250 specimens *examined* annually, with no downward deviation permitted.

Two other performance criteria have been introduced since 1973 as supplements to the basic criterion. One of these requires that at least 66% of the 250 minimum be patient specimens examined to assist in diagnosis, while the remainder may be studied for research purposes. The other criterion requires that at least 51% of the minimum number of examinations be on specimens from VA patients.

The performance criteria have been well accepted by the directors of the EM units and have proved most helpful in the evaluation process. The various criteria may not be directly applicable elsewhere, but their successful use in the VA suggests that with appropriate modification they may have wider potential.

Evaluation of the EM Programs

The evaluation process measures performance of each EM unit in relation to the criteria and also assesses its overall effectiveness in meeting the goals of the program.

The basis for evaluation is the semiannual report prepared by each program director, using a standardized format. The semiannual reports include the following information:

- The number of specimens accessioned and how many of these were examined by EM, by thick section only, or embedded and retained but not studied during the reporting period.
- The distribution of specimens by source, such as kidney, liver, skin, nervous system, and so forth.
- The names and grades of personnel assigned to the EM unit, with the number of hours per week each spends in the unit.
- The number of written reports issued for inclusion in the patients' medical records.
- The use of immunofluorescence.
- The use of scanning electron microscopy.
- The number, location, and activity of any sharing agreements.
- The training and teaching activities.
- The publications and presentations at professional meetings resulting from EM studies during the reporting period.
- The significant contributions to diagnosis.

In addition, several representative patient reports with attached electron micrographs from each program are forwarded for review.

The semiannual reports are analyzed both individually and programmatically by staff in the pathology service. The data from all the reports are arrayed so that the total program can be assessed and changes and trends observed. The consolidated data also allow ready comparison of the performance of each unit with the total VA picture. Individual letters are sent to the directors of all EM units in response to the semiannual reports. These letters discuss the program in some detail, comment on particular strengths, and provide constructive suggestions for correction of weaknesses.

Ratings are assigned semiannually to each EM unit, based upon the analysis of the reports, supplemented as needed by the advice of experts and findings at a site visit. The final ratings are either satisfactory or unsatisfactory. The latter rating leads to a unit being placed on probation. Failure to meet the performance criteria is the most common reason for assigning an unsatisfactory rating. A program given such a rating has a grace period of 12 months in which to overcome the deficiencies. During this period, there are several interchanges between the EM unit director and the staff of the pathology service in the central office that are designed to improve the program, and a site visit is generally made. These various measures have proved helpful, and, so far, only two programs have been terminated.

During its regular annual meeting, the EM advisory group reviews the reports from all EM units in detail and gives particular consideration to the quality and effectiveness of both the diagnostic and training activities. At the same time, the group gains an overview of the total program, and this sometimes leads to suggestions and recommendations for changes.

PROCEDURES FOR IDENTIFYING THE NEED FOR EM UNITS

Deciding how many diagnostic EM units are needed in the VA is difficult. The provision of such facilities in every hospital with graduate medical education programs is not practical because resources are limited and the work load would not be sufficient to meet the performance criteria. Since the major contributions of EM are in surgical pathology (including percutaneous biopsy materials), it was decided to use the numbers of surgical specimens to estimate the need for EM facilities.

In 1972, a total of 264,000 surgical specimens was examined in all VA hospitals, and the 23 diagnostic EM units operational at that time studied some 6,000 specimens. However, the value of ultrastructural examination was not then widely appreciated, and the majority of the VA hospitals did not have ready access to diagnostic electron microscopy. It was decided that a better estimate of the potential requirements could be obtained by sending a questionnaire to the chiefs of staff at each of the VA's general hospitals asking for their estimates of EM needs. In our experience, such questionnaires provide helpful information, but the judgments of perceived need are usually low when facilities are not readily available. Responses to the questionnaire showed the need for an estimated 3,500 additional ultrastructural examinations. It was concluded, therefore, that there were about 10,000 surgical specimens (including those from kidney and liver biopsies) on which the diagnosis would probably have been improved by ultrastructural study. On this basis, a preliminary decision was reached that EM studies were probably indicated on 10,000, or almost 4% of the 264,000 surgical specimens in the VA.

Recent data show that EM studies are performed on about 4% of all VA surgical specimens. Recently, on a VA-wide basis, the number of patient specimens examined by the 42 diagnostic EM units has amounted to 3.5% of all surgical specimens in the VA (Table 2). These units probably account for about 85% of all diagnostic EM studies in the VA, and the addition of the other studies carried out in research facilities or referred elsewhere would raise the figure to about the 4% level. Higher percentages are examined in hospitals with EM units, and the probable explanation for this finding is discussed in a later section. Further experience with these trends or the discovery of new applications for EM in diagnosis may lead the VA to revise the percentage upward or downward or develop new forecasting procedures.

With information available on the number of surgical specimens, the requirements for diagnostic electron microscopy can be estimated. As discussed earlier, the minimum performance criterion for an EM unit is 250 specimens annually. Therefore 4% of the surgical specimens divided by 250 equals the appropriate number of EM units. For instance, a hospital with 4,000 surgical specimens annually would be expected to generate only 160 specimens for EM study, a number insufficient to support a unit. On the other hand, an institution with 7,000 surgical specimens should generate 280 EM specimens, which is just enough to justify a unit. A medical district with seven hospitals and a total of 13,000 surgical accessions annually could be expected to have need for 520 EM examinations and two EM units, while a district with eight hospitals and 25,000 surgical accessions could provide sufficient material for four EM units.

Although this method of estimating EM requirements has proved helpful in the VA, it probably cannot be applied unmodified to other settings. The VA hospitals serve predominantly men, and the majority of the patients are more

Table 2. Experience With the VA Diagnostic EM Program

	Fiscal Year			
Activity	1974	1975	1976	1977
No. of EM units	41	39 [a]	42	42
Total accessions	11,766	15,455	18,657	19,876
Accessions per unit (mean)	287	396	444	473
Total patient accessions	9,768	11,635	15,511	15,737
VA patient accessions	7,352	9,779	12,510	12,615
Total specimens examined	NA	12,459	15,854	18,165
Specimens examined per unit (mean)	NA	319	377	432
Total patient specimens examined	NA	9,182	13,588	14,000
VA patient specimens as percent of surgicals				
Hospitals with EM units	6.5%	6.0%	8.0%	7.0%
All VA hospitals	3.2%	3.5%	3.6%	3.5%

[a] Two EM programs discontinued in FY 1975.

than 55 years old. In hospitals with different patient populations, the requirements for EM studies may constitute a different percentage of the surgical accessions.

SHARING AGREEMENTS FOR DIAGNOSTIC ELECTRON MICROSCOPY

As indicated earlier, the VA has statutory authority to share available capacity in specialized patient care resources with approved non federal medical institutions and to receive reimbursement for the costs. Diagnostic EM is considered to be such a specialized resource, and there are a number of agreements whereby diagnostic EM studies are provided by the VA for other hospitals. At this time some 10 VA hospitals are providing diagnostic EM studies for about 46 community hospitals. On the other hand, the VA has a few agreements with other institutions to provide EM studies for facilities lacking readily accessible resources. The number of specimens examined through such agreements is small and is not included in the data presented on VA experiences.

EM UNIT DESIGN

The diagnostic EM units in the VA are placed, whenever possible, physically within the laboratory service of the hospital to insure a close working relationship with the rest of the pathology program. A factor, however, which is usually decisive in determining the exact location of the EM unit is the need for freedom from undue vibration. This consideration frequently requires, particularly in newer hospitals with a wide-span steel supporting structure, placement of the EM unit close to a supporting vertical column. Proximity, both horizontal and vertical, to elevators and heavy machinery must be avoided because of the possibility of disturbing vibrations. As a result, it is sometimes necessary to place the EM unit on a basement level of the hospital.

The basic space allowance for an EM unit in the VA is 970 net square feet. The distribution of the space is shown in Table 3. Such facilities have proved satisfactory for a diagnostic EM unit with one pathologist and two technologists who use

Table 3. Basic Space Allowance for an EM Unit in the VA

Rooms	Net Square Feet [a]
Electron microscope room	150
Dark room	50
Preparation room	200
Cutting room	160
Developing, printing, and enlarging room	120
Finishing room	120
Office for pathologist	170
TOTAL	970

[a] Net square feet approximates 60% of gross square feet.

a single high resolution transmission instrument. Programs with larger staffs or a greater variety of instruments may require additional space and design considerations.

THE DIAGNOSTIC EM EXPERIENCE IN THE VA (1974–1977)

Much of the experience described is statistical, represents only the activities in diagnostic EM units, and has been collected from the semiannual reports of the local program directors. Although the data are interesting and unique, they are not fully reflective of the program. The many important contributions to patient care are not readily captured from the reports, and yet they are the real source of the growth and vitality. Since the EM units are located in VA hospitals with medical school affiliations and linkages with community hospitals, their experiences are reasonably indicative of the usage and applications of electron microscopy in diagnosis for a considerable segment of American medicine. The information, therefore, is of general interest.

The annual experience with the numbers of specimens accessioned and examined is shown in Table 2. Patient material is specifically identified as a component of the annual totals. The category of "Specimens Examined" includes both the majority studied by EM and the minority by thick section only. An annual increase has occurred each year throughout the four-year period. In 1977, however, there was a smaller increase than previously both for the total numbers of accessions and for the numbers of patient specimens accessioned and examined. The small increase in patient specimens is partly attributable to the sharper delineation made in this year's reports between specimens received for diagnosis and those received for clinical investigation.

The data suggest that by 1977 substantial fulfillment of the diagnostic needs had occurred at the hospitals with EM units. There may not be further significant increases in the numbers of patient studies unless there is more regional use of existing facilities or additional units are opened to meet presently unrealized needs elsewhere. The current demand may possibly be augmented also by the emergence of new applications for diagnostic EM.

Throughout the fiscal years 1974 through 1977, patient specimens have constituted most of the activities, denoting appropriate emphasis upon diagnostic studies. The differences between the total accessions and examinations and those on patient specimens are accounted for by the specimens processed for teaching and research. The difference between accessions and examinations represents specimens accumulated for later use in teaching programs.

The sources and types of patient specimens studied in the EM units are shown by percentage distribution in Table 4. Over the four-year period, specimens from the kidney, the hematopoietic system (which includes lymph nodes and spleen), and tumors have constituted the majority of all specimens. Although the percentage of kidney specimens dropped in 1977, the absolute numbers showed an increase. The kidney is the single largest source of specimens received through sharing agreements. The category labeled "Other" in Table 4 consists of specimens coming in small numbers from a variety of sources, including myocardium, salivary gland, and body fluids. An increasing number of tissues and exudates are being examined for viral particles. However, no separate count is available for this group. Viral studies on tissues, such as the liver or central nervous system are included in the specific categories while those from exudates are included in the category designated as "Other."

The pathology service has a continuing interest in the actual percentage of surgical specimens that are examined by EM because of the 4% level used in determining the requirement for EM units (see above). Therefore, the numbers of EM specimens have been calculated annually as percentages of all surgical specimens VA-wide and of those in hospitals with EM units. The EM accessions have shown a percentage increase almost four times greater than that of the surgical accessions for each year except 1977, when the increases were approximately the same. The findings for 1974 through 1977 are shown in Table 2. The percentages within each group over this period have varied only slightly from year to year, but consistently the percentage in hospitals with EM units has been approximately twice that for the entire system. In our view, the primary reasons for the higher percentages in hospitals with EM units are the ready availability of the resource and the greater appreciation of its value in diagnosis. Another reason may be the location of the units in teaching hospitals with tertiary level treatment services. The higher percentages may be also slightly misleading because of the inclusion in the EM numbers of some material from autopsies and body fluids that are not counted as surgical accessions. However, even when the differences are considered, the percentage in hospitals with EM units remains significantly greater.

Immunofluorescence procedures are available to most EM units. Scanning electron microscopes, because of the present limited applications in diagnostic electron microscopy, are not generally acquired but some program directors have access to a scanning EM to meet particular needs.

Training and education are part of the activities of all theEM units. Ultrastructural findings are discussed and demonstrated at general and specialty hospital conferences, and instruction in the techniques of electron microscopy is included in many of the graduate medical education programs. At the VA Hos-

Table 4. Percentage Distribution of EM Specimens for Diagnosis by Source

	Fiscal Year			
Source	1974	1975	1976	1977
Kidney	22	16	27	14
Liver	15	15	8	9
Skin	3	5	4	7
Muscle	4	3	5	4
Nervous system	4	3	5	4
Respiratory system	2	3	4	6
GI system	2	3	3	4
Hematopoietic system	13	19	10	18
Tumors [a]	20	22	27	24
Other	15	11	7	11
TOTAL	100	100	100	100

[a] Tumors from sites other than the hematopoietic system.

pital in Albany, New York, there is a formal one-year training program for EM technicians that is part of the baccalaureate program of a local college. At most of the hospitals with EM units, on-the-job training is provided for some of the technical staff.

The professional staffs of the electron microscopy units make presentations on their work at local, national, and international medical meetings and publish in the medical literature. These contributions are too numerous and varied for inclusion in this chapter.

The experiences described are in reality only the skeleton of the VA's diagnostic EM program and do not reflect the extent of its contributions. A more vital picture can be obtained by reference to presentations at the three VA conferences held in 1973 (1–11), 1975 (12), and 1977 (13), at which the program directors describe their diagnostic achievements and investigations. These two-day conferences are organized by the pathology service in the central office with assistance from the EM advisory group and some program directors. They are attended by all the EM program directors and some invited speakers. The sessions include formal presentations, workshops, and poster sessions.

The titles of the papers given at the 1977 conference by the VA program directors provide some indication of the range of activities: EM examination of tumor invasion of human skin by Dr. Scott McNutt, VA Hospital, San Francisco, CA; Role of EM in evaluating the presence of "mycosis cell" in biopsy specimens by Dr. John Guccion and associates, VA Hospital, Washington, DC; Usefulness of immunofluorescence and EM studies in dermatopathology by Dr. Ferenc Gyorkey, VA Hospital, Houston, TX; Combined use of direct immunofluorescence and ultrastructure in the diagnosis of lymphomas and allied conditions by Dr. Monica Bishop, VA Hospital, Albany, NY; Ultrastructural and immunologic changes of the synovium in hypertrophic pulmonary osteoarthropathy by Dr. Victor Pardo and associates, VA Hospital, Miami, FL; Electron, light microscopy and immunofluorescence microscopy evaluation of kidney biopsy material from 104 patients with systemic lupus erythematosus by Dr. Jesus M. Vasquez, VA Hospital, San Juan, PR; EM as a diagnostic tool in viral diseases of the central nervous system by Dr. Suzanne Mirra, VA Hospital, Atlanta, GA; A unified concept of lung cancer by Dr. Raymond Yesner, VA Hospital, West Haven, CT; an Ultrastructural study of lung tumors by Dr. N. M. Kandawalla and associates, VA Hospital, Tampa, FL; Use of Epon-embedded tissue for X-ray microanalysis by Dr. John D. Shelburne, VA Hospital, Durham, NC; Differential diagnosis of non-neoplastic liver diseases by EM by Dr. Oscar Iseri, VA Hospital, Baltimore, MD; Diagnosis of alcoholic hepatitis by EM by Dr. Itaru Watanabe, VA Hospital, Kansas City, MO; EM studies on micronodular cirrhosis of the liver by Dr. Ernesto Hoffmann and associates, VA Hospital, New Orleans, LA. The conferences are very popular and provide a forum for information exchange and mutual reinforcement in the multiple uses and applications of EM in diagnostic pathology.

SUMMARY

The VA experiences with diagnostic electron microscopy over more than a decade demonstrate its vitality and growing contributions to patient care.

The VA's program is characterized by centralized planning, site selection, funding, and evaluation, while the local program directors have considerable autonomy for daily operations. The level of activity has increased progressively as the services have become more widely available, and appreciation of the value and applications of ultrastructural study has grown. Statistical data on activities for the period 1974 through 1977 are presented and discussed.

The various procedures developed for management of the diagnostic EM program in the VA are considered to have general interest and possible applicability in other settings after appropriate adjustments to recognize different patient populations, levels of funding, and staffing.

The VA's experiences are indicative of the uses of diagnostic EM in a considerable segment of medical practice because of the geographical distribution of the units, their location in teaching hospitals with medical school affiliations, and their linkages with community hospitals through sharing agreements.

REFERENCES

1. Williams MJ: Electron microscopy in the Veterans Administration for service pathology: Some administrative aspects of the program. *Hum Pathol* 6:399–400, 1975.
2. Bloodworth JMB Jr, Azar HA, Yodaiken RE: Symposium on electron microscopy in diagnostic pathology. *Hum Pathol* 6:403, 1975.
3. Spargo BH: Practical use of electron microscopy for the diagnosis of glomerular disease. *Hum Pathol* 6:405–420, 1975.
4. Gyorkey F, Min KW, Krisko I, Gyorkey P: The usefulness of electron microscopy in the diagnosis of human tumors. *Hum Pathol* 6:421–441, 1975.
5. Morningstar WA: Whipple's disease. *Hum Pathol* 6:443–454, 1975.
6. Yodaiken RE, Pardo V: Diabetic capillaropathy. *Hum Pathol* 6:455–465, 1975.
7. Burns WA, Zimmerman HJ, Hammond J, Howatson A, Katz A, White J: The clinician's view of diagnostic electron microscopy. 6:467–487, 1975.
8. Azar HA: Significance of the Reed-Sternberg cell. *Hum Pathol* 6:479–484, 1975.
9. Azar HA, Morningstar WA, Vracko R, and White HJ: Problems in the development of a clinically oriented program in diagnostic electron microscopy. *Hum Pathol* 6:485–489, 1975.
10. Bloodworth JMB Jr, Burns WA, Daoud AS, Vracko R, Waldo ED: Education and training in electron microscopy. *Hum Pathol* 6:491–497, 1975.
11. Trump BF, Valigorsky JM, Jones RT, Mergner WJ, Garcia JH, Cowley RA: The applications of electron microscopy and cellular biochemistry to the autopsy. *Hum Pathol* 6:499–516, 1975.
12. Veterans Administration, 1B-11-53: Proceedings of the Veterans Administration Second Electron Microscopy Conference, May 13–14, 1975.
13. Veterans Administration: Proceedings of the Veterans Administration Third Electron Microscopy Conference, May 10–11, 1977, in preparation.

2
Liver Pathology

Kyuichi Tanikawa, M.D., Ph.D.
Professor, Department of Medicine
Chief of Electron Microscopy Laboratory
Kurume University School of Medicine
Kurume, Japan

With the development of electron microscopy, fine structural observations have become very important, not only for pure morphological studies but also for biochemical correlations that reveal the close relationship between function and structure.

From the clinical point of view, the remarkable development of new procedures such as endoscopic examinations has made biopsy materials from various organs much easier to obtain. As far as the liver is concerned, liver biopsy is a relatively easy technique, and numerous observations and studies of biopsy materials have already been made. However, because of its complex functions and the numerous disorders affecting this organ, the relationships between functional deteriorations and fine structural changes are still not fully established. For this reason, relatively few accounts (1,2,3) have been published on the systematic application of electron microscopy to the clinical diagnosis of liver disease.

On the basis of our studies of more than 1,000 liver biopsy specimens under the electron microscope, the following seem to be fruitful areas for electron microscopy:

Discovery of Infectious Agents. In various infectious diseases caused by viruses, protozoa, bacteria, etc., the infectious agents can be identified by electron microscopic observations. In differentiating various infectious hepatic diseases, hepatitis A and B viruses, infectious mononucleosis virus, and *Leishmania donovani* have been demonstrated by electron microscopy.

Studies of Pathogenesis. Electron microscopy is an important tool for the study of pathogenesis and the fine structural characteristics of each hepatic disorder. Trump and his associates have demonstrated fine structural alterations characteristic of lipid peroxidation in carbon tetrachloride-induced liver injury in rats. In human fatty liver, electron microscopy may help in the elucidation of the mechanism of fat accumulation.

Diagnosis of Diseases with Characteristic Changes in Organelles.   Any liver disease exhibits more or less characteristic alterations in various organelles of the hepatocyte. Since many of these alterations are not disease-specific, a definite diagnosis of certain hepatic diseases is difficult. However, electron microscopy may provide some data to make a differential diagnosis, for example, between viral hepatitis and alcoholic hepatitis, since the latter shows more characteristic mitochondrial alterations and alcoholic hyalin. In the future, quantitative stereology may permit correlation with liver function tests.

Organelles in the hepatocyte show various changes in pathological states. Some of them may be specific to certain liver diseases, and some may be merely reflections of whole hepatocytic alterations. At present, numerous forms of organellar alterations have been identified by electron microscopy, although their causes or meanings have not been fully understood. From the clinical standpoint, some of these organellar changes could be diagnostically helpful.

Recently, in specific disease an entity that can be diagnosed by changes in organelles and is the cause of cryptogenic cirrhosis has been defined by Sharp et al. in homozygous PiZZ children (4). A pattern of change can be either macro- or micronodular, but the distinctive finding is the presence of intracytoplasmic eosinophilic inclusions in the periphery of lobules or pseudolobules that are stained with PAS technique, and remain stained after diastase digestion. Conclusions are easily missed in routine hematoxylin- and eosin-stained sections. By electron microscopy, the inclusions consist of finely granular homogeneous material within smooth and rough endoplasmic reticulum. The material is closely related to alpha-1-antitrypsin.

Diagnosis of Deposit or Storage Diseases.   Metals such as iron have a characteristic fine structure and can also be defined by X-ray analysis. Stored materials in various storage diseases also have their own characteristic fine structural features. Thus, electron microscopic study is useful in the diagnosis of these disorders.

Identification of Cell Origin or Cell Type.   It is sometimes difficult to distinguish cell type or cell origin by light microscopy. However, electron microscopy often makes it possible to clarify the cell origin or cell type, especially in malignant tissue or cell infiltrates.

Evaluation of Liver Response to Drugs.   Recently, many new drugs have been introduced for clinical use. In the evaluation of the reactions in the liver to these drugs, mainly through morphological investigation, the electron microscope is an essential tool.

BASIC PROBLEMS IN FINE STRUCTURAL OBSERVATION OF THE LIVER

Before discussing the subjects described above, we must mention some basic problems in fine structural interpretations of the liver.

Variations Within the Lobule

In the normal lobule, there are considerable variations in cellular structure, such as shape, size, or number of hepatocytic organelles, as well as various histochemically demonstrated enzymes within the lobule. Although the significance of such variations has not been fully elucidated, they must be closely related to the physiology of the liver. The intralobular variations of organelles in normal human livers (5) are briefly summarized in the Table 1.

In pathological states, such intralobular variations may become more obvious. For instance, intrahepatic cholestasis occurs primarily in the centrilobular area, and fatty metamorphosis appears to be different in severity with localization within the lobule, depending upon the causative factors. Therefore, before examining specimens under the electron microsope, one must decide what part of the lobule is to be examined, by first checking a thick section of the specimens stained with toluidine blue (or other stains) by light microscopy (see Chapter 4, Vol. 1, of this series for an explanation of this technique).

Morphometric Analysis

Electron microscopy has been extremely valuable for qualitative interpretations of normal or pathological conditions. However, for quantitative studies of fine structural changes, morphometric analysis has inevitably become important, especially in relationship to functional or biochemical changes. For instance, so-called vesiculation of smooth endoplasmic reticulum (SER) has been subjectively interpreted as an increase or proliferation of SER. However, our recent morphometric study of alcohol-treated rats (6) has shown no increase in the amount of SER in the hepatocyte, contrary to our previous assumptions. In Dubin-Johnson syndrome, the reported bile canalicular alterations have not

Table 1. Differences Between Hepatocytic Organelles Within the Lobule of Human Liver

Organelles	Centrolobular	Peripheral
Bile canaliculus		More numerous microvilli Wider lumen
RER		More numerous
SER	More abundant	
Mitochondria	Round or oval in shape	More numerous Larger Oval or oblong in shape
Peroxisomes	More numerous	
Golgi complexes		More numerous Larger Often distended with VLDL
Lysosomes		Larger

been completely accepted. However, our morphometric studies (7,8) revealed definite changes of the canaliculus in comparison with the normal.

Recently, morphometric methods have become well established (9,10) and in the future will probably become more important in fine structural studies. However, because they are complex and time-consuming, they may have limited clinical application.

Considerations of Age, Sex, and Social Habits

Age changes of the liver in animals have been studied at the ultrastructural level, although most of the studies have not been analyzed by morphometric techniques. In human liver, a few morphometric studies of age changes have been made. Morphometric analysis by Tauchi and Sato (11) has revealed that mitochondria decrease in number but increase in size, especially after 60 years of age. Giant mitochondria (megamitochondria) also are frequently found in the aged group. From my experience, megamitochondria are frequently observed in biopsy samples from older patients, as compared with samples from younger ones. In addition, peroxisomes and lipofuscin granules appear to be more numerous in the aged groups. Thus, in interpreting fine structural changes of the liver, one should take into consideration aging changes of the liver.

So far, no extensive studies have been made on sex differences in the fine structure of the liver. However, as we know, biotransformation of female sex hormones is performed mainly by the liver, and reactions of the liver to drugs are quite different between male and female in animals. Thus, some fine structural differences in the liver might be found between sexes in humans, although we know little about them at present.

When we evaluate fine structural alterations of biopsy materials, the nutritional and social habits (e.g., smoking, alcohol consumption, or medication) of the patients should be checked, since they affect the fine structural changes of hepatocytic organelles, especially the amount of SER.

BASIC CHANGES OF THE LIVER

Fine Structural Characteristics of Reversible or Irreversible State of the Hepatocyte

One type of necrosis occurs when the hepatocytic contents become coagulated. Such coagulation necrosis, as exemplified by the acidophilic body by light microscopy, is shown by electron microscopy as an entire cell that appears to be smaller than normal and remarkably dense (Fig. 1). This body loses its surface microvilli and becomes detached from neighboring cells. Glycogen particles are readily observed. However, the endoplasmic reticulum and mitochondria are recognizable, with few alterations in the cytoplasm of such a cell (2,12,13). In general, the degenerating hepatocyte appears to be remarkably electron-dense and smaller, with a shrunken nucleus, irregularly shaped mitochondria, and fat droplets, and is often surrounded by macrophages, lymphocytes, or fibroblasts (Fig. 2).

Trump and his group have studied the ultrastructural changes following cell injury and for the sake of discussion have designated them stages of cell injury.

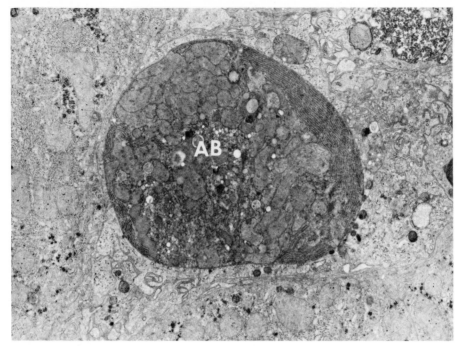

Figure 1. ACIDOPHILIC BODY
The body (AB), detached from the hepatocyte cord, appears to be electron-dense and small, compared with normal hepatocytes. The mitochondria and endoplasmic reticulum are recognizable. (×5,400)

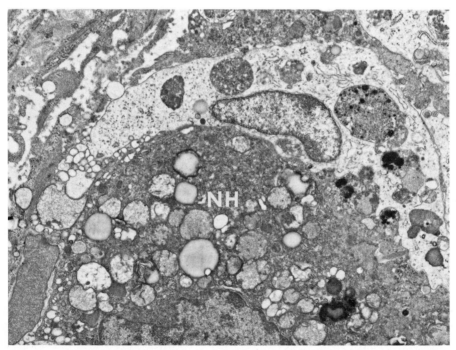

Figure 2. DEGENERATING HEPATOCYTE IN CHRONIC ACTIVE HEPATITIS
The hepatocyte (NH) undergoing necrosis has a shrunken nucleus and irregularly shaped, swollen mitochondria, and is partly surrounded by a macrophage. (×6,000)

19

These are numbered from stage 1 (the normal cell) to stage 7 (the necrotic cell). For a discussion of this subject, see Chapter 1, Volume 1, of this series.

Injuries to any organelle may produce changes in the plasma membrane. For example, injuries to mitochondria may result in deficient energy supply to the plasma membrane, leading to functional impairment of the plasma membrane. Consequently, the intracellular sodium to potassium ratio increases with increased water content in the cytoplasm, resulting in swelling of the hepatocyte. Such cells have been recognized as "ballooned cells" in acute hepatitis (Fig. 3).

Occasionally, the plasma membrane forms large blebs, and free organelles are observed in the tissue spaces (Fig. 4). Although those changes may be fixation artifacts or other processes, the fact that these findings are frequently observed in pathological states, such as in hepatitis, is suggestive of the susceptibility of plasma membranes to injury. Those changes of the plasma membrane associated with organellar alterations may be reversible, but may also result in liquefaction necrosis of the hepatocyte (13).

In severe cholestasis, dense bile materials may occupy most of the cytoplasm of some hepatocytes and may be indicative of bile necrosis (Fig. 5).

Fine Structure of Cholestasis

One of the major pathological processes in the liver is cholestasis. Under the electron microscope, cholestatic changes are seen in the hepatocyte, bile canaliculus, and the epithelial cell of the bile duct or ductule. Hypertrophy of the SER, associated with a decrease of its enzymic activities (hypoactive hypertrophy), has been considered, in intrahepatic cholestasis, as a primary lesion, and in extrahepatic cholestasis, as a secondary effect due to the detergent action of accumulated bile acids in the hepatocyte (14,15). The mitochondria appear to be generally enlarged, with occasional curling of cristae, or inclusions in the matrix (16). Such morphological alterations are probably associated with inhibition of oxidative phosphorylation in cholestasis (17). Under the electron microscope, bile materials accumulated in the hepatocyte appear as aggregates of fine granular and fibrillar materials, surrounded or not surrounded by a membrane (18,19) (Fig. 6).

The bile canaliculus is usually dilated, with a loss or stunting of its microvilli, and often contains bile thrombi. The pericanalicular ectoplasm is widened. The bile thrombus appears as an accumulation of dense materials in the dilated bile canaliculus. These may be finely granular, crystalline, lamellar, or whorled in shape (16,18) (Fig. 7). Such bile materials can also be seen in the cytoplasm of bile ductular epithelial cells or between two ductular epithelial cells. The lysosomes typically contain such whorled lamellar material.

As described above, cholestasis has some characteristic fine structural features in the liver. However, at present, it is difficult to differentiate intrahepatic from extrahepatic cholestasis by electron microscopy, because no remarkable differences in fine structure have been noted. In the future, differences may be found, especially in early stages of cholestasis, between intrahepatic and extrahepatic origins, or among various types of intrahepatic cholestasis with different etiologies. Recent interest has been evoked in the role of cell filament contraction failure in cholestasis.

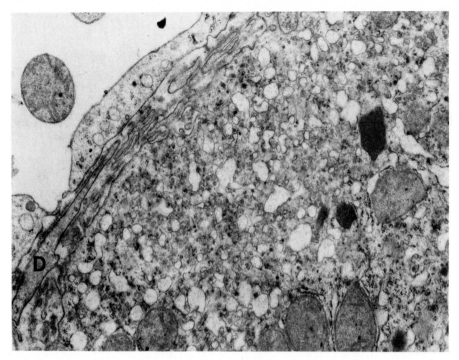

Figure 3. "BALLOONED CELL" IN ACUTE VIRAL HEPATITIS
A flattened plasma membrane facing the space of Disse (D) is noted, with numerous vesicles and vacuoles in the cytoplasm of the hepatocyte. (×18,000)

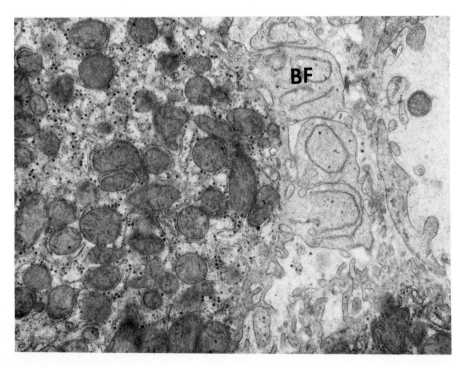

Figure 4. BLEB FORMATION OF THE PLASMA MEMBRANE IN WEIL'S DISEASE
The plasma membrane of the hepatocyte facing the space of Disse shows several bleb formations (BF) in which organelles are absent. (×10,000)

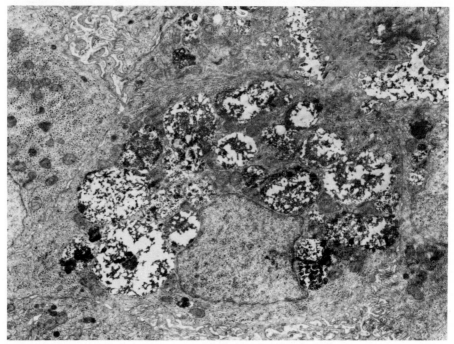

Figure 5. BILE NECROSIS IN OBSTRUCTIVE JAUNDICE
The cytoplasm of the hepatocyte (center) is mostly occupied by electron-dense materials with an irregularly shaped nucleus. Glycogen particles are, however, fairly well preserved. This is probably in a state of bile necrosis. (×2,700)

Fine Structure of Fibrosis

In normal liver, the fibroblast has been recognized only in the portal area. Thus, one aspect of fibrosis occurring within the lobule has been considered to be the result of the collapse of preexisting fibrous stroma after the necrosis of hepatocytes. However, fibrogenesis within the lobule has been more easily understood since the fat-storing cell, first described by Ito and Nemoto in 1952 (20), has been accepted as a resting fibroblast or a cell having a fiber-forming capacity (21,22). The fat-storing cell, localized in the space of Disse or between two hepatocytes in normal liver, is hypertrophied and increased in number in the immature connective tissue located around degenerating hepatocytes. Those fat-storing cells have proliferated rough-surfaced endoplasmic reticulum (RER) with faint flocculent material in their dilated cisternae, suggesting active protein production, probably protocollagen (22) (Fig. 8).

Mature connective tissue septa, as seen in cirrhosis, appear as numerous interlaced collagen fiber bundles with a few fibroblasts and containing small blood vessels (Fig. 9). Electron microscopy is apparently helpful sometimes in the diagnosis of macronodular, so-called "cryptogenic" cirrhosis. This pattern is seen in liver biopsy specimens, sometimes without the patient's having a clear history of chronic alcoholism, or after a period of abstinence. Among the electron microscopic criteria are Mallory's alcoholic hyalin, which can be easily missed by light microscopy, and centrilobular hyalin, which is considered specific for alcoholic

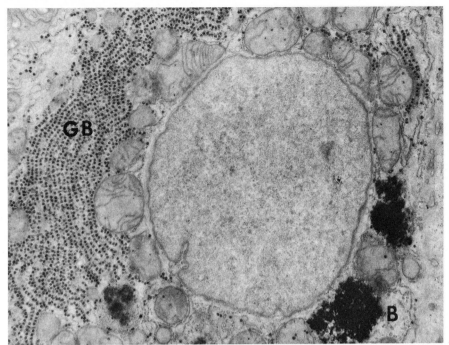

Figure 6. THE HEPATOCYTE IN CHOLESTASIS
Electron-dense aggregates (B), probably bile materials, are noted in the cytoplasm. The mitochondria have some peculiar cristae. Glycogen bodies (GB) are often seen in cholestasis. (×12,000)

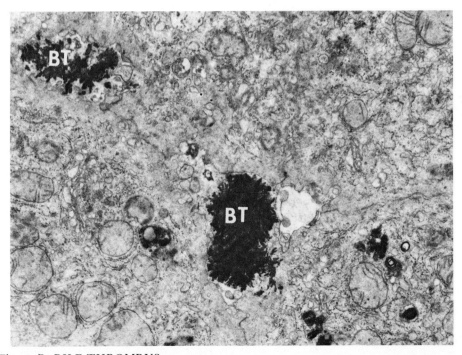

Figure 7. BILE THROMBUS
Bile thrombi (BT) are noted in dilated bile canaliculi that lose their microvilli. The pericanalicular ectoplasm is widened. (×8,000)

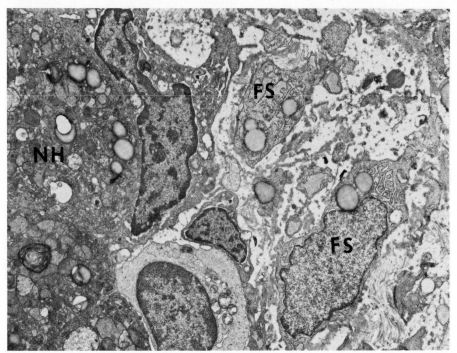

Figure 8. FAT-STORING CELL AND FIBROSIS
Fat-storing cells (FS) are recognized in the immature connective tissue around the necrotizing hepatocyte (NH) in active chronic hepatitis. (×4,800)

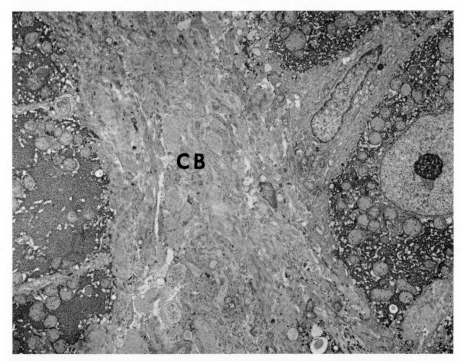

Figure 9. MATURE CONNECTIVE TISSUE SEPTUM IN CIRRHOSIS
The septum is characterized by numerous interlaced collagen bundles (CB) with few cell components. (×3,000)

liver disease. Giant mitochondria, especially those that are spherical, are characteristic of chronic alcoholism, and sometimes are present when hyalin is not. Also characteristic are paracrystalline cytoplasmic inclusions in the mitochondria.

The formation of a basement membrane in the space of Disse, with continuous or proliferated sinusoidal lining cells, has been noted in cirrhosis. Such capillarization of the sinusoid, first described by Schaffner and Popper (23), would be expected to inhibit the supply of nutrients and oxygen from the sinusoid to the hepatocyte, resulting in further hepatocytic injury.

In chronic hepatitis, cirrhosis, or obstructive jaundice, bile ductules are often proliferated and are accompanied with fiber formation around them (24) (Fig. 10). Such ductular proliferation has been considered to be closely related to fiber formation, although the mechanism is not clear.

Sometimes the diagnosis of primary biliary cirrhosis by light microscopy is extraordinarily difficult, the usual differential being extrahepatic obstruction, chronic and regressive hepatitis, or certain drug reactions in liver disease secondary to ulcerative colitis. Chedid et al. have described bundles of fibrils and biliary epithelial cells in disruption of basal lamina, with occasional multilayering and loss of continuity of the basement membrane. Such changes were not found by those authors in extrahepatic obstruction, chronic aggressive hepatitis, or postnecrotic cirrhosis (25).

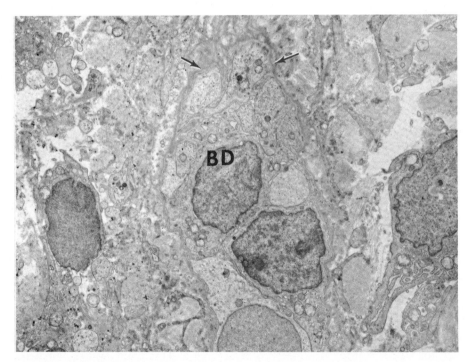

Figure 10. PROLIFERATED BILE DUCTULE WITH FIBROSIS
The proliferated bile ductule (BD) is surrounded by abundant collagen fibers and fibroblasts. A basement membrane (arrows) of the bile ductule is irregularly duplicated. (×3,000)

ELECTRON MICROSCOPIC DIAGNOSIS OF LIVER DISEASE

Discovery of Infectious Agents

Among various infectious diseases of the liver, viral hepatitis is the most common. Following the discovery of hepatitis B surface antigen and Dane particles (probably whole virus particles of hepatitis B) in the serum under the electron microscope, similar particles have been demonstrated in the hepatocyte. Spherical particles about 20 nm in diameter have been observed mainly in the nucleus and occasionally in the cytoplasm of the hepatocyte in HBs antigen carriers and in patients with acute or chronic hepatitis and cirrhosis with HBs antigenemia (26,27). Larger particles approximately 34 to 45 nm in diameter have also been demonstrated in membrane-bound vesicles in the cytoplasm of the hepatocyte (28,29). In addition, spherical and tubular structures or filaments approximately 20 to 35 nm in diameter are found in the cisternae of the endoplasmic reticulum (ER) (30,31,32).

The relation of these particles or tubular structures, seen in the thin sections of the liver tissue, to the three types of particles observed in the serum is uncertain. However, it seems likely that the 20-nm particles and larger particles approximately 35 to 45 nm in diameter in thin-sectioned materials correlate well with the 20-nm paricles of HBs antigen and the 42-nm double-shelled Dane particles in the serum, respectively; and tubular or filamentous structures observed in the cisternae of the ER seem to be closely related to the tubular form of HBs antigen in the serum. The Shikta orcein stain for HBs antigen has been shown to be specific for the hepatitis antigen and can be demonstrated in formalin-fixed paraffin-embedded tissue, although orcein-positive cells correspond to the so-called ground-glass hepatocytes. The latter can be very difficult to find in routine hematoxylin- and eosir-stained sections. Thus, discovery of such particles under the electron microscope would be diagnostic for hepatitis B and also helpful in the differential diagnosis of viral hepatitis from alcoholic or drug-induced hepatitis. Unfortunately, however, only a few specimens obtained from acute hepatitis B patients show such particles in the liver by electron microscopy, and, at present, the clinical significance of the presence of such particles in the tissue is limited.

Although 27-nm particles have so far been found only in feces of hepatitis A patients (32) and in the hepatocyte of marmosets infected with hepatitis A virus (33), in the near future, hepatitis A-associated antigens may be demonstrated in human liver tissue under the electron microscope.

Recently, virus particles have been demonstrated in the liver of patients with infectious mononucleosis (34). The mature particles with enveloping membrane measuring about 110 to 150 nm and the immature particles about 80 nm in diameter are noted in the nucleus and the cytoplasm of the hepatocyte. The mature particles contain a dense central nucleoid approximately 45 nm in diameter.

Kala-azar is one of the commonest protozoal diseases involving the liver, and is essentially a disease of the reticuloendothelial system. Leishman bodies stay in the liver for years. Under the electron microscope, Leishman bodies, measuring about 1.5 to 2.0 μm in length and 1 μm in width, are readily and clearly identified in Kupffer cells from their fine structural detail (2,35,36) (Fig. 11). In malaria, similar infectious agents have been found under the electron microscope (36) (Fig. 12). These protozoa are not well demonstrated and diagnosed by conventional

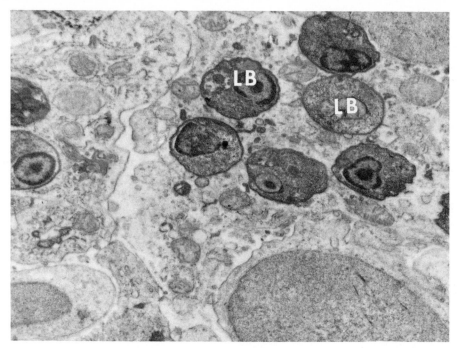

Figure 11. LEISHMAN BODIES IN KALA-AZAR
Leishman bodies (LB) are recognized in the Kupffer cells with detailed structures. (×11,000)

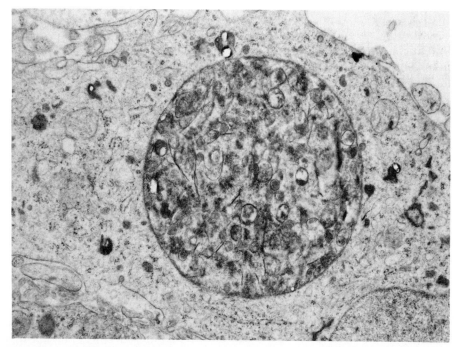

Figure 12. LIVER IN MALARIA
A large, round body, probably a schizont of the malaria parasite, is seen in the Kupffer
cell of a patient with tropical malaria. (×11,000)

27

light microscopy, but electron microscopy is very valuable and is the best diagnostic means.

Study of Pathogenesis

It has been suggested by Recknagel and others (37,38) that lipid peroxidation is a main causative factor in liver injury in carbon tetrachloride poisoning. Smuckler and his group have studied this lesion extensively (39). Trump and his associates have revealed by in vivo and in vitro studies that a large aggregate consisting of clumping of vesicles associated with dense material accumulation is a characteristic morphological change in membrane peroxidation (40,41,42) (Fig. 13). Similar lesions in the liver have also been described in phosphorus and iodoform intoxications in which lipid peroxidation has been suggested as a causative factor.

Such morphological changes are not usually observed in human livers except following accidental events. In carbon tetrachloride-induced liver injury, the RER is markedly altered with dilation of the cisternae associated with detachment of ribosomes. Disturbed lipoprotein synthesis due to such alterations of the RER has been implicated in the triglyceride accumulation in the liver. Similar findings have been observed by us in the livers of patients accidentally exposed to vanasium vapor (Fig. 14). In those cases, the remarkably fatty livers were assumed to be caused by alterations of the RER with subsequent reduction of lipoprotein synthesis similar to that in carbon tetrachloride intoxication.

In the alcoholic fatty liver, on the other hand, morphological changes of the RER, such as detachment of ribosomes, are thought to be rare even in severe cases. Sometimes, however, dilation of SER with flocculent materials in the cisternae is noted (43). The mechanism of triglyceride accumulation in the alcoholic fatty liver has been extensively studied by Lieber and his group, and it currently appears that a combination of factors is responsible, including increased mobilization from peripheral tissues, decreased fatty acid oxidation, increased synthesis of fatty acids, triglycerides, and decreased secretion from the liver (44). Thus, the mechanism of triglyceride accumulation seems to be different from the above-described carbon tetrachloride- or vanasium-induced fatty liver.

In Reye's syndrome, numerous small fat droplets are observed in the hepatocyte. Because of the small size of the fat droplets, they are occasionally overlooked by light microscopy of hematoxylin- and eosin-stained sections. However, under the electron microscope, they are clearly demonstrated and are about 1 to 2 μm in diameter (Fig. 15). Mitochondrial alterations have been described that are thought to be characteristic of this disorder (45,46).

In Reye's syndrome, unfortunately, we still do not understand the biochemical basis of the fatty liver. However, there have been fairly extensive studies on mitochondrial function, and it may well be that mitochondrial functional alterations are in some way related, for example, by decreased fatty acid oxidation.

More recently, other types of abnormalities in the experimental animal have been associated with fatty livers, and these should be studied in humans, e.g., in patients receiving chemotherapy for cancer in the form of agents such as vinblastine and other vinca alkaloids. In the experimental animal, those agents that interfere with microtubule function and cause depolymerization of microtubules are associated with fatty livers that have been related in the literature to decreased

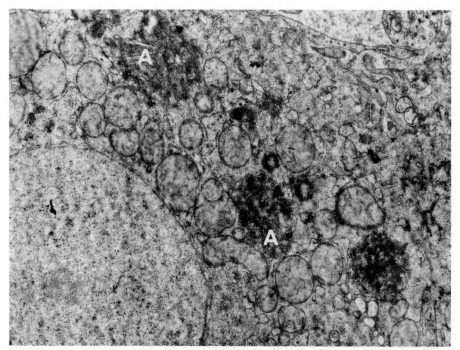

Figure 13. CARBON TETRACHLORIDE-INDUCED LIVER INJURY IN THE RAT
Membranous aggregates with electron-dense materials (A) are characteristic. These represent a state of lipid peroxidation. (×10,000)

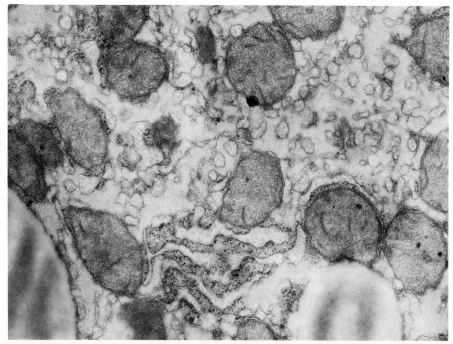

Figure 14. VANASIUM-INDUCED FATTY LIVER
RER is partially disrupted with fat droplet formation. (×20,000)

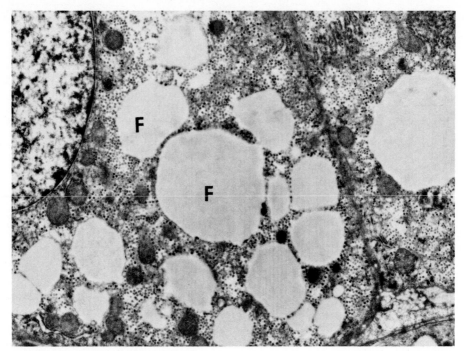

Figure 15. REYE'S SYNDROME
Numerous fat droplets (F) fairly similar in size are noted in the cytoplasm of the hepato-cyte. (×10,000)

secretion of lipoproteins; the microtubules are believed to be important in the secretory process. Furthermore, in experimental animals vinblastine and related alkaloids also result in massive increases in the number of autophagic vacuoles (49).

Diagnosis of Constitutional Hyperbilirubinemia

Constitutional hyperbilirubinemia, such as Dubin-Johnson syndrome, Rotor syndrome, or Gilbert's disease, has become a relatively commonly diagnosed icteric state due to the recent developments in liver function tests and biopsy techniques. Recently, fine structural features of these congenital disorders have been fairly well established by morphometric analysis (8), although some controversy still exists. Dubin-Johnson syndrome can be diagnosed histologically by numerous yellowish-brown granules in the hepatocyte. However, when the granules are not numerous, it becomes difficult to diagnose this disease. The granules in the hepatocyte of Dubin-Johnson syndrome have fine structural features different from lipofuscin granules or other granules such as hemosiderin. The former granules are round or oval and measure about 0.5 to 2.5 μm in diameter, and are most frequently observed surrounding the bile canaliculus (Fig. 16). They are delimited by a single membrane, and their matrix contains a great many ferritin-like fine grains with larger dense particles. Most of them resemble the structure of lipofuscin bodies, but are morphologically distinguishable from the latter when they have a prominent lipid component (2,50). The granules in Dubin-Johnson

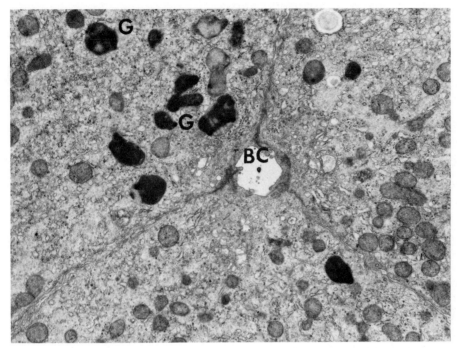

Figure 16. DUBIN-JOHNSON SYNDROME
Electron-dense granules (G), characteristic of the syndrome, are seen around the bile canaliculus (BC), which is slightly dilated, with loss of its microvilli. (×7,000)

syndrome are also easily distinguishable from the hemosiderin or copper-containing granules in Wilson's disease.

In addition, in Dubin-Johnson syndrome, characteristic alterations are seen in the bile calaliculi. They are often slightly dilated, with stunting or loss of microvilli, and in some instances there are electron-dense granules in the lumen (7,8) (Fig. 16). These bile canalicular alterations are not only of interest for their etiological considerations, but also they are diagnostic for this disorder when they occur along with the fine structural characteristics of the granules.

In Gilbert's disease, no remarkable changes are found in the liver by light microscopy. However, electron microscopy reveals characteristic alterations, such as flattening of the hepatocytic plasma membrane facing the space of Disse and changes of the SER such as an increase in amount and vesiculation or vacuolation (2,7,8) (Fig. 17). Thus, these changes provide the morphological basis for the diagnosis of this disease. At present, Gilbert's disease, clinically manifested by unconjugated hyperbilirubinemia, can be divided into two types. One variety of Gilbert's disease is considered to be a disturbance of bilirubin uptake at the plasma membrane facing the space of Disse. The other variety is due to a partial defect of bilirubin conjugation in which glucuronyl transferase is deficient. However, fine structural differences in the liver between the two types have not been established.

In Rotor's syndrome, light microscopy also shows normal liver architecture. However, by electron microscopy, the bile canalicular changes as seen in Dubin-Johnson syndrome and alterations of plasma membrane facing the space of Disse

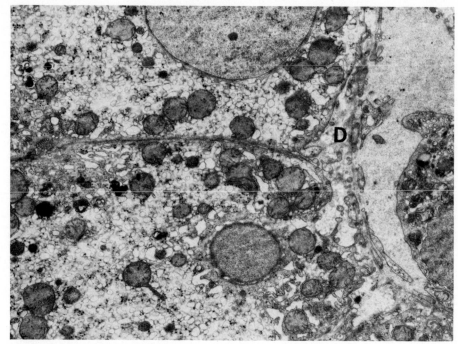

Figure 17. GILBERT'S DISEASE
The plasma membrane of the hepatocyte facing the space of Disse (D) is flattened, and its cytoplasm is mostly occupied by vesiculated SER. (×6,000)

or seen in Gilbert's disease have been found (2,8) (Fig. 18). Although these changes are not prominent, Rotor's syndrome seems to be situated morphologically between Dubin-Johnson syndrome and Gilbert's disease (8).

Diagnosis of Deposit or Storage Diseases

Excess accumulation of iron in the liver occurs in hemochromatosis and hemosiderosis. Ferritin particles are about 120 Å in diameter; however, they are demonstrated in micrographs as electron-dense particles about 70 Å in diameter because only part of the ion micelle is seen under the electron microscope (51). These particles are scattered throughout the cytoplasm or concentrated within lysomes as hemosiderin granules in the hepatocyte. Hemosiderin granules are delimited by a single membrane and contain numerous ferritin particles, usually without any other components (Fig. 19). In iron deposit disease, numerous ferritin particles or hemosiderin granules are observed. Thus, morphologically, such iron deposits can be easily identified by electron microscopy. Recently, extensive experimental investigations of the pathophysiology involved in iron overload in the liver and other cells have been reported in several reviews.

In Wilson's disease, it is well known that copper accumulation occurs in the liver. In the symptomatic stage, the hepatic copper concentration is higher than in the later symptomatic stage. In the former stage, copper particles are diffusely distributed in the cytoplasm of the hepatocyte, but unlike ferritin particles, they are too

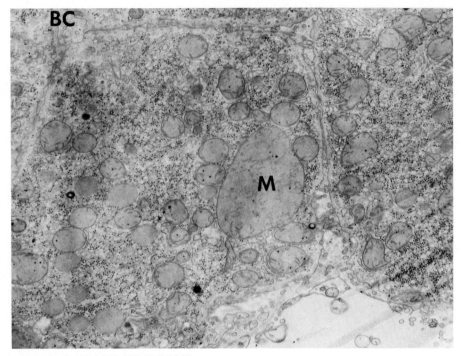

Figure 18. ROTOR'S SYNDROME
The plasma membrane of the hepatocyte facing the space of Disse is slightly flattened, with dilated intercellular space. The bile canaliculus (BC) is mildly dilated, with loss of its microvilli. A megamitochondrion (M) is seen in the cytoplasm of the hepatocyte. (×8,000)

small to be identified under the electron microscope. However, characteristic changes of mitochondria can be noted, such as increase in density of the matrix or alterations of the mitochondrial membrane (52,53) (Fig. 20). Thus, such mitochondrial alterations aid in the diagnosis of the asymptomatic stage of Wilson's disease.

In the symptomatic stage of this disease, hepatocellular copper is concentrated in lysosomes. Under the electron microscope, electron-dense granules characteristic of this disease appear fairly round, often multivacuolated, and measure about 1 to 2 μm in diameter (2) (Fig. 21). By using an energy-dispersive x-ray spectrometer combined with a scanning divice attached to a conventional transmission electron microscope, these granules were shown to have a high concentration of copper. By this procedure, the hemosiderin granules, naturally, showed a high content of iron (54). In the future, such x-ray analysis at the ultrastructural level will undoubtedly be applied more often and become an important diagnostic technique.

To date, a considerable number of storage diseases have been identified, most of which involved the liver, since it is the locus of various metabolic pathways. Recently, fine structural changes of the liver in these disorders have been extensively studied and reviewed (55,56,57).

In glycogen storage diseases, the hepatocytic cytoplasm is crowded with excess

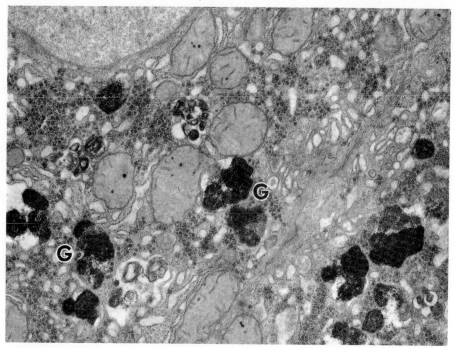

Figure 19. IRON DEPOSIT DISEASE
Hemosiderin granules (G) are numerous in the cytoplasm of the hepatocyte. They are delimited by a single membrane and contain many ferritin particles. (×13,000)

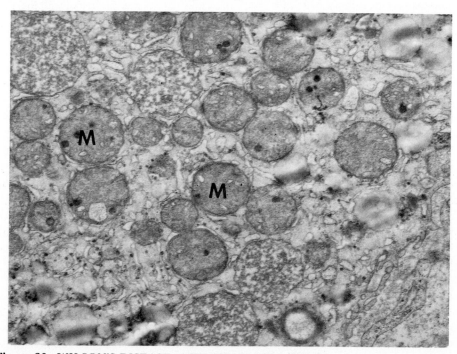

Figure 20. WILSON'S DISEASE; ASYMPTOMATIC STAGE
Mitochondria (M) with increased density of the matrix are noted. Several mitochondria are markedly altered. (×10,000)

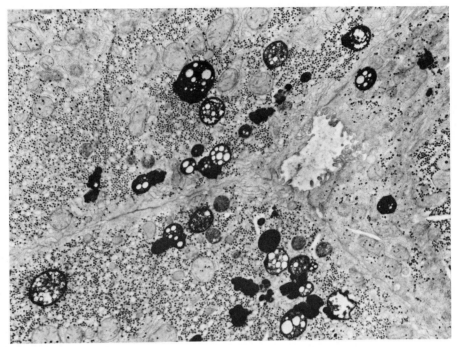

Figure 21. WILSON'S DISEASE; SYMPTOMATIC STAGE
Numerous electron-dense granules are noted, especially around the bile canaliculus. Multivacuolated granules seem to be specific in this disorder. (×6,000)

glycogen particles, and in type II glycogenosis, abnormal lysosomes filled with glycogen particles are characteristic (58,59). In lipidosis, excess lipid accumulation occurs in the Kupffer cells, and sometimes in the hepatocyte and other cell components in the liver. The different kinds of lipidoses have their own fine structural features (60). For instance, in Niemann- Pick disease, numerous round or oval vacuoles filled with loosely arranged, sometimes concentric or parallel osmophilic membranes are seen in hepatocytes, Niemann-Pick cells, and Kupffer cells. In Gaucher's disease (61), the Gaucher cell contains many Gaucher bodies containing membrane-enclosed cytoplasmic bodies filled with tubular structures.

In acid mucopolysaccharidosis (62), numerous electron-lucid vacuoles, bounded by a single membrane and containing several small electron-dense granules, are found in the hepatocytes (Fig. 22), Kupffer cells and epithelial cells of the bile duct. These vacuoles have been shown to contain mucopolysaccharide. Thus, from fine structural characteristics of accumulated materials in the liver, the diagnosis can be easily made in some storage diseases.

Differential Diagnosis of Hepatitis

In the differential diagnosis of viral hepatitis from alcoholic hepatitis, electron microscopy can be helpful to some extent.

In viral hepatitis, type B, particles immunologically identified as hepatitis B

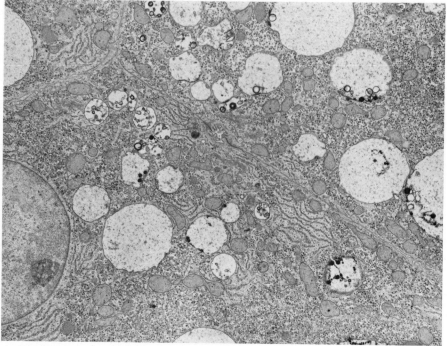

Figure 22. ACID MUCOPOLYSACCHARIDOSIS
Numerous vacuoles with electron-dense particles are seen in the cytoplasm of the hepatocyte. These vacuoles have been proved to contain acid mucopolysaccharides. (×5,000)

antigen or its core or surface antigens have been demonstrated in the nuclei or in
cytoplasms of the hepatocytes, but in only very rare cases. In alcoholic hepatitis,
alcoholic hyalin bodies are found as fibrillar deposits in the cytoplasm of the
hepatocytes (63,64,65,66,67,68). The frequency of these bodies seen by electron
microscopy may be limited.

In viral hepatitis, mainly the endoplasmic reticulum seems to be injured (Fig.
23). On the other hand, the mitochondrial changes are remarkable in alcoholic
hepatitis with minimal changes of the RER (69) (Fig. 24). Thus, such organellar
changes are of help for differentiation of these major liver diseases.

In summary, differentiation between alcoholic and viral hepatitis can be
difficult, especially if viral hepatitis occurs in alcoholics. In chronic alcoholics or
some cases of acute alcoholic hepatitis, the presence of numerous megamitochondria and abundant alcoholic hyalin is of help. Furthermore, the changes in the ER
are minimal as compared with the massive dilation that occurs in viral hepatitis.
Viral particles are not very helpful in evaluating the acute stages; they are too
quickly absent.

Insofar as drug-induced liver injuries are concerned, cholestatic liver injuries
show bile canalicular changes, such as dilation and loss or stunting of microvilli.
Various forms of bile thrombi have also been recognized by electron microscopy.
However, these changes are indistiguishable from cholestatic viral hepatitis. In
halothane hepatitis, mitochondrial alteration may be specific (70).

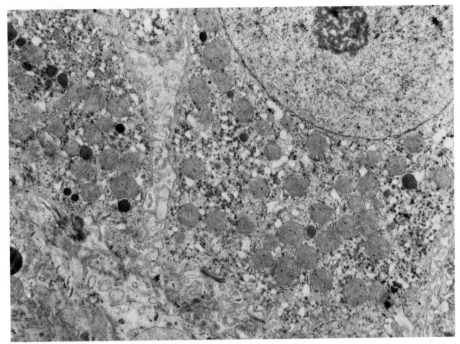

Figure 23. ACUTE VIRAL HEPATITIS
Alterations of endoplasmic reticulum appear to be more remarkable in comparison with other organellar changes. (×6,000)

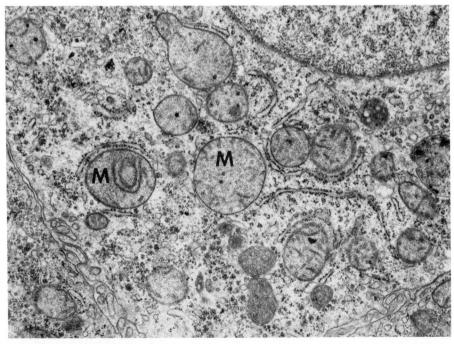

Figure 24. ALCOHOLIC HEPATITIS
Mitochondria (M) appear to be swollen and have peculiarly shaped cristae. RER is fairly well preserved. (×10,000)

Evaluation of Liver Response to Drugs and Other Agents

Numerous new and old drugs in clinical use and many additives in food preparations are metabolized mainly in the liver. The liver responds in variable degrees to these agents and may show identifiable changes in fine structure. Therefore, electron microscopy has become very important for the morphological evaluation of the liver response to these agents (71). At the present, however, few ultrastructural studies have been incorporated into testing programs for new drugs.

In Japan, a phospholipidosis, induced by a coronary vasodilator that had been widely used throughout the world, was discovered by electron microscopic observations of liver biopsy materials (72,73) (Fig. 25). This drug-induced liver effect would have been found before distribution of the drug if there had been electron microscopic studies during the testing stage. Thus, for the future, it may well be that electron microscopy will be included as an important tool for testing new drugs.

Many drugs lead to a poliferation of SER, the site of biotransformation, in association with increased cytochrome P-450 and its associated microsomal enzymes. This may be considered as an adaptive change of the liver to these drugs. This phenomenon is striking in the liver of man and experimental animals following phenobarbital treatment (Fig. 26). Although such a condition has been called a hypertrophic and hyperactive state of the SER, it may be replaced by a state of decompensation in which the SER is shown to be tightly clumped. The latter has been described as hypoactive hypertrophy because of an associated decrease of enzyme activites. Thus, such tightly clumped SER may be a morphological hallmark. In general, it may be difficult to decide whether these changes should be considered as a sign of adaptation or injury in individual cases.

An increase in the number of peroxisomes has been noted after the administration of clofibrate (74) (Fig. 27). Whether or not this change is adaptive or a pathological change is unknown, because the function of this organelle remains to be elucidated.

In a study of chronic toxication by chlorobiphenyls in man, increased amounts of SER were predominant, and there was also an increased number of lipofuscin-like granules (75) (Fig. 28). The increase of lipofuscin-like granules in the hepatocyte may be taken as an indication of mild chronic toxicity because such granules are considered to be residual bodies. At the present time, the border between adaptation to drugs and injuries induced by drugs has not been fully established at the ultrastructural level. However, accumulating data should clarify such problems in the future.

Identification of Cell Origin or Cell Type

It is occasionally difficult to differentiate hepatoma from metastatic carcinoma of the liver by light microscopic observations of liver biopsy specimens because specimens are frequently too small for adequate evaluation. However, under the electron microscope, hepatoma cells are generally very similar to normal hepatocytes with the presence of α-type glycogen particles and bile canaliculus formation (2) (Fig. 29).

In acute or chronic hepatitis, a considerable number of mononuclear cells are noted infiltrating, mostly in the portal tract. Most of them are lymphocytes and

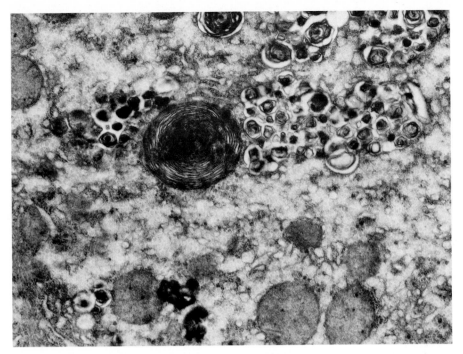

Figure 25. DRUG-INDUCED PHOSPHOLIPIDOSIS
Numerous laminated, electron-dense granules are deposited in the cytoplasm of the hepatocyte. (×11,000)

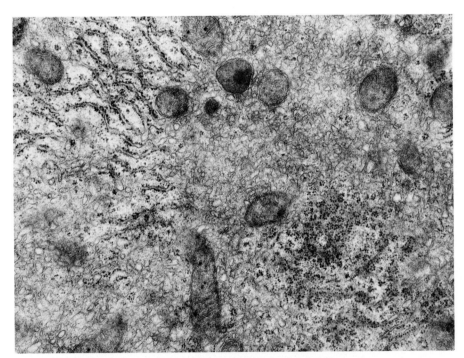

Figure 26. LIVER OF PHENOBARBITAL-TREATED RAT
SER is proliferated and is partially continuous with RER. (×18,000)

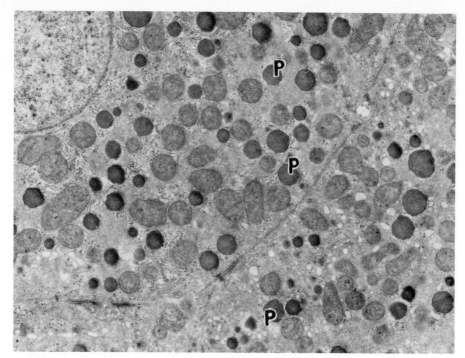

Figure 27. LIVER OF CLOFIBRATE-TREATED RAT
Peroxisomes (P) are observed in the cytoplasm of the hepatocyte. (×6,400)

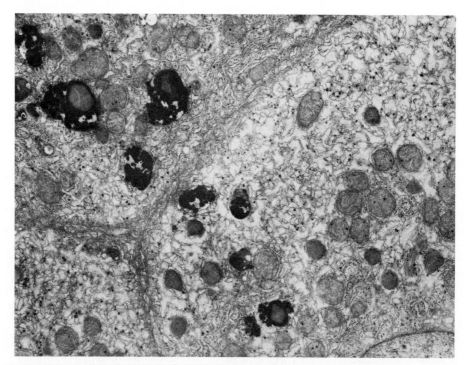

Figure 28. CHLOROBIPHENYL INTOXICATION
SER is proliferated, with an increase of lipofuscin granules. (×7,000)

40

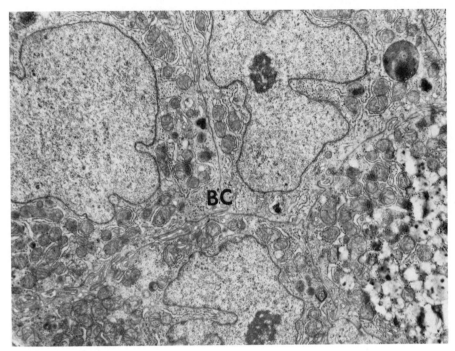

Figure 29. HEPATOMA
Hepatoma cells have irregularly shaped, large nuclei. However, the bile canaliculi (BC) are incompletely formed between hepatoma cells. α-Glycogen particles and fat droplets are noted in their cytoplasm. (×5,000)

macrophages. However, though we describe them merely as mononuclear cells, the cell type or origin is usually not completely clear. "Oval cells" have been identified as ductular epithelial cells by electron microscopy (76). It has been a surprise for me to find that a considerable number of mononuclear cells infiltrating the portal area are of ductular cell origin when seen under the electron microscope (Fig. 30). "Small hepatocytes" have also been identified as accompanying proliferated bile ductules in our recent studies in active chronic hepatitis and cirrhosis (77) (Fig. 31). In infectious mononucleosis, numerous, large, irregular mononuclear cells are present within the lumen of the sinusoid. In viral hepatitis such a finding is not common.

Metastatic tumors of the liver are often difficult to diagnose when the primary focus is unknown. The use of electron microscopy has helped considerably in identifying the type of tumor and the primary site. For example, carcinomas can usually be clearly distinguished from sarcomas by the presence of well-defined cell junctions and other specializations in the former. Specific secretory granules serve to identify both endocrine and exocrine tumors, as in the case of pancreas adenocarcinomas. Intra- and intercellular alveoli are important characteristics of metastatic adenocarcinomas from almost any site. Recently, there has been a report of formation of cell junctions between matastatic oat cell carcinoma and normal uninvolved hepatic parenchymal cells (78).

Thus, electron microscopy is valuable for identification of cell type and origin.

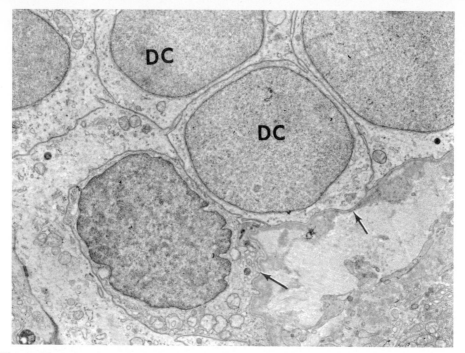

Figure 30. PROLIFERATED BILE DUCTULAR CELLS
Proliferated ductular cells (DC) are surrounded by a basement membrane (arrows).
These cells are difficult to recognize as ductular cells by light microscopy. (×5,000)

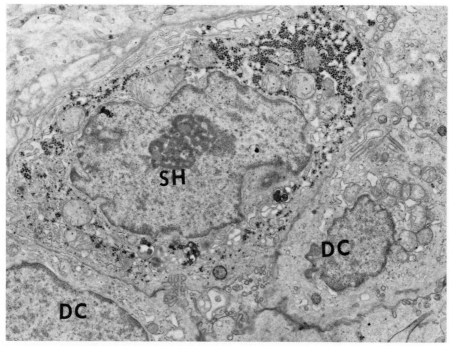

Figure 31. "SMALL HEPATOCYTE"
A "small hepatocyte" (SH) is occasionally noted with proliferated bile ductules. DC=ductular cell. (×8,000)

42

REFERENCES

1. Tanikawa, K, Okuda K: Electron microscopy in the diagnosis of liver disease, in *Recent Advances in Gastroenterology. Proceedings of the Third World Congress of Gastroenterology.* Tokyo, 1967, vol 3, p 385.

2. Tanikawa K: *Ultrastructural Aspects of the Liver and Its Disorders.* Berlin, Springer-Verlag, 1968, p 234.

3. Popper H: Electron microscopy in the diagnosis of liver disease, in Read AE (ed): *The Liver.* London, Butterworths, 1967, p 47.

4. Sharp HL, Bridges RA, Krivit W, et al: Cirrhosis associated with alpha-1-antitrypsin deficiency: A previously unrecognized inherited disorder. *J Lab Clin Med* 73:934, 1969.

5. Ma MH, Biempica L: The normal liver cell. *Am J Pathol* 62:353, 1971.

6. Ikejiri N, Tanikawa K: Morphometric analysis of fine structural changes of the liver in acute and chronic alcoholic intoxications. *Kurume Med J,* in press.

7. Tanikawa K, Abe H, Miyakoda U, et al: Electron microscopic findings of the liver in constitutional hyperbilirubinemia. *Acta Hepatol Jap* 12:160, 1970.

8. Miyakoda U: Morphometric analysis of fine structural changes of the liver in patients with constitutional hyperbilirubinemia. *Acta Hepatol Jap* 16:121, 1975.

9. Loud AV: A quantitative stereological description of the ultrastructure of normal rat liver parenchymal cells. *J Cell Biol* 37:27, 1968.

10. Weibel ER: Stereological principles for morphometry in electron microscopic cytology. *Int Rev Cytol* 26:235, 1969.

11. Tauchi H, Sato T: Age changes in size and number of mitochondria of human hepatic cells. *J Gerontol* 23:454, 1968.

12. Klion FM, Schaffner F: The ultrastructure of acidophilic "Councilman-like" bodies in the liver. *Am J Pathol* 48:755, 1966.

13. Schaffner F: Some unsolved ultrastructural problems encountered in the study of the liver and its diseases, in Schaffner F, Sherlock S, Leevy CM (eds): *The Liver and Its Diseases.* New york, Intercontinental Med Corp, 1974, p 7.

14. Popper H, Schaffner F: The pathophysiology of cholestasis. *Hum Pathol* 1:1, 1970.

15. Schaffner F, Popper H: Cholestasis is the result of hypoactive hypertrophic smooth endoplasmic reticulum in the hepatocyte. *Lancet* 2:355, 1969.

16. Desmet VJ: Morphologic and histochemical aspects of cholestasis, in Popper H, Schaffner F (eds): *Progress in Liver Diseases.* New York, Grune & Stratton, 1972, Vol. 4, p 97.

17. Lee MJ, Whitehouse MW: Inhibition of electron transport and coupled phosphorylation in liver mitochondria by cholanic acids and their conjugates. *Biochim Biophys Acta* 100:317, 1965.

18. Biava C: Studies on cholestasis. II. The fine structure and morphogenesis of hepatocellular and canalicular bile pigment. *Lab Invest* 13:1099, 1964.

19. Hubner G: Die Rolle der Lysosomen beim Ikterus der Leber und bei der Pigmentablagernung in den Leberzellen, in Beck K (ed): *Ikterus.* Stuttgart, New York, Schattauer, 1968, p 115.

20. Ito T, Nemoto M: Uber die kupfferschen Sternzellen und die "fettspeicherungs Zellen" ["fat-storing cells"] in der Blutkapillarenwand der menschlichen Leber. *Okajimas Folia Anat Jap* 24:243, 1952.

21. Popper H, Undenfriend S: Hepatic fibrosis. Correlation of biochemical and morphologic investigations. *Am J Med* 49:707, 1970.

22. Tanikawa K: Ultrastructure of hepatic fibrosis and fat-storing cells, in Popper H, Becker K (eds): *Collagen Metabolism in the Liver.* New York, Stratton Intercontinental Medical Book, 1975, p 93.

23. Schaffner F, Popper H: Capillarization of hepatic sinusoids in man. *Gastroenterology* 44:239, 1963.

24. Popper H, Paronetto F, Schaffner, F., et al: Studies on hepatic fibrosis. *Lab Invest* 10:265, 1961.

25. Chedid A, Spellberg MA, DeBeer RA: Ultrastructural aspects of primary biliary cirrhosis and other types of cholestatic liver disease. *Gastroenterology* 67:858, 1974.

26. Ahmed MN, Huang S, Spence L: Australia antigen and hepatitis. An electron microscopic study. *Arch Pathol* 92:66, 1971.

27. Krawczynski K, Nazarewicz T, Brzosko WJ, Nowoslawski A: Cellular localization of hepatitis-associated antigen in liver of patients with different forms of hepatitis. *J Infect Dis* 126:372, 1972.

28. Huang S: Hepatitis-associated antigen hepatitis. An electron microscopic study of virus-like particles in liver cells. *Am J Pathol* 64:483, 1971.

29. Haynes ME, Haynes DWG, Williams R: Cytoplasmic particles in hepatocytes of patients with Australia antigen-positive disease. *J Clin Pathol* 25:403, 1972.

30. Stein O, Fainnaru M, Stein Y: Visualization of virus-like particles in endoplasmic reticulum of hepatocytes of Australia antigen carriers. *Lab Invest* 26:262, 1972.

31. Gerber MA, Paronetto F: Hepatitis B antigen in human tissue, in Schaffner F, Sherlock S, Leevy C (eds): *The Liver and Its Diseases*. New York, Intercontinental Med Corp, 1974, p 54.

32. Feinstone SM, Kapikian AZ, Purcell RH: Hepatitis A: Detection by immune electron microscopy of a virus-like antigen associated with acute illness. *Science* 182:1026, 1973.

33. Provost PJ, Wolanski BS, Miller WJ, et al: Physical, chemical, and morphologic dimensions of human hepatitis A virus strain CR326 (38578). *Proc Soc Exp Biol Med* 148:532, 1975.

34. Chang MY, Campbell WG: Fatal infectious mononucleosis. Association with liver necrosis and herpes-like virus particles. *Arch Pathol* 99:185, 1975.

35. Tanikawa K, Hojiro O: Electron microscopic observation of the liver in Kala azar. *Kurume Med J* 12:148, 1965.

36. Miwa S, Tanikawa K: Electron microscopic observation of the liver in malaria and Kala azar. *Rev Int Hepatol* 15:489, 1965.

37. Recknagel RO, Glende EA: Lipid peroxidation in acute carbon tetrachloride liver injury, in Brown H, Hardwick DF (eds): *Intermediary Metabolism of the Liver*. Springfield, Ill, Charles C Thomas, 1973, p 23.

38. Recknagel RO, Ugazio G, Koch RR, Glende EA: New perspectives in the study of experimental carbon tetrachloride liver injury, in Gall EA, Mostofi FK (eds): *The Liver*. Baltimore, Williams & Wilkins, 1973, p 150.

39. Smuckler EA: Structural and functional changes in acute liver injury. *Environ. Health Prospect* 15:13, 1976.

40. Trump BF, Arstila AU: Cellular and subcellular reactions of cells to injury, in LaVia N, Hill R (eds): *Principles of Pathobiology*. New York, Oxford University Press, 1971, p 9.

41. Arstila AU, Smith M, Trump BF: Microsomal lipid peroxidation; morphological characterization. *Science* 175:530, 1972.

42. Trump, BF, Dees JH, Sherburne JS: The ultrastructure of the human liver cell and its common patterns of reaction to injury, in Gall EA, Mostofi fk)eds): *The Liver*. Baltimore, Williams & Wilkins, 1973, p 80.

43. Tanikawa K, Miyakoda U: Étude au microscope électronique des mécanismes de dépôt et de disparition des lipides hépatiques dans la stéatose ethylique. *Ann Gastroenterol Hepatol* 9:411, 1973.

44. Lieber CS: Liver disease and alcohol: fatty liver, alcoholic hepatitis, cirrhosis and their relationship. *Ann NY Acad Sci* 252:63, 1975.

45. Partin JC, Schubert WK, Partin JS: Mitochondrial ultrastructure in Reye's syndrome (encephalopathy and fatty degeneration of the viscera). *N Engl J Med* 285:1339, 1971.

46. Schubert WK, Partin JC, Partin JS: Encephalopathy and fatty liver (Reye's syndrome), in Popper H, Schaffner F (eds): *Progress in Liver Diseases*. New York, Grune & Stratton, 1972, vol 4, p 489.

47. Snodgrass PJ, DeLong GR: Urea-cycle enzyme deficiencies and an increased hydrogen load producing hyperammonemia in Reye's syndrome. *N Engl J Med* 294:55–60, 1976.

48. Brown T, Hug G, Lansky L, et al: Transiently reduced activity of carbamoyl phosphate synthetase and ornithine transfer of amylase in liver of children with Reye's syndrome. *N Engl J Med* 294:861–867, 1976.

49. Hirsimäki Y, Arstila AU, Trump BF: Autophagocytosis: In vitro induction by microtubule poisons. *Exp Cell Res* 9:11, 1975.

50. Toker C, Trevino N: Hepatic ultrastructure in chronic idiopathic jaundice. *Arch Pathol* 80:453, 1965.

51. Crichton RR: Ferritin: structure, synthesis and function. *N Engl J Med* 284 (25):1413, 1971.

52. Sternlieb I: Mitochondrial and fatty changes in hepatocytes of patients with Wilson's disease. *Gastroenterology* 55:354, 1968.

53. Sternlieb I: Evolution of the hepatic lesion in Wilson's disease (hepatolenticular degeneration), in Popper H, Schaffner F (eds): *Progress in Liver Diseases.* New York, Grune & Stratton, 1972, vol 4, p 511.

54. Tanikawa K, Abe H, Miyakoda U: Electron microscopic observation and X-ray analysis of deposit granules in the hepatocyte. *J Clin Electron Microscopy* 6:332, 1973.

55. Hers HG, Van Hoof F: The genetic pathology of lysosomes, in Popper H, Schaffner F (eds): *Progress in Liver Diseases.* New York, Grune & Stratton, 1970, vol 3, p 185.

56. Volk BH, Wellmann KF, Wallace BJ: Hepatic changes in various lipidoses: electron microscopic and histochemical studies, in Popper, H, Schaffner F (eds): *Progress in Liver Diseases.* New York, Grune & Stratton, 1970, vol 3, p 206.

57. Hug G: Nonbilirubin genetic disorders of the liver, in Gall EA, Mostofi FK (eds): *The Liver.* Baltimore, Williams & Wilkins, 1973, p 21.

58. Baudhuin P, Hers HG, Loeb H: An electron microscopic and biochemical study of type II glycogenosis. *Lab Invest* 13:1139, 1964.

59. Hug G, Garancis JC, Schubert, WK, et al: Glycogen storage disease, type II, III, VIII, and IX. *Am J Dis Child* 111:457, 1966.

60. Volk BW, Wallace BJ: The liver in lipidosis. An electron microscopic and histochemical study. *Am J Pathol* 49:203, 1966.

61. Fisher ER, Reidbord H: Gaucher's disease: pathogenetic considerations based on electron microscopic and histochemical observations. *Am J Pathol* 41:679, 1962.

62. Callahan WP, Lorincz AE: Hepatic ultrastructure in the Hurler syndrome. *Am J Pathol* 48:227, 1966.

63. Biava C: Mallory alcoholic hyaline: a heretofore unique lesion of hepatocellular ergastoplasm. *Lab Invest* 13:301, 1964.

64. Flax MH, Tisdale WA: An electron microscopic study of alcoholic hyalin. *Am J Pathol* 44:441, 1964.

65. Smuckler EA: Ultrastructure of human alcoholic hyalin. *Am J Clin Pathol* 49:790, 1968.

66. Iseri OA, Gottlieb LS: Alcoholic hyalin and megamitochondria as separate and distant entities in liver disease associated with alcoholism. *Gastroenterology* 60:1027, 1971.

67. Yokoo H, Minick OT, Batti F, et al: Morphologic variants of alcoholic hyalin. *Am J Pathol* 69:25, 1972.

68. Gerber MA, Orr W, Denk H, et al: Hepatocellular hyalin in cholestasis and cirrhosis: its diagnostic significance. *Gastroenterology* 64:89, 1973.

69. Schaffner F: Electron microscopy of virus, alcohol and drug-induced hepatitides, in Saunders SJ, Terblanche J (eds): *Liver. Proceedings of an International Liver Conference with Special Reference to Africa.* London, Pitman Medical, 1973, p 46.

70. Klion FM, Schaffner F, Popper H: Hepatitis after exposure to halothane. *Ann Intern Med* 71:467, 1969.

71. Jezequel AM, Orlandi F: Fine morphology of the human liver as a tool in clinical pharmacology, in Orlandi F, Jezequel AM (eds): *Liver and Drug.* New york, Academic Press, 1972, p 145.

72. Oda T, Shikata T, Suzuki H, et al: Phospholipidosis of the liver cell: A report of three cases with a new type of fatty liver. *Acta Hepatol Jap* 10:530, 1969.

73. De la Iglesia FA, Feuer G, Takada A, et al: Morphologic studies on secondary phospholipidosis in human liver. *Lab Invest* 30:539, 1974.

74. Svoboda D, Grady H, Azarnoff D: Microbodies in experimentally altered cell. *J Cell Biol* 35:127, 1967.

75. Hirayama C, Irisa T, Yamamoto T: Fine structural changes of the liver in a patient with chlorobiphenyls intoxication. *Fukuoka Acta Med* 60:455, 1969.

76. Grisham JW, Hartroft WS: Morphologic identification by electron microscopy of "oval" cells in experimental hepatic degeneration. *Lab Invest* 10:317, 1961.

77. Tanikawa K, Abe H, Miyakoda U, et al: Hepatic fibrosis and bile ductular proliferation. *The 10th Annual Meeting of Japanese Hepatology Society,* Kyoto, 1974, p 55.

78. Jesudason ML, Iseri OA: Host-tumor cellular junctions: An ultrastructural study of hepatic metastases of bronchogenic oat cell carcinoma. *Hum Pathol,* in press.

3

The Hematopoietic System

Henry A. Azar, M.D.

Chief, Laboratory Service
Veterans Administration Hospital
Professor of Pathology
University of South Florida
College of Medicine
Tampa, Florida

INTRODUCTION

Electron microscopy has been gradually adapted as an adjunct to the diagnosis of a variety of hematological conditions and the elucidation of their pathogenesis. There are already a number of excellent atlases and treatises dealing partly or exclusively with electron microscopy of blood cells (1–4). In our own diagnostic electron microscopy unit at the Laboratory Service of the Tampa Veterans Administration Hospital, hematological tissues (i.e., bone marrow, peripheral blood (buffy coat), lymph nodes, and spleen) constitute the largest group of tissue samples processed for electron microscopy (5). The other two large groups are kidney biopsy samples and neoplasms, with the exclusion of hematopoietic and lymphoreticular malignancies (Table 1). Our experience is entirely based on transmission electron microscopy. The reader should be alerted, however, about the vast possibilities of applying scanning electron microscopy and X-ray microanalysis to the study of blood cells (6). Refinements in histochemical techniques applicable to electron microscopy may some day sharpen our ability to distinguish between closely related cells or single out early pathological changes, thus helping us in the differentiation of the "unclassifiable" leukemias (7).

In dealing with electron microscopy as an adjunct to the diagnosis of hematological conditions, it is important to approach ultrastructural studies as a natural extension of routine examinations that begin with a good clinical workup, a complete blood count with study of the peripheral blood smear, and, whenever indicated, a thorough study of a bone marrow aspirate or bone section, or a histological study of a lymph node. In other words, the electron microscopist dealing with hematopoietic cells should also be ideally a good hematologist and a well-trained hematopathologist.

47

Table 1. Types of Specimens Submitted for Electron Microscopy, Tampa VA Hospital (October 1973 to January 1977)

Year	Total	Bone Marrow	Buffy Coat	Lymph Nodes and Spleen	Kidney	Research	Others	
1973 (3 mos)	43	5		3	2	22	Tumor misc.	6
							Liver	2
							Others	3
							Total	11
1974	192	71	13	6	9	22	Tumor misc.	27
							Liver	6
							Nervous system	13
							Endocrine gland	5
							Others	20
							Total	71
1975	328	60	27	10	58	78	Tumor misc.	49
							Liver	4
							Nervous system	3
							Endocrine gland	11
							Others	28
							Total	95
1976	352	111	22	17	78	14	Tumor misc.	72
							Liver	6
							Nervous system	6
							Endocrine gland	7
							Others	19
							Total	110

The first half of this chapter will deal mainly with normal and abnormal ultrastructural features of the three major hematopoietic cell lines, that is, erythroid cells, granulocytes, and megakaryocytes. The second half will discuss lymphoreticular cells, including lymphocytes, plasma cells, macrophages, and monocytes. In both parts, we have given preference to aspects of electron microscopy that, in our hands, have helped in establishing or confirming the diagnosis in difficult hematological conditions and, whenever possible, in relating altered function to ultrastructural changes.

I. ERYTHROCYTES, GRANULOCYTES, AND MEGAKARYOCYTES

HEMATOPOIETIC TISSUES

Among the various hematopoietic tissues, the one that is the most complex, yet the most frequently dealt with, is the bone marrow. The would-be electron microscopist needs to get acquainted with the fine structure not only of the major hematopoietic cell lines but also of supporting stromal and vascular elements.

There are at least three elusive areas in studies of marrow ultrastructure. One concerns the hematopoietic stem cells and early precursors and how to recognize them. Another is suggested by the persistent problem of how to recognize monocytes and their precursors, and distinguish them from precursors of neutrophilic granulocytes. A third subject that continues to intrigue morphologists is the sinusoidal system of the marrow and the mode of exit of maturing cells into the bloodstream.

The ultrastructure of hematopoietic cells reinforces some known or suspected intimate functional interrelationships of different cell lines: hemosiderin-laden macrophage and normoblasts in bone marrow, macrophage and lymphocytes in the cortex of the lymph node, thymic epithelial cells and lymphocytes.

Divisions and General Considerations

The hematopoietic system may be defined for our purposes, albeit simplistically, as a far-flung, structurally and functionally complex organ that supplies the blood with its cell constituents. In this complex and intertwined system, one recognizes three convenient but arbitrary divisions:

1. *The bone marrow,* a major source of erythrocytes, granulocytes, and platelets, all presumably derived from common hematopoietic stem cells.
2. *The lymphoid tissues,* including the lymphoepithelial organs (thymus, tonsils, and adenoids), the lymph nodes and bowel-associated lymphoid follicles, the lymphoid aggregates and nodules of the marrow, and the white pulp of the spleen.
3. *The reticuloendothelial system,* a widely distributed network of fixed and wandering macrophages, including potential macrophages, the monocytes of marrow origin.

Functionally, hematopoietic cells are concerned principally with oxygen transport (erythrocytes), homeostasis and coagulation (thrombocytes), phagocytosis, metabolic degradation of particulate objects as well as micromolecules, and transformation of antigens into effective immunogens (macrophages and granulocytes), and cell-borne and humoral immunity (T and B lymphocytes). Ultrastructurally, differentiated hematopoietic cells are recognized as cells primarily engaged in protein synthesis for intracellular storage (normoblasts) or for export (activated B lymphocytes and plasma cells). The formation of lysosomal-type granules is particularly well developed in all three types of granulocytes, and in monocytes. Platelets are particularly suited for aggregation and plastic conformation to defective vascular walls. These are some considerations that will be discussed in subsequent sections. The lymphoreticular cells will be discussed in the second half of the chapter.

Procedures Relevant to Hematopathology

In processing hematological specimens, we have favored glutaraldehyde fixation, osmiation, and *en bloc* staining with uranyl acetate. Epon-embedded 1μ-thick sections stained with methylene blue polychrome are used for selection of representative areas and for relating ultrathin sections to light microscopic preparations. Ultrathin sections are also stained with 2.5% lead citrate. A disadvantage of *en bloc* staining with uranyl acetate is the loss of glycogen granules.

Blood cells can be readily studied by means of a *buffy coat* preparation (Fig. 1). A carefully prepared and well-oriented buffy coat sample yields three major layers: platelets, being the most superficial one, mononuclear cells (lymphocytes, monocytes, and "band" cells), and polymorphonuclear leukocytes next to the red cell zone (Fig. 2). A buffy coat preparation is particularly useful for the study of platelets in various functional disorders and leukemias, in the study of acute leukemias, and in the evaluation of patients with pancytopenias in whom hairy cell leukemia is suspected.

A sample of *bone marrow fragments,* consisting of well-selected red marrow tissue without blood clots, is routinely set aside in buffered (pH 7.4) 2.5% glutaraldehyde at 4°C, until marrow smears and routine paraffin sections are examined. Should the initial hematological or histological study disclose a diagnostic problem that requires verification or further study, the glutaraldehyde-fixed marrow sample can then be further processed for electron microscopy.

Lymph node and *spleen* samples are treated in the same manner as bone marrow samples: "precautionary" fixation in 2.5% buffered glutaraldehyde and further processing for electron microscopy should the histological examination warrant it.

Histochemical studies can be done in all hematological samples at the level of both light and electron microscopy. Among the reactions that are particularly useful in hematology are acid phosphatase (with tartrate resistance) for hairy cell leukemia (8), and peroxidase, periodic acid-Schiff (PAS), and esterase reactions for the study of acute leukemias and myelosarcoma (7,9,10).

Finally, the *electron microscopy report* should receive great attention. For examination of hematopoietic tissues, the electron microscopist should carefully word his diagnosis in the light of clinical and histological findings. It could be dangerous to rely solely on electron microscopic findings, particularly should these clash with clinical judgment or traditional laboratory data. It is obvious that the electron microscopist should resist operating in a detached scientific vacuum when dealing with a minute tissue sample.

ERYTHROID CELLS

With the exception of the situation in neonatal life and in extramedullary hematopoiesis, erythropoietic precursors are almost entirely confined to the marrow, where they are considerably outnumbered in the normal state by granulocytic precursors, or where they tend often to gather in erythroid islets surrounding a hemosiderin-laden macrophage or nursing cell (1,3).

The Erythroid Series

Since one cannot at this time deal in electron microscopy with shades of basophilia and hemoglobinization, the interpreter has to learn to identify erythroid precursors as such by negative as well as positive attributes: absence of granular inclusions that characterize both myeloid cells and megakaryocytes,

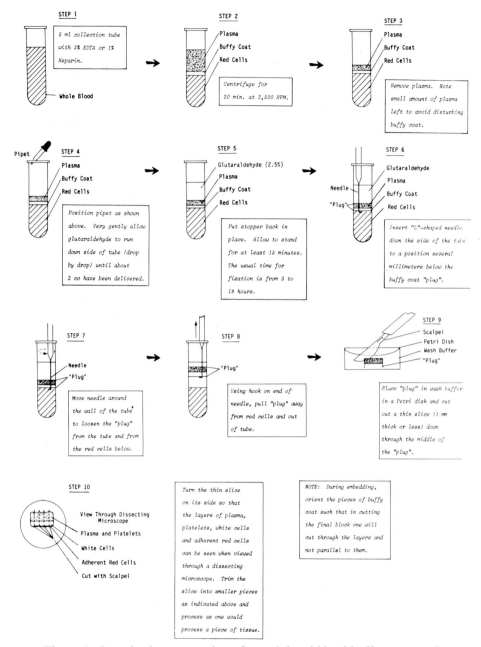

Figure 1. Steps in the preparation of a peripheral blood buffy coat sample.

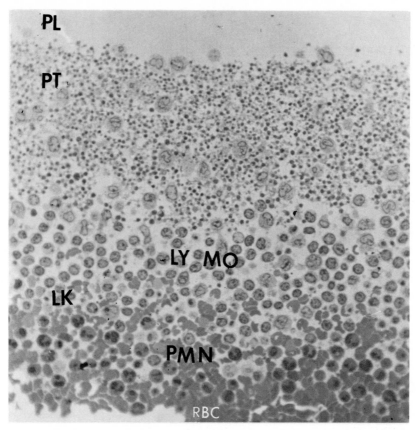

Figure 2. Cross section of a buffy coat "plug" showing (from top to bottom) plasma layer (PL), platelets (PT), leukocytes (LK)—with upper layer of lymphocytes (LY) and monocytes (MO) and lower layer of polymorphonuclear leukocytes (PMN)—and bottom RBC layer. (1μ-thick section, ×250)

absence of significant amounts of smooth or rough endoplasmic reticulum, and presence of polyribosomal granules that fade during the maturation process while the cytoplasmic background becomes increasingly electron-dense as hemoglobinization progresses from early to late normoblasts (Fig. 3). More specific attributes of erythroid cells include their close proximity to cytoplasmic processes of macrophages and development of rhopheocytotic invaginations and vacuoles leading to the formation of siderosomes, or ferritin-rich and membrane-bound cytoplasmic inclusions (Figs. 4 and 5).

Anemias

One seldom needs to study by electron microscopy blood cells or marrow samples of patients with anemias, with the rare exceptions of ringed sideroblastosis of refractory or unexplained anemias, anemias with bizarre inclusions such as Heinz bodies, and the even rarer dyserythropoietic anemias.

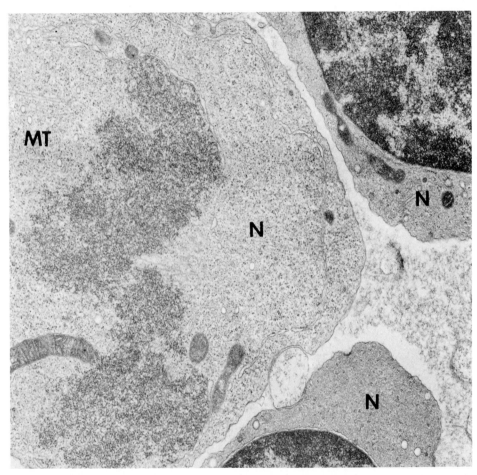

Figure 3. Electron micrograph of field of bone marrow showing three normoblasts (N) in various stages of hemoglobinization. The largest and earliest normoblast is in mitosis (MT) and shows the least dense cytoplasmic background, suggesting the least hemoglobinization. (×16,744) *(This and subsequent preparations were, unless otherwise indicated, stained with uranyl acetate and lead citrate.)*

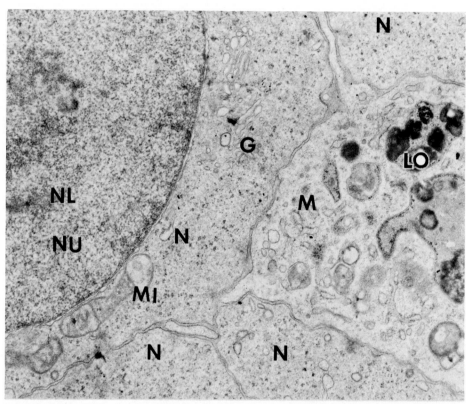

Figure 4. Electron micrograph showing an erythroid islet with macrophage (M), containing lysosomal complexes (LO) and neighboring normoblasts (N). Note the Golgi region (G), mitochondrion (MI), nucleus (NU), and a nucleolus (NL) of an early normoblast. (×16,350)

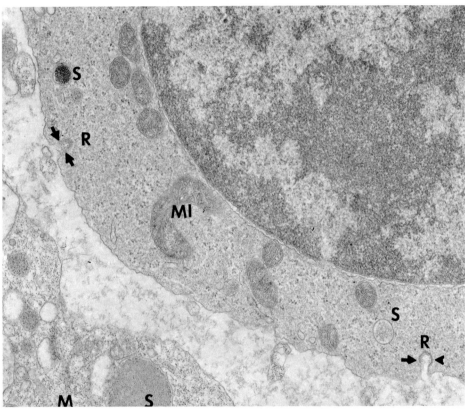

Figure 5. Portion of a normoblast containing mitochondrion (MI) and siderosomes (S). There is evidence of rhopheocytosis (R) between the arrows. Also represented in this field is a macrophage (M) containing siderosomal granules (S). (×31,900)

In *sideroblastic anemias,* it is important to distinguish between the classical "ringed" sideroblasts with dense, amorphous intramitochondrial and intercristal deposits (Figs. 6 and 7) and the presumably less serious and slightly more frequent "ringed" sideroblasts due to increased numbers of siderosomes and often with increased rhopheocytotic activity (Figs. 8 and 9). These two patterns of sideroblastosis cannot be distinguished by light microscopy alone. It is apparent that sideroblastic anemias constitute a heterogeneous group of etiologically unrelated disorders. It is probable that the two patterns of "ringed" sideroblasts represent at least two different pathophysiological processes, each with its own prognostic significance (11).

The megaloblastosis of *pernicious anemia* or *folate deficiency* offers the expected picture of a hemoglobinized cytoplasm, the nucleus being relatively arrested in its maturation process (Fig. 10). This is applicable not only to erythroid cells but less noticeably to granulocytic and megakaryocytic cell lines, as well as to nonhematopoietic cells.

An interesting observation that is particularly applicable to severe *anemias of renal failure* is the development of cytoplasmic processes or pedicles in normoblasts. These processes are characterized by their content of parallel, aligned microtubules, often juxtaposed with histiocytic processes (Fig. 11). Similar findings are observed in severe *iron deficiency anemia,* which characteristically lacks siderosomes in normoblasts and ferritin-rich phagolysosomes in macrophages (Fig. 12).

Sickle cells have been observed to contain crystalloid structures, 150 Å across, arranged parallel to the axis of sickled cells provided that osmiation is omitted, osmic acid being a strong oxidizing agent (12). Erythroid cells in *thalassemia,* on the other hand, contain dispersed ferritin and dense cytoplasmic inclusions, thought to represent precipitates of either excess hemoglobin peptide chain or denatured hemoglobin (13). Still denser and often angular inclusions are observed in *Heinz bodies* (3).

Polycythemia Vera

The normoblasts in polycythemia vera, though greatly increased in number, are as a rule shockingly normal in their appearance (Fig. 13). In the evolution of polycythemia vera, the marrow may increasingly show a marked granulopoiesis and megakaryocytosis.

Erythremia and Di Guglielmo's Disease

Erythremias are considered to be neoplastic proliferations of erythroid cells as a single or predominantly single cell line, or as an erythroleukemia or classical Di Guglielmo's disease. In both situations, erythroid precursors may appear obviously pathological. Giant and multinucleated pronormoblasts are noted. There is an exaggerated megaloblastic tendency (Fig. 14). Although nonspecific, PAS-positive and glycogen-rich deposits are often present within the cytoplasm of these abnormal erythroid cells (14).

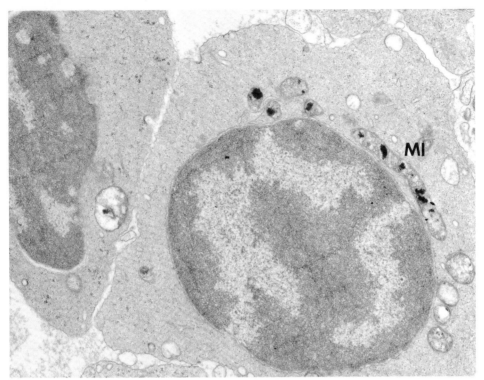

Figure 6. Ringed sideroblasts with mitochondria (MI) containing dense siderophilic deposits. (×20,240)

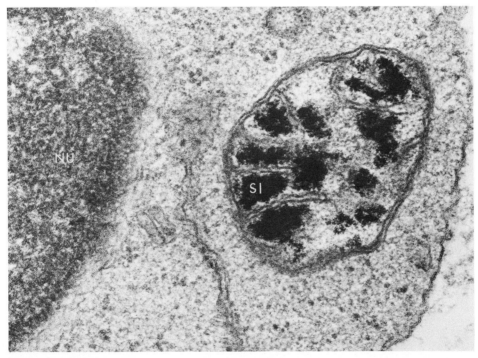

Figure 7. Higher magnification of a field in ringed sideroblast showing mitochondrion with siderophilic deposits (SI) between mitochondrial cristae. (×98,600)

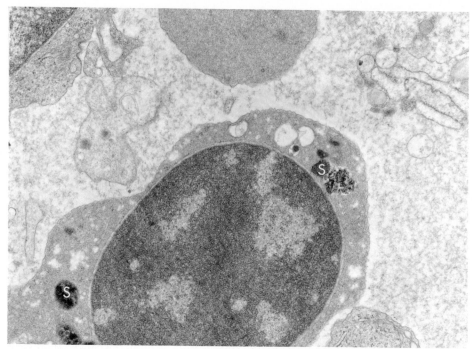

Figure 8. Ringed sideroblast with dense deposits within siderosomes (S), and not within mitochondria. (×20,240)

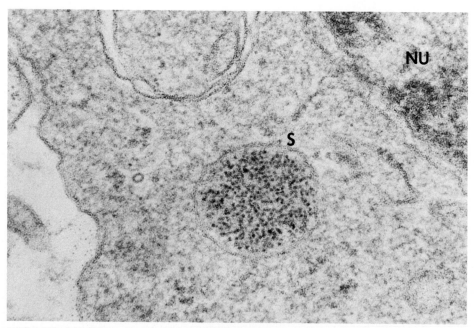

Figure 9. Higher magnification of a field of ringed sideroblasts with siderosomal deposits of powdery ferritin-like deposits (S). (×165,300)

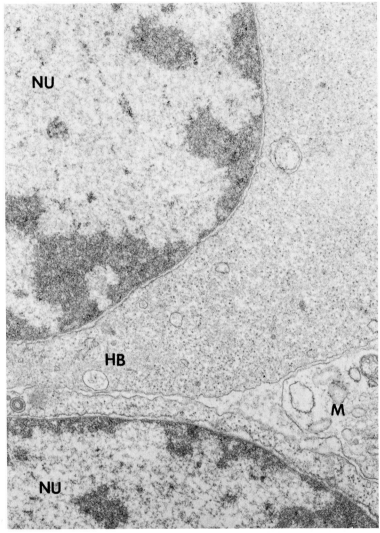

Figure 10. Megaloblast with immature nucleus showing dispersed chromatin, and a cytoplasm with considerable hemoglobinization (HB) for this stage of development. Also represented is a portion of a macrophage (M). (×31,900)

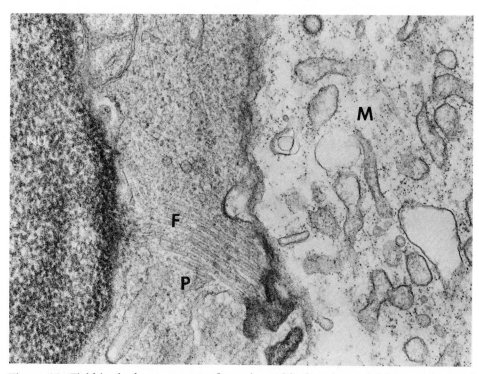

Figure 11. Field in the bone marrow of a patient with chronic renal failure and treated with hemodialysis. The patient was recently injected with an iron dextran preparation. A normoblast to the left shows a pseudopod formation (P) containing parallel microtubules. A macrophage (M) contains numerous fine iron dextran particles. (×62,500)

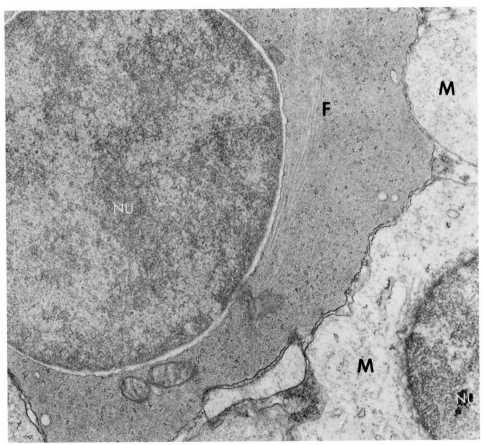

Figure 12. Normoblast with irregular borders and containing fibrils (F). Note absence of dense iron particles in either normoblast or adjacent macrophages (M). No demonstrable iron was found in the paraffin section stained with prussian blue reaction. (×27,600)

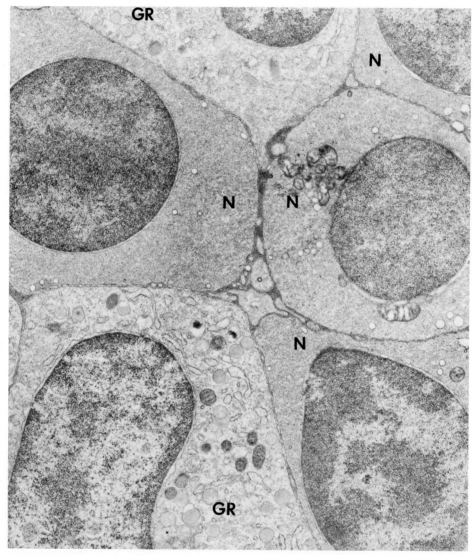

Figure 13. Field of bone marrow showing four normoblasts (N) together with two neutrophilic granulocytic precursors (GR). From a patient with polycythemia vera. (×13,500)

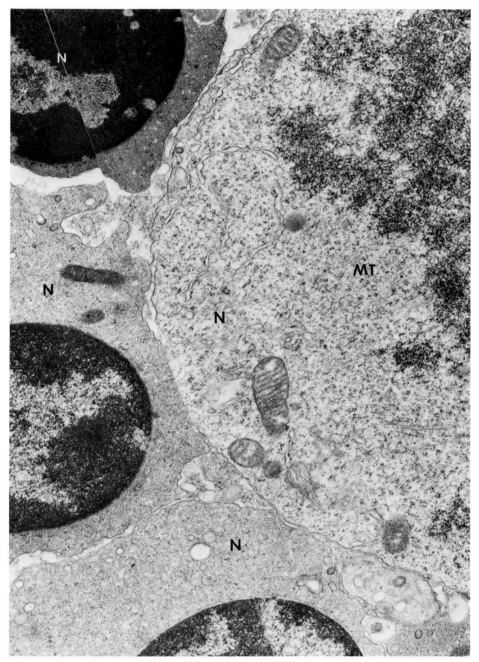

Figure 14. Field of bone marrow showing four normoblasts (N) in various stages of development; the largest normoblast is in mitosis (MT). From a patient with Di Gugliel-mo's disease. (×26,390)

NEUTROPHILIC GRANULOCYTES

Neutrophilic granulocytes constitute a large proportion of marrow cells, and they are the commonest leukocytes observed in the peripheral blood of normal adults. Human marrow and peripheral blood buffy coat preparations appear quite suitable for electron microscopic as well as enzymatic studies; yet our current information on neutrophilic granulocytes and their granules continues to be largely based on observations made on rabbit polymorphonuclear leukocytes of peritoneal exudate (15) and rabbit bone marrow (16).

The major single indication for the ultrastructural study of human leukocytes in the blood or marrow from a practical diagnostic standpoint has been the determination of the type of leukemia, particularly in acute and seemingly unclassifiable forms. In order to contribute effectively to this difficult field, the electron microscopist should be well acquainted with the process of granulopoiesis, as well as with the fine structure and histochemical attributes of granulocytes, monocytes, and the various lymphocytes.

It is doubted at present that all three granulocytic cell lines (i.e., neutrophils, eosinophils, and basophils) derive from a common ancestor, an indeterminate myeloblast or promyelocyte in which azurophilic or primary "nonspecific" granules differentiate in later forms into "specific" or secondary granules. It appears more probable that as soon as granules are detected in an early granulocytic precursor, there has already been a commitment on the part of that cell to differentiate irrevocably along the path of neutrophilic, eosinophilic, or basophilic cell lines. In man, there has been, to our knowledge, no documented observation of a "hybrid" granulocyte, such a hypothetical "eosino-neutrophil," or a "baso-eosinophil." On the other hand, there is considerable merit for the theory that granulocytes, erythroid cells, and megakaryocytes share a common pluripotential hemopoietic stem cell. This theory is supported by the frequent observation in chronic myeloid leukemia and in a variety of myeloproliferative disorders of a proliferative activity of all three major hemopoietic cell lines. The Philadelphia or Ph[1] chromosome marker, when present, has been observed not only in granulocytic leukemic cells but in erythroid and megakaryocytic precursors (17). Lymphoid and fibroblastic cells do not, however, appear to share the Ph[1] abnormality commonly observed in chronic myeloid leukemia (18,19).

The Granulocytic Series

Let us now focus on neutrophilic granulopoiesis and review the classical observations of Bainton and Farquhar (16,20,21) in rabbit bone marrow polymorphonuclear leukocytes (PMNs). Two distinct types of rabbit PMN granules were characterized by light and electron microscopy, and cytochemistry. These were azurophilic and specific granules that differed from each other on the basis of size, density, origin in relation to the Golgi region, and chemical content. The large, dense *azurophilic granules* measured about 800 mμ and appeared early in the promyelocyte from the inner or concave face of the Golgi complex (16). These correspond to a type of primary lysosome rich in peroxidase as well as other lysosomal enzymes (20, 21). The smaller and less dense *specific granules* are produced at a later stage, and are formed from the outer or convex surface of the Golgi region. They contain alkaline phosphatase and lack lysosomal enzymes (20,21). Azurophilic granules cease to be formed after the myelocyte stage and are reduced in number by mitotic divisions, whereas specific granules continue to accumulate during PMN maturation. The mature rabbit PMN contains thus 75% specific granules and 25% azurophilic granules. The possibility of the presence of a third type of granule in rabbit PMNs has also been raised.

It has also been pointed out by Watanabe et al. (22), Scott and Horn (23), and Spitznagel (24), among others, that the granules of human neutrophils may be considerably more heterogeneous than rabbit PMNs in their fine structure as well as in their chemical constituents. In dealing with human neutrophils, one needs to be particularly careful in associating type of granule with specific enzyme activity. There is also a general awareness that the classification of human neutrophilic granules into azurophilic (primary or type A) and specific (secondary) granules may be too simplistic. An additional form of neutrophilic granule with layered or crystalloid structure has been also described (22,25). The fine structure of granules may also differ according to the technique of fixation and staining used; thus, more elements of confusion are added.

The developmental stages of neutrophilic granulocytes as studied by light microscopy in preparations stained with Romanovsky dyes need not be reviewed here. The general indices of nuclear size and shape, chromatin pattern, and prominence of nucleoli, as well as characteristics of cytoplasmic granules and other organelles serve also in the analysis of granulocytes by electron microscopy. The maturation sequence does not obviously "leap" from stage to stage but progresses subtly from a poorly defined myeloblast to a highly differentiated and easily recognized mature segmented neutrophil. Because of this subtle maturation, and in view of the inherent difficulty in ultrathin sections in obtaining complete hemispheric sections passing through both Golgi area and nucleoli of a sizable percentage of cells, it is desirable to classify neutrophilic granulocytes into three major divisions, each with its own nuclear and cytoplasmic features (Table 2). The same approach may be used with other hematopoietic cell lines.

Early neutrophils comprise myeloblasts and promyelocytes. Traditionally, a myeloblast would be considered by light microscopy an undifferentiated hemopoietic cell with no evidence of granule formation, yet acceptable as a granulocytic precursor because of its association with differentiated granulocytes. By electron microscopy, what is considered conventionally to be a myeloblast may demonstrate a number of granules of varying size and density (dense, intermediate light; compact or flocculent) within or in proximity to the Golgi area, and less often elsewhere in the cytoplasm (Fig. 15). Promyelocytes exhibit an obvious granulation throughout the cytoplasm (Fig. 16). Although this interpretation is derived partly from observations in cases of acute myeloid leukemia and a case of granulocytic sarcoma, findings in normal marrow and in granulocytic hyperplasia suggest that granules that are missed on light microscopy, even in 1μ-thick sections, may be readily observed in early neutrophils by electron microscopy. Other features of early neutrophils that help to set them apart from early normoblasts include a tendency to slight invagination or flattening of the nucleus in the region of a well-developed Golgi area, presence of a prominent nucleolus or nucleoli away from the nuclear membrane, and segments of rough endoplasmic reticulum.

Table 2. Classification of Neutrophilic Granulocytes

Stages of Development	Nucleus	Nucleoli	Golgi Area	Dense Granules [a]	Light Granules [b]	Other Organelles
Early granulocytes (myeloblasts, promyelocytes)	Euchromatin, or dispersed chromatin	Prominent, complex, often several	Prominent, site of first granules	Absent or rare in "blasts," to many in promyelocytes; usually large (800 mμ)	Absent or rare in "blasts," to few in promyelocytes; variable in size	Polyribosomes abundant; granular reticulum gradually increases in promyelocytes; mitochondria prominent
Intermediate granulocytes (myelocytes, metamyelocytes)	Gradual chromatin condensation, and indentation	Less prominent	Less prominent	Diminish with cell division	Increase in number (300–500 mμ)	Polyribosomes and granular reticulum diminish; mitochondria relatively few and small
Late granulocytes (bands, segmented neutrophils)	Progressive lobation and increased chromatin condensation	Absent	Rudimentary	Minority	Majority; with maturation, granules condense	Polyribosomes and granular reticulum disappear; mitochondria rare; glycogen accumulates

[a] Dense or primary granules correspond to azurophilic granules. In the rabbit, these are rich in myeloperoxidase (21).
[b] Light or secondary granules correspond to specific neutrophilic granules. In the rabbit, these are rich in alkaline phosphatase (21).

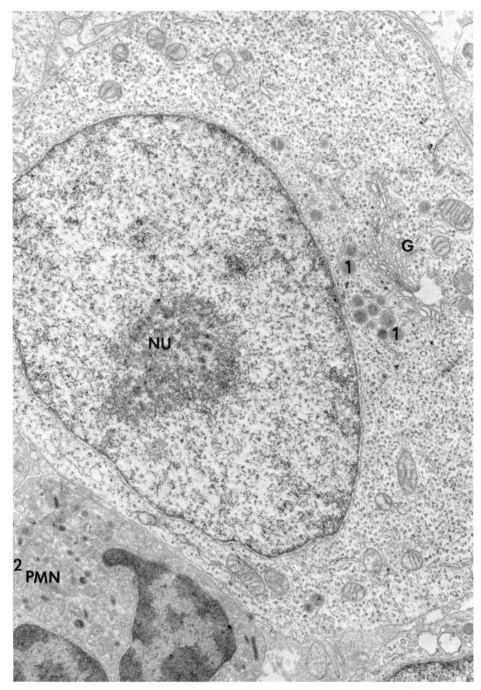

Figure 15. Early granulocytic precursor with Golgi region (G) and presence of primary or azurophilic granules (1) located mainly within the narrow zone between Golgi region and nucleus. Also note polymorphonuclear leukocytes (PMN) with beginning formation of secondary or specific neutrophilic granules (2). (×12,125)

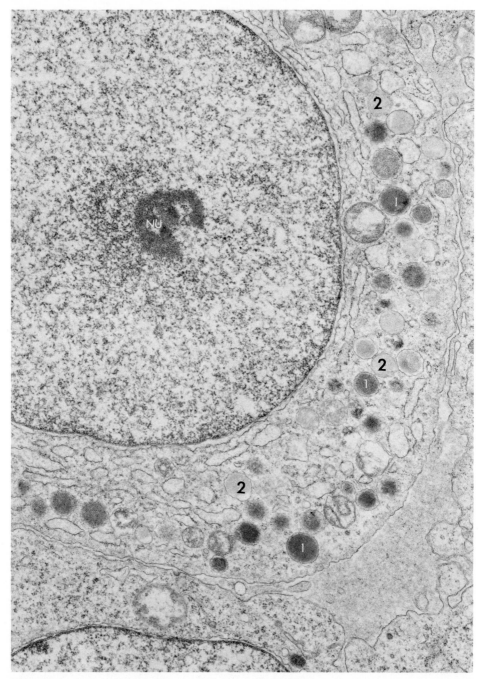

Figure 16. Early granulocyte precursor, probably a promyelocyte, showing both primary dense staining granules (1) and secondary or specific neutrophilic granules (2). (×25,390)

Intermediate neutrophils include myelocytes and metamyelocytes. In these, the characteristic euchromatin or the dispersed "open" chromatin pattern of early granulocytes begins to condense and the nucleoli become less prominent. Mitochondria decrease in number. Secondary or specific granules, recognizable by their smaller size and low density, some with a crystalloid structure, become more evident and gradually predominate over the larger and often denser primary granules (Figs. 17 and 18). The latter are markedly reduced in number and probably stop being formed in metamyelocytes. We have had considerable difficulty in distinguishing monocytes from partly degranulated and vesiculated metamyelocytes (see below).

Late neutrophils include nonsegmented (bands) and segmented neutrophils. The process of chromatin condensation and margination continues. The Golgi area atrophies. Secondary granules of low density predominate over dense granules. Both light and dense granules tend to be smaller than their counterparts in early granulocytes. Dumbbell and oblong dense granules, some with a crystalloid structure, may be observed. (Figs. 19 and 20). The nuclear segments or lobules are interconnected with narrow bridges composed of thin chromatin strands lined by nuclear membrane.

Congenital Anomalies

The various types of inherited variations in leukocytes have been reviewed by Davidson (26). Some of these anomalies have been the subject of electron microscopic observations and will be briefly mentioned here.

The *Pelger-Huët anomaly* is characterized by a failure of morphogenesis of nuclear lobes of neutrophils with no apparent functional impairment (26). Homozygotes have mature neutrophils with a rounded nucleus containing coarsely clumped chromatin. The neutrophils of heterozygotes have the characteristic pince-nez bilobed nucleus.

The *May-Hegglin anomaly* is associated with the triad of large intracytoplasmic, RNA-rich bodies known as Döhle bodies, giant platelets, and thrombocytopenia with hemorrhagic manifestations (27). The Döhle bodies consist of clusters of granules of about 200 Å together with free fibrils of 50 Å in diameter (27,28).

The *Alder-Reilley anomaly* of neutrophils and other leukocytes is one of several conditions resulting from a disturbance of polysaccharide metabolism and associated with "giant" granules (29).

The rare *Chediak-Higashi syndrome* is often fatal at an early age and is usually associated with partial occulocutaneous albinism, susceptibility to viral and bacterial infection, and a high risk of leukemia and lymphoma (30,31). Circulating neutrophils and other leukocytes, as well as a variety of somatic cells, contain giant lysosomal granules that appear to have adequate bactericidal activity but are more stable than normal granules during phagocytosis (32,33).

Acquired or Functional Changes

The primary function of neutrophils is phagocytosis, which appears to take place in an orderly sequence of events (34): *a)* Pseudopods form and surround the phagocytosed particle; *b)* A phagocytic vacuole or phagosome is thus created; *c)*

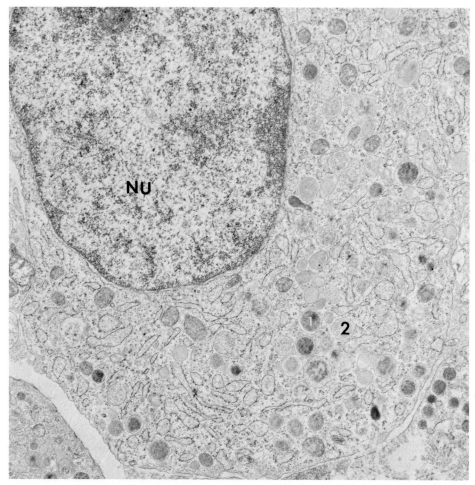

Figure 17. Intermediate granulocyte precursor, probably a myelocyte, showing presence of primary- and secondary-type granules together with a fair amount of rough endoplasmic reticulum. (×16,530)

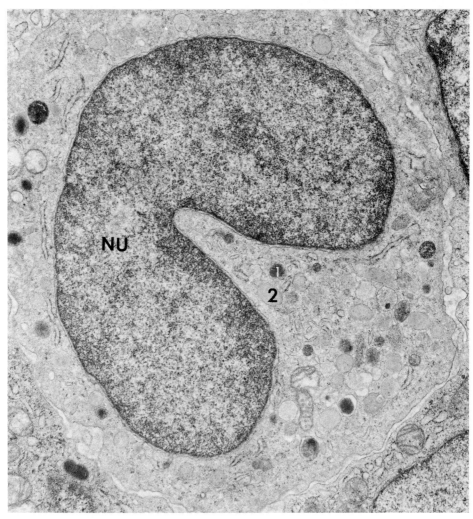

Figure 18. Intermediate granulocyte precursor, a metamyelocyte, showing predominance of light-staining secondary-type granules (2). (×16,530)

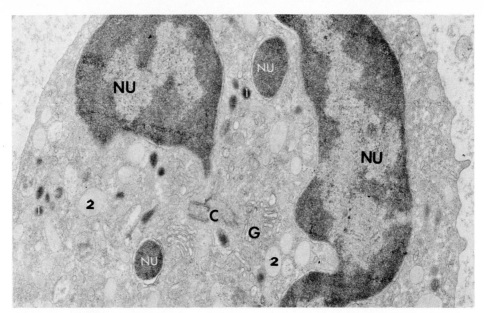

Figure 19. A late granulocyte, a polymorphonuclear leukocyte, showing a well-developed Golgi region (G) and centriole (C). There are persistent primary or azurophilic granules (1). These are much smaller than in earlier or intermediate granulocytes. The predominant granules are of the specific neutrophilic type (2). (×26,390)

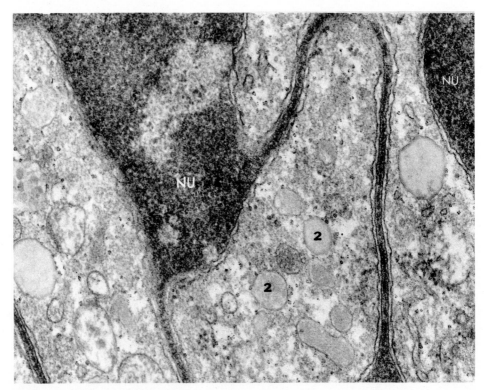

Figure 20. Higher magnification of a segmented neutrophil showing long, tortuous, and slender communicating bands between nuclear lobes. Also note the presence of specific neutrophilic granules (2). (×72,500)

A process of degranulation occurs as lysosomal granules meet with phagocytic vacuoles and discharge their content of hydrolytic enzymes within the vacuoles; *d)* Coalescence takes place not only between lysosomes and phagosomes, resulting in phagolysosomes, but also between newly formed phagolysosomes. Residual bodies or complex bodies are then formed prior to cell death.

These sequences of phagocytosis are readily observed in man in the phenomenon of platelet satellitosis or platelet neutrophil adherence (35,36), which is followed by pseudopod formation, engulfment of platelets, degranulation of the phagocytic neutrophil or monocyte by formation of phagolysosomes and eventual digestion of phagocytosed platelets (Figs. 21 to 24).

In another form of degranulation, seen where neutrophils are exposed to bacterial toxins, there is random rupture of lysosomes and direct release of lysosomal contents into the cytoplasm in a manner of "suicide bags" (37).

The "toxic" granules seen under the light microscope have been interpreted as phagocytosed material (34), or as large or immature azurophilic granules (38). Döhle bodies are known to contain RNA and were found to contain aggregates of rough endoplasmic reticulum (38) which presumably persist in neutrophils when myelocytes have been exposed to toxic agents.

Patients with bacteremia tend to have in the marrow immature neutrophils with numerous dense primary (azurophilic) granules and aggregates of rough endoplasmic reticulum (38). In the peripheral blood, the "shift to the left" is characterized by an outpouring of younger neutrophils, some with Döhle bodies, "toxic" granules, and vacuoles, the latter probably representing phagocytic vacuoles.

Leukemias and Myeloproliferative Disorders

It is generally assumed, but far more proved, that leukemic cells are metabolically and morphologically transformed cells that should be recognized as deviants from their normal counterparts. Another approach, less orthodox but perhaps as tenable as the former, is that leukemia results from an accumulation of a hemopoietic subpopulation that reflects a maturation arrest or a lack of response to normal regulatory mechanisms. Regardless of views on the nature of leukemias, these are still empirically classified by means of cytological methods based on Romanovsky dyes. With the availability of an increasing number of therapeutic regimens "tailored" for specific morphological types and even subtypes of leukemia, the pathologist is under great pressure to provide a classification that is biologically relevant or that could be related to a statistically predictable pattern of response to therapy. Thus, it is becoming increasingly important to define the commoner leukemias as acute or chronic, and as myeloid, monocytoid, or lymphoid. This can be resolved by traditional cytological techniques in the vast majority of chronic leukemias. "Second level" methods, including histochemical, electron microscopic, chromosomal, and immunochemical studies, are now resorted to only in acute or "unclassifiable" leukemias and in certain chronic leukemias presenting unusual problems of clinical management or with morphologically "atypical" forms (i.e., hairy cells, cerebriform Sézary cells, immunoblasts). A partial listing of the more widely used "second level" methods appears in Table 3.

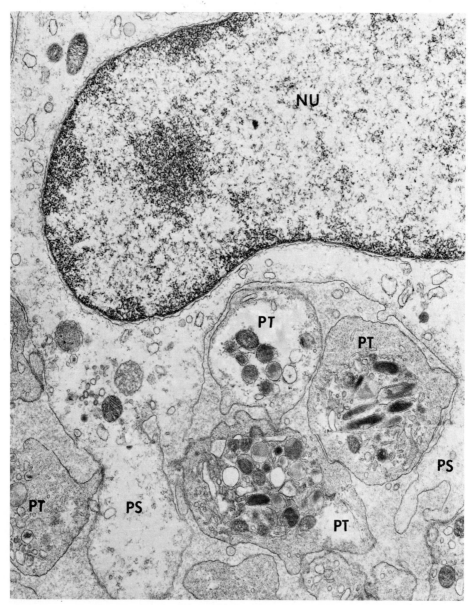

Figure 21. Buffy coat preparation showing neutrophil with two pseudopods (PS) engulfing three platelets (PT). Another platelet is adherent to the left side of the neutrophil. Note paucity of neutrophilic granules. (×26,390)

74

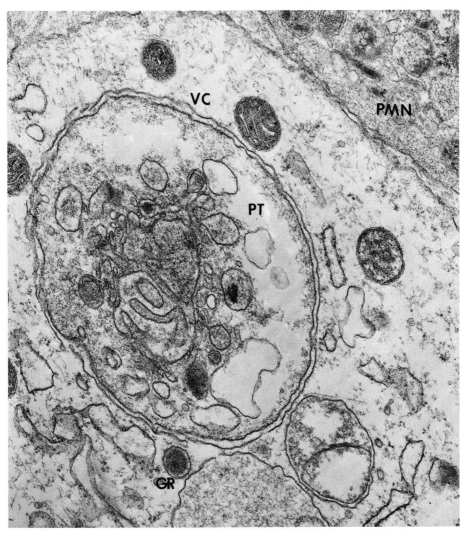

Figure 22. Portion of neutrophil with phagocytic vacuoles (VC) containing engulfed platelet (PT). Note proximity of lysomal granule (GR). Another polymorphonuclear leukocyte (PMN) is seen in the right upper quadrant . (×55,100)

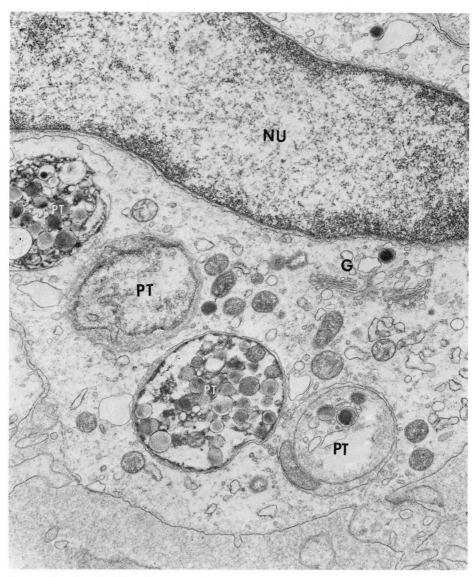

Figure 23. Neutrophil engulfing four platelets (PT) in various stages of degeneration. Note Golgi region (G). (26,390)

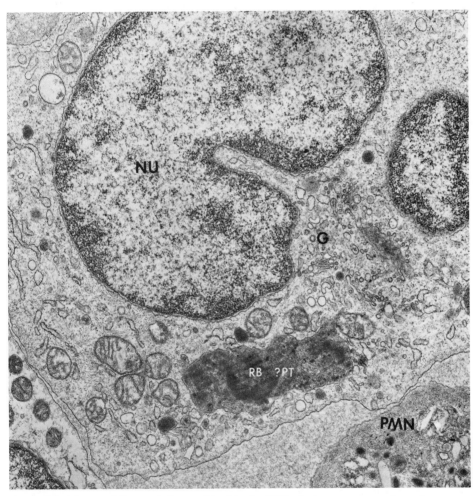

Figure 24. Neutrophil containing residual body (RB) of what appears to be a degenerating platelet (PT). (×26,390)

Table 3. Methods Suggested in the Classification of Acute Leukemias (7, 9)

	ALL	AML	CML (Blast Crisis)	APL	AMoL
Cytochemistry					
Peroxidase	–	+	++	++	±
Naphthol AS-D chloroacetate	–	+	++	++	–
Naphthol AS-D acetate	±	+	+	++	++
Naphthol AS-D acetate; NaF.	no inhibition	no inhibition	no inhibition	no inhibition	inhibition
PAS	0–+ (> 50%)	±	±	±	±
Electron microscopy (granules)	–	(1°, ? 2°) Auer rods	1° and 2°	(1° and 2°) Auer rods	(?1° only) Auer rods
Ph¹ chromosome			+ (85%)		
Lysozyme (urine, serum)			±		++
B₁₂ binding capacity (transcobalamin I)		±	+	+	?

Key: ALL = acute lymphoid leukemia; AML = acute myeloid leukemia; CML = chronic myeloid leukemia; APL = acute promyelocytic leukemia; AMoL = acute monocytoid leukemia.

In dealing with the difficult problem of classification of *acute leukemias,* some laboratories have favored cytochemical reactions (9,39). We have had more experience with electron microscopy, which, in an ongoing unit, can provide in our opinion results that are easier to interpret and faster to obtain than by cytochemistry. The key question raised in relation to chemotherapy of acute leukemias, especially in adults, is simply whether the leukemia is myeloid (granulocytic), or nonmyeloid and, therefore, presumably lymphoblastic. In our hands, this question can be answered with a considerable degree of certainty by electron microscopic studies of buffy coat preparations and of bone marrow samples. The presence of an appreciable number of lysomal granules within the Golgi area and the rest of the cytoplasm is strong indication of granulocytic differentiation (Figs. 25 and 26). A tendency to nuclear indentation in the Golgi area, a moderate number of granules of similar size and electron density, and the presence of tell-tale Auer bodies add more support to this interpretation. It is doubted whether other methods, such as cytogenetics and cell culture, offer much in the classification of acute leukemias. The use of xenogeneic innoculations of human leukemic cells in athymic "nude" (40,41) or athymic-asplenic mice (42), maintained under specific pathogen-free conditions, may prove to be useful in characterizing the dominant or more aggressive leukemic population. "Nude" mice bearing xenogeneic innoculates of leukemic cells are also convenient hosts for therapeutic trials.

The distinction between *acute myeloid leukemia* (AML) and *acute monocytoid leukemia* (AMoL) may be impossible to resolve by morphological means alone. Ideally, both cytochemical and electron microscopic methods should be applied to this problem, but this is costly and probably of limited practical value. Chemical markers such as serum vitamin B12 binding capacity, mainly transcobalamin I, which is largely derived from granlocytic cells (43), and serum or urine lysozyme or muramidase (44) may be conveniently used in differentiating myeloid from monocytoid leukemias (45,46).

Acute promyelocytic leukemia (APL) is characterized by an abundance of early granulocytes filled with primary-type (azurophilic) granules and occasional Auer bodies (47,48). Rod-shaped granules and splinter forms progressing into Auer bodies have to be described in this type of leukemia (47). It is important to single out APL from AML because of the known association of APL with disseminated intravascular coagulation and a poor response to chemotherapy.

Auer bodies are intracytoplasmic needle-like inclusions with either a lamellar structure or a form of periodicity. Like azurophilic or primary granules, they display a generally strong peroxidase activity. Auer bodies are encountered most commonly in acute promyelocytic leukemia, but they have been also described in acute myeloid and acute monocytoid leukemia (48).

The ultrastructural features of *chronic myeloid leukemia* (CML) permit us to review cell characteristics that were already broached in discussing the granulocytic series. Both bone marrow and peripheral blood contain a vast number of predominantly neutrophilic granulocytes undergoing a seemingly orderly maturation, but resulting in a preponderance of early granulocytes rather than nonsegmented and segmented neutrophils (Figs. 27 and 28). The overall impression is a "shift to the left" but without a background of sepsis or known toxicity. That CML cells are genetically transformed cells is, however,

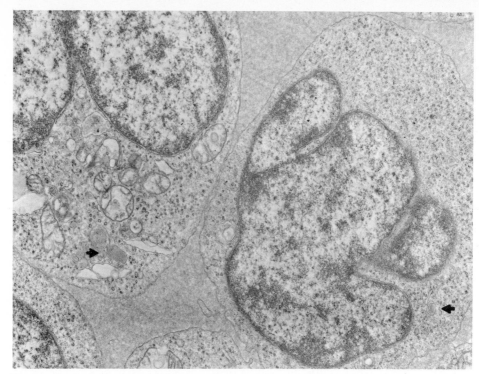

Figure 25. Peripheral blood buffy coat preparation showing two "blast" cells with early granulocytic differentiation. Granules are shown at arrows. From a patient with acute leukemia showing early myeloid differentiation. (×13,064)

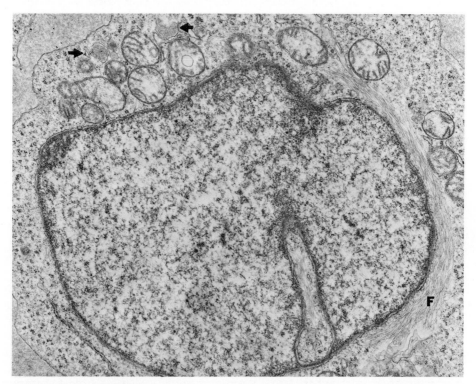

Figure 26. Same preparation as above. Note granules at arrow. A bundle of filaments is also shown (F). (×20,240)

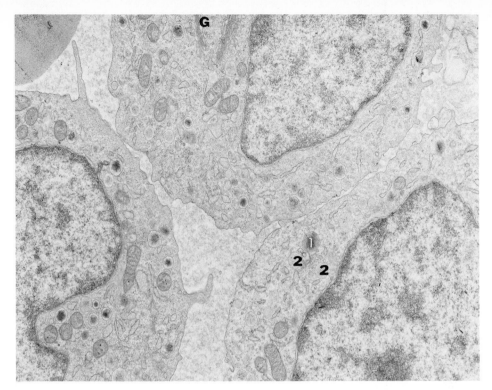

Figure 27. Bone marrow field showing three intermediate neutrophilic granulocytes. Note primary (1)- and secondary (2)-type granules. G represents the Golgi region. From a patient with chronic myeloid leukemia. (×13,064)

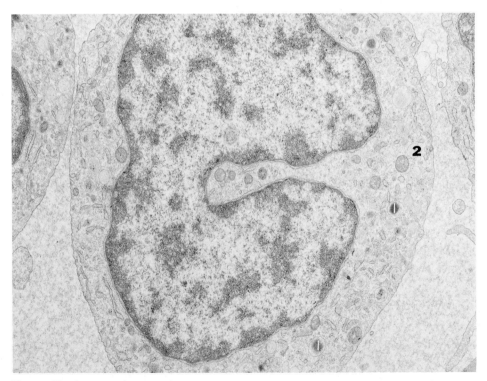

Figure 28. Same patient as above. A metamyelocyte with primary-and secondary-type granules. (×13,064)

suggested by the presence of Ph[1] chromosome marker in myeloid cells of about 85% of cases (49). CML cells exhibit a strong peroxidase reaction that is shared by most normal and leukemic granulocytes. Their alkaline phosphatase reaction is, however, weak, reflecting perhaps a decreased number of late granulocytes with secondary-type granules (50).

During the natural course of CML, there often develops an accelerated phase characterized by blastoid transformation and bone marrow failure. This *blast cell crisis* of CML may be difficult to distinguish from forms of acute leukemia, particularly in the absence of previous laboratory and clinical data. In general, the blastic phase of CML continues to exhibit features of granulocytic differentiation (Fig. 28); however, there have been instances in which totally undifferentiated cells or cells resembling lymphoblasts have been described (51).

Granulocytic sarcomas (myelosarcomas, chloromas) may or may not be associated with a leukemic component. Some granulocytic sarcomas have been associated with osteolysis (52,53). In one instance, because of the finding of lytic lesions and a collapse of L1, and poor interpretation of bone marrow sections and smears, the diagnosis of "nonsecretory" multiple myeloma was entertained until a marrow sample was studied by electron microscopy. Many "blasts" were readily recognized as granulocytes; others were undifferentiated (Fig. 29 and 30). The granulocytes were peroxidase-positive (Fig. 31). The Leder stain, done on paraffin sections, was likewise positive.

Myeloproliferative disorders encompass a constellation of conditions that includes polycythemia vera, chronic myeloid leukemia, essential thrombocythemia, myelofibrosis and osteosclerosis, and agnogenic myeloid metaplasia (54). All these conditions share a proliferative tendency of a multipotential hemopoietic cell that differentiates into a predominant cell line or multiple cell lines, including marrow stromal elements. Unlike the situation in most cases of chronic myeloid leukemia, the granulocytes in other forms of myeloproliferative disorders lack the Ph[1] chromosome marker and show an increased alkaline phosphatase activity. The evolution of polycythemia vera into a predominantly granulocytic myeloproliferative disease has been mentioned earlier (Fig. 32).

EOSINOPHILS

Eosinophils are granulocytes with coarse and distinctive granules. When viewed by electron microscopy, these granules are characterized in man by the presence of a densely staining transverse crystalloid bar, *the internum,* which is surrounded by a lighter matrix, the *externum* (Figs. 33 and 34). The crystalloid bars are more sharply delineated in maturer cells than in eosinophilic myelocytes. Still earlier eosinophilic precursors are difficult to recognize, but are suspected because of their large, dense, "azurophilic" primary granules showing a beginning of central condensation. *Charcot-Leyden crystals* are thought to be derived from degraded and recombined eosinophilic crystalloids (55). Eosinophilic granules are rich in lysomal enzymes; they are particularly rich in a peroxidase enzyme that differs antigenically and spectrophotometrically from that of neutrophils. Little is known about the properties or significance of the crystalloid bars.

Eosinophils are phagocytes and are capable of ingesting a variety of particles, including bacteria. Their most characteristic attribute seems to be their attraction

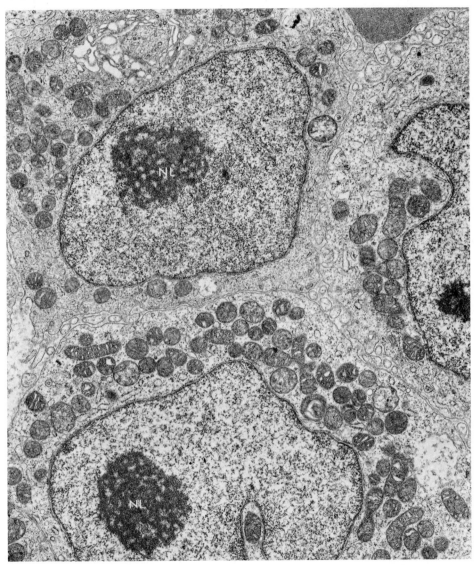

Figure 29. Bone marrow field of granulocytic sarcoma with three undifferentiated blast cells with no appreciable granulocytic differentiation. Note well-developed nucleolus (NL) and abundance of mitochondria. (×13,050)

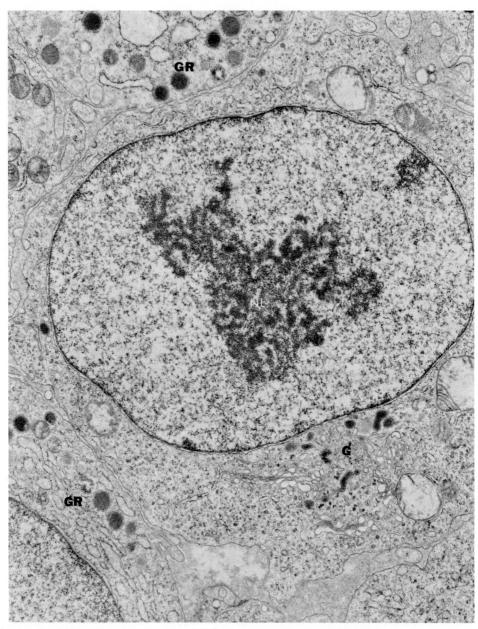

Figure 30. Same bone marrow preparation as above. This is an early granulocytic precursor, probably at the level of a myeloblast or early promyelocyte, showing primary type of granules in the region of Golgi (G). (×12,750)

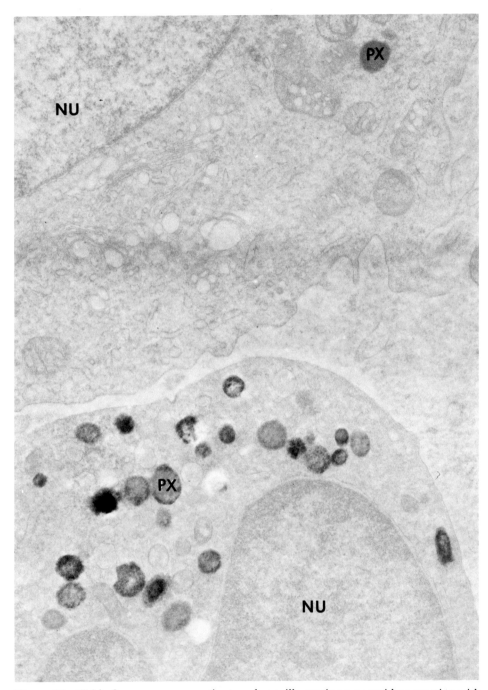

Figure 31. Field of marrow, same patient as above, illustrating a peroxidase reaction with one positive granule in the upper cell and multiple positive granules in the lower cell. The peroxidase-positive granules are marked PX. (×31,900)

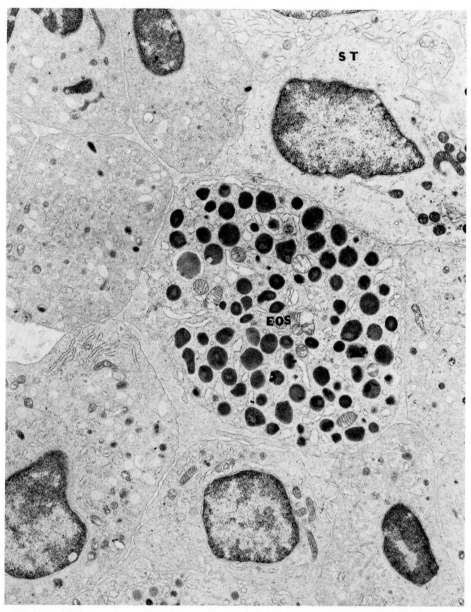

Figure 32. A bone marrow field showing crowded granulocytic cells that are all polymorphonuclear leukocytes except for an eosinophilic precursor (EOS) in the center and a stromal cell (ST). From a patient with myeloproliferative disorder. (×7,500)

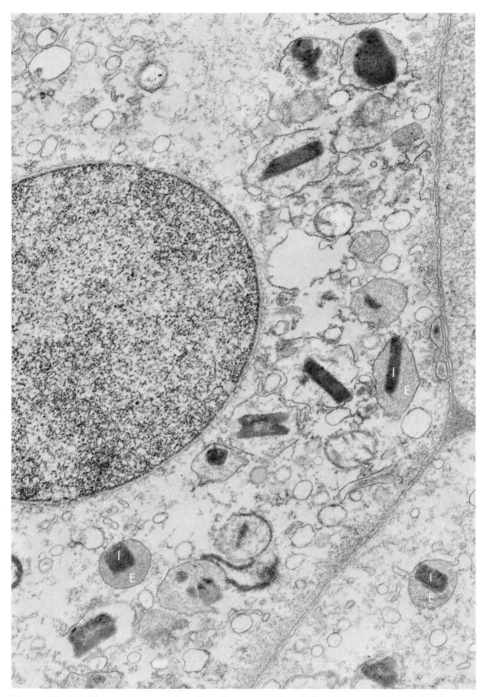

Figure 33. Buffy coat preparation showing two eosinophils. The specific grnules demonstrate a central crystalloid bar (I) and an external less dense matrix (E). From a patient with "hypereosinophilic syndrome." (×39,900)

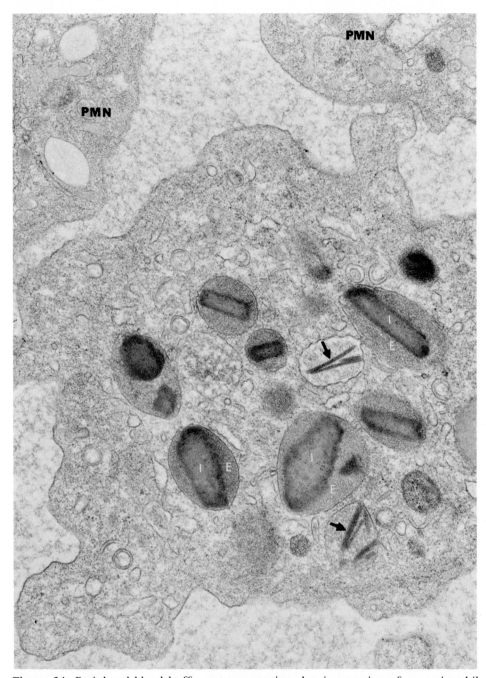

Figure 34. Peripheral blood buffy coat preparation showing portion of an eosinophil with formation of crystalloid bars (I) in various stages of development, including early splinter forms (arrows). From another patient with "hypereosinophilic syndrome." (×43,500)

to and phagocytosis of antigen-antibody complexes (56). It also appears that certain types of antibody responses, particularly those involving reaginic antibody and IgE, have a strong tendency to elicit an eosinophilic response. The reader is referred to Zucker-Franklin's review on eosinophilic function and disorders (57).

The conditions commonly associated with marked eosinophilia are drug reactions and a variety of other allergic conditions, and parasitic infestations. Other conditions associated with significant eosinophilia include bacterial infections, autoimmune diseases, and malignancies, particularly Hodgkin's disease. Eosinophilia can be a significant component of chronic myeloid leukemia and other myeloproliferative disorders (58).

A syndrome of progressive eosinophilia without identifiable underlying cause and bordering on leukemia has received considerable attention in recent years (57,59). This primary *"hypereosinophilic" syndrome* is frequently accompanied by endomyocardial fibrosis with superimposed mural thrombosis. It is suspected that eosinophils play a role in producing this and other organ damage. Zucker-Franklin (57) appears to accept the progressive eosinophilic syndromes as a manifestation of eosinophilic leukemia. She stresses, however, that these rare conditions should have no underlying disease and, unlike the majority of eosinophilic conditions, should not respond to steroids or antihistamines. The eosinophils in progressive eosinophilic syndromes are mature, but their granules may display considerable variation (57).

Eosinophilic leukemia is extremely rare and has been the subject of controversy, principally because it appears to overlap with "collagen" diseases with eosinophilia, Loeffler's fibroblastic endocarditis with eosinophilia, and other hypereosinophilic states (57,58). It is also important to separate "pure" eosinophilic leukemias from chronic myeloid leukemia with a large eosinophilic component. The latter would be expected to have a positive Ph[1] chromosome marker. It should be noted that the eosinophils seen in cases of eosinophilic leukemia, if these are indeed leukemic, are usually mature, although their granules vary considerably in size, shape, and distribution (57). This observer has not to date felt secure enough to make a diagnosis of "pure" eosinophilic leukemia. He has, however, observed conditions of marked eosinophilia in both chronic myeloid leukemia and what may be called a primary hypereosinophilic syndrome (Figs. 33 and 34).

BASOPHILS AND MAST CELLS

There are considerable morphological and functional similarities between the basophils of the blood and marrow and the fixed tissue mast cells. In man, both have strikingly similar coarse metachromatic granules, both are rich in histamine, heparin, and 5-hydroxytryptophan, and both play a role in inflammatory and immune reactions. including anaphylaxis (60,61,62).

Basophils are filled with large granules which on Romanovsky-stained smears appear purplish or dark blue, and often completely cover the nucleus. The metachromasia of these granules is imparted by their acid mucopolysaccharides, mainly by heparin and hyaluronic acid. Extensive cytochemical and biochemical analyses of basophilic granules are not yet available. Basophils are thought to be derived principally, if not solely, from the marrow.

By electron microscopy, basophilic granules tend to be homogeneously dense and have a closely fitting limiting membrane. Occasionally, they are found to have a powdery or finely lamellated internal structure (Fig. 35). A banded or a finely fibrillar pattern is found in granules of both normal basophils and those of chronic myeloid leukemia (22,60). Terry et al. (63) have described, in guinea pig basophils, granules with banded, honeycomb or hexagonal, and rectangular lattice patterns. The functions of basophils and mast cells are poorly understood. They are capable of some sluggish phagocytic activity, and they degranulate with release of histamine and heparin in response to certain stimuli, particularly some antigens and immune complexes (64). IgE appears to sensitize tissues by fixing itself on cells through the Fc portion of its molecules. Sensitized mast cells and basophils release histamine upon addition of specific antigen or antibody to IgE (65,66).

Disease states associated with basophilia include a variety of chronic inflammations, hypersensitivity, and myxedema. High levels of basophils are noted in myeloproliferative disorders, particularly in chronic myeloid leukemia (67). Numerous basophils and mast cells have been identified in tissues involved with Hodgkin's disease (68). We have been impressed by the presence of basophils in the marrow in fields infiltrated by chronic lymphocytic leukemia cells and in lymphomatous infiltrates.

Mast cells of man are readily observed in lymph nodes and in a variety of inflammatory conditions. They are usually located near blood vessels. Ultrastructurally, they are characterized by peripheral filiform processes and large granules that show considerable variation in their internal structure. Scroll-like formations or "curlicues" are frequently found in these granules alongside homogeneously dense or lamellated granules (Figs. 36 and 37), thus raising in our mind some doubt as to the validity of a sharp distinction between tissue mast cell and basophilic granules. Mast cells are markedly increased in urticaria pigmentosa (69) and in systemic mast cell disease (70,71). Two cases of leukemic manifestations of systemic mastocytosis, or mast cell leukemia, have been reported (72,73).

MEGAKARYOCYTES AND PLATELETS

The study of platelets and megakaryocytes has received considerable attention during the past decade, particularly because of the increased and effective use of platelet transfusions, and the possibility of storing platelets by freezing. The reader is referred to the monograph of Baldini and Ebbe (74) and the review of White (75) for a background on newer concepts in platelet function and morphology.

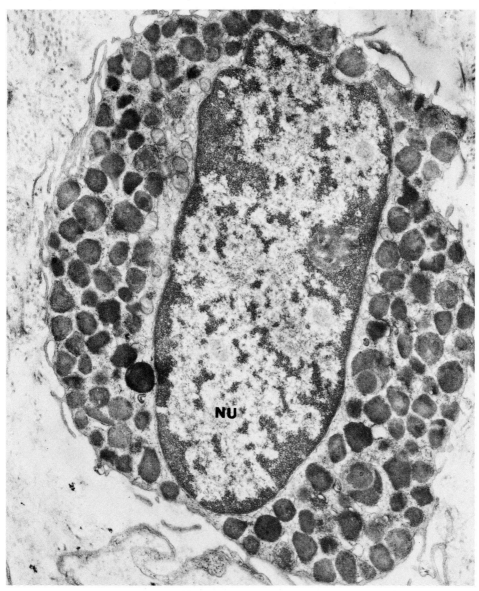

Figure 35. Basophil in lymph node. Note the dense staining granules, which show a crowded powdery appearance with some lamination. (×16,800)

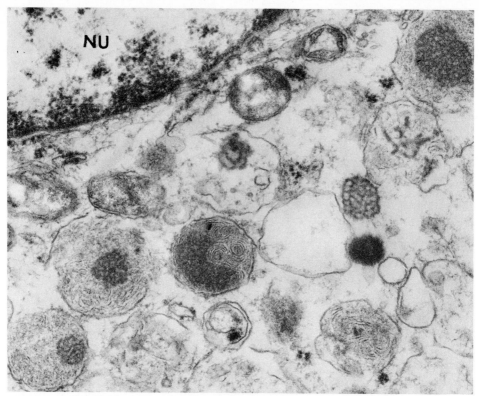

Figure 36. Portion of a mast cell showing granules contining scroll-like formations. Note the tremendous variation in size and internal structure. (×47,000)

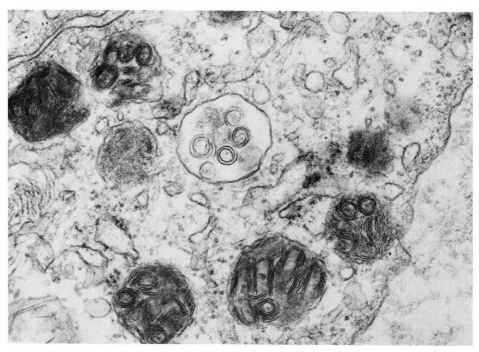

Figure 37. Portion of a mast cell demonstrating the scroll-like and intricate internal structure. (×72,500)

Megakaryocytes of adults are observed principally in the bone marrow, and far less frequently in pulmonary alveolar capillaries, renal glomeruli, splenic red pulp, and liver sinusoids.

Megakaryocytes undergo a unique maturation process in which multiple cycles of DNA synthesis take place without cell division. When megakaryoblasts reach the level of 8N, they become recognizable from other hemopoietic cells. In addition to their marked polyploidy (up to 32N), megakaryocytes are characterized by an abundance of cytoplasmic volume, the presence of a large number of specific "azurophilic" granules, and an extensive agranular endoplasmic system with demarcation lines that will form the boundaries of future platelets. Megakaryocytic granules frequently contain a dense core or nucleoid that gives the appearance of a "bull's eye" (Fig. 38). There appears to be no need to classify these granules further. Mechanisms of platelet release seem to involve both detachment of peripheral cytoplasmic blebs and disintegration of the cytoplasm along vesicles and cleavage lines formed by the elaborate endoplasmic tubular system.

Platelets or thrombocytes are cytoplasmic fragments of megakaryocytes. Ultrastructurally, they display the same "bull's eye"-type granules, although these tend to vary in number, size, and density. In addition, their cytoplasm contains vesicles, a few mitochondria, glycogen granules, segments of smooth endoplasmic reticulum, and microtubules (Fig. 39). Microfibrils have been isolated in platelets, and these are thought to contain a contractile protein (76). The primary function of platelets appears to be the effective plugging of fine vascular defects and a contribution to clot formation. During platelet aggregation and clot formation, platelets tend to group their granules in their center, thus forming a peripheral clear zone or hyalomere.

Most pathological disorders involving megakaryocytes and platelets appear to be functional, without any known morphological basis. Ultrastructural observations are beginning to accumulate in thrombopathic conditions in which platelet functional abnormalities are characterized by defective release mechanisms. In one 70-year-old man with a mild bleeding diathesis and a chronic, mild nonimmunological thrombocytopenia, the platelets were giant, bizarre, and vacuolated. Half of the daughter's platelets were very large but not vacuolated (77). Giant platelets have been also described in patients with May-Hegglin abnormality (27), and have been observed in children suffering from renal insufficiency, hearing loss, and hemorrhagic tendencies (78).

The phenomenon of platelet satellitism or platelet to leukocyte adherence is one of several causes of spurious thrombocytopenia, and has been mentioned in relation to platelet phagocytosis by neutrophils (see in section on Neutrophils). In one patient, this phenomenon could be transferred to normal blood by means of an IgG fraction obtained from the serum of a patient with platelet satellitosis (70).

Thrombocytosis and megakaryocytic proliferation are well known to occur in the myeloproliferative disorders, including chronic myeloid leukemia. Maldonado et al. (80) observed morphological abnormalities in at least some of the platelets of all five patients with myeloproliferative diseases. These "dysplastic" changes included mainly giant forms, paucity or absence of granulation, disorganization and scarcity of microtubules, and hypertrophy of the tubular and

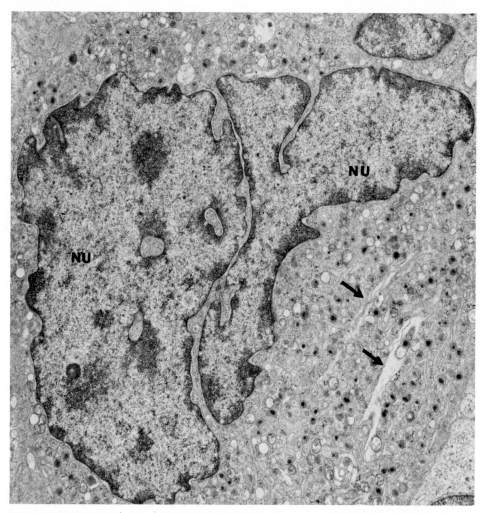

Figure 38. Portion of a megakaryocyte showing intricate or lobulated segmented nucleus (NU). The cytoplasm is filled with specific megakaryocytic granules with dense internal core. The arrows point at cleavage lines that will form demarcation lines of future platelets. (×20,590)

canalibular systems. Small circulating megakaryocytes were also found in four of these patients (81). Abnormal morphological features of megakaryocytes of the marrow in refractory anemia and myelomonocytic leukemia included micromegakaryocytes, immature megakaryocytes with nucleocytoplasmic asynchronism, and a variety of forms of abnormal thrombocytopoiesis (82).

Rare cases of megakaryoblastic leukemia, or variants of chronic myeloid leukemia with a large megakaryoblastic component, have been reported. In one case report, the majority of marrow-derived dividing cells obtained in one sample were megakaryoblasts, and 100% of the metaphases were Ph[1]-positive (83).

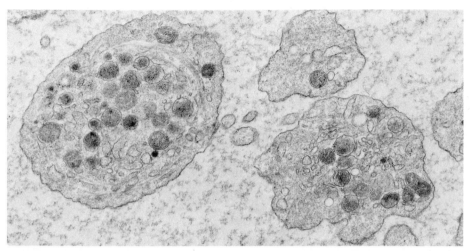

Figure 39. Peripheral blood preparation showing platelets with granules crowded in the center. Note presence of microtubules. (×26,390)

II. LYMPHORETICULAR CELLS

LYMPHORETICULAR SYSTEM

The lymphoreticular system, also known as the lymphatic, lymphoid, and reticuloendothelial system, may be defined as a heterogeneous group of organs, tissues, and widely scattered cells that are ultimately involved in phagocytic and immune functions.

Among the various lymphoreticular cells, the small lymphocytes seem to occupy a pivotal role. First thought to be end cells, some small lymphocytes may be regarded as embryonal cells capable of successive stages of transformation until a stage of irreversible differentiation is reached. The ultimate form of differentiation may be a plasma cell or another form of B cell, a T cell, or a macrophage.

For our purposes, we have conveniently divided lymphoreticular cells into lymphoid cells (i.e., lymphocytes and related forms), plasma cells and other immunoglobulin-forming cells, and histiocytes or macrophages. Monocytes are generally considered to be precursors of macrophages.

LYMPHOID CELLS

Lymphocytes and Lymphocytosis

Our current knowledge about lymphocytes has been greatly influenced by a number of observations made during the past two decades which have helped us discern the functional and morphological diversity of these cells. Among these observations, one may single out:

1. The ability of small, "mature-looking" lymphocytes to transform in vitro, and probably in vivo, under the influence of mitogens into much larger blast cells and then undergo mitosis (84,85)

2. The ability of lymphocytes to recirculate from the blood back into the lymph via postcapillary venules of lymph nodes (86)

3. The existence of at least two populations of lymphocytes from the standpoint of longevity: a minor group that is short-lived and a larger group that may persist in the circulation for years (87,88)

4. The functional attributes of lymphocytes as T and B cells, null cells, and possibly M cells, together with subdivisions of these functional groups (89,90)

From a strictly morphological viewpoint, lymphocytes display a marked heterogeneity, which can be appreciated in routinely prepared blood smears or tissue sections of lymph nodes. This morphological heterogeneity is evidenced by *a)* size of cells: small, medium, large; *b)* nuclear chromatin pattern: clumped (heterochromatin) or dispersed (euchromatin); *c)* nuclear configuration: cleaved and noncleaved, cerebriform, and convoluted; *d)* presence or absence of cytoplasmic features such polyribosomes, rough endoplasmic reticulum, cytoplasmic inclusions, peripheral "hairy" processes.

As observed under the electron microscope, small lymphocytes have a dense nucleus with a small nucleolus. The cytoplasm is rather scant and empty except for a Golgi region, a few mitochondria, and free ribosomal granules (Fig. 40). In contrast, large lymphocytes have a nucleus with dispersed chromatin, one or several prominent nucleoli, and a cytoplasm rich in polyribosomes. There may be segments of rough endoplasmic reticulum. The latter description fits the appearance of many germinal center cells or centrocytes (Fig. 41). Forms intermediate between small and large lymphocytes exist. Although lymphocytes are frequently described as round, the marked plasticity and mobility of live lymphocytes can be readily visualized with the phase microscope. Fine cytoplasmic processes, pseudopods, and other deviations from the spherical state can be appreciated particularly in fields of inflammation. Desmosomes, of the *macula adherens* type, have been observed in the dendritic reticulum cells of germinal follicles (90a). Desmosome - associated dendritic reticulum cells were also found in cases of nodular (follicular) lymphoma (90b).

The small lymphocyte with scant and relatively empty cytoplasm and a dense nucleus has the morphological appearance of a genetically inactive, nonfunctioning "reserve" cell. Burnet (91) has called it "a mobile repository of genetic information." If unprovoked, this is also a long-lived cell. This apparent inactivity is, however, deceptive, for the potential of the small lymphocyte to grow, divide, and differentiate is remarkable. The large lymphocyte appears often to be a protein-synthesizing cell. Those of germinal follicular origin show evidence of immunoglobulin synthesis (92). Functionally and morphologically "activated" large B cells, of follicular or nonfollicular origin, represent transitional stages between lymphocytes and plasma cells.

Considerable attention has been given in recent years to immunological and functional markers of lymphoid cells that help in differentiating them as T cells (or thymus-derived), B cells (or immunoglobulin-synthesizing cells of bone mar-

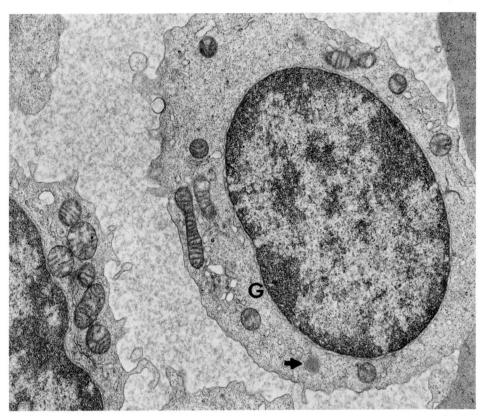

Figure 40. Peripheral blood buffy coat preparation showing two small lymphocytes. Note paucity of organelles. G represents a diminutive Golgi region; arrow points to a rare lysosomal (primary, azurophilic) granule. (×13,050)

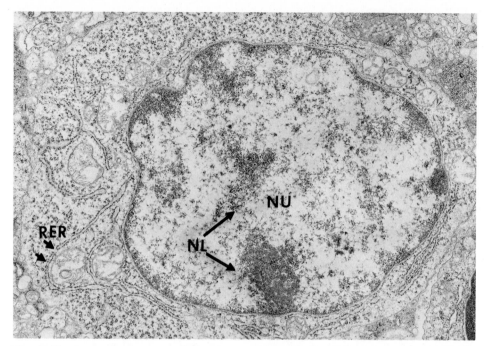

Figure 41. A germinal center cell showing a moderately developed nucleolus (NL). Note the beginning of a formation of rough endoplasmic reticulum (RER) at arrows. (×13,050)

row or bursa-equivalent tissues), M cells (or potential macrophages or monocytoid cells), and N or null cells, that is, uncommitted or undifferentiated lymphocytes with no recognizable markers (89,90). Attempts to recognize these cells by transmission or scanning electron microscopy have not yielded so far any practical advantage. Lymphocyte markers have also been used in introducing a functional classification of the lymphomas (93). More recently, human lympho-blastoid cell lines characterised as T or B cells by various markers were shown to have distinct differences in nuclear morphology, amount of cytoplasm, pyroninophilia, and PAS staining. Transmission electron microscopy generally confirmed these light microscopic observations (93a).

Reactive lymphocytosis in man, particularly in relation to infectious monocytosis and other viral infections, may be likened to the in vitro blastoid transformation of lymphocytes when exposed to a variety of mitogens. This in vitro reaction, first described by Nowell (84) in 1960 in cultures of human lymphocytes exposed to phytohemagglutinin, can be readily observed with the phase microscope. Small lymphocytes enlarge within 24 hours into blast forms containing at least one prominent nucleolus (Fig. 42). Mitoses can be seen at 48 to 72 hours. In later cultures, the cells acquire more and more granules and vacuoles; they become gradually more ovoid and assume the appearance of phagocytic cells with numerous refractile granules (Fig. 43). Electron microscopic studies of phytohemagglutinin-induced blast cells reveal an increased content of polyribosomes with a poorly developed rough endoplasmic reticulum (Fig. 44). In later cultures, the cells are characterized by the presence of membrane-bound granules, numerous vesicles, phagosomes, and residual bodies (Figs. 45 and 46). These changes, which are now thought to represent features of T cells, also suggested the possibility of the eventual transformation of some lymphocytes into macrophages (94). The lymphoid cells of infectious mononucleosis (Fig. 47) and the similarly transformed cells in influenza-type upper respiratory infection (Fig. 48) bear a striking resemblance to the blastoid forms of lymphocytes exposed to mitogens.

Lymphoid Leukemias

Lymphoid leukemias may be classified, as for other leukemias, according to level of development of the predominant leukemic cell population, as well as the mode of clinical presentation. One hesitates to allude to "mature" and "immature" lymphocytes, particularly since it has been mentioned earlier that normal small, so-called "mature" lymphocytes are dormant preblast cells.

Because of prognostic and therapeutic considerations, it has become important to document as much as possible by means of histochemical and electron microscopic means the various cytological types of leukemias. A distinction between acute lymphoid leukemia of childhood and early adulthood, and acute myeloid or monocytoid leukemia is particularly important from the standpoint of immediate management. This may even call for a "stat" electron microscopic examination of blood or marrow cells, since different therapeutic protocols have been established for different types of leukemia.

The 1976 WHO nomenclature of hematopoietic and lymphoid neoplasms (95) recognizes a variety of acute and chronic lymphoid leukemias without describing their ultrastructural features.

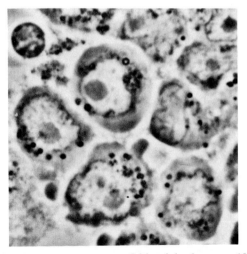

Figure 42. Phase microscopy appearance of blood leukocytes 48 hours after in vitro incubation with phytohemagglutinin. Blastoid transformation of lymphocyte is marked. Nuclear chromatin is dispersed, and the nucleolus is prominent. Dark refractile granules are present principally in the region of the Golgi apparatus. (×1,200) *(Courtesy of Dr. Farid Khouri; unpublished data)*

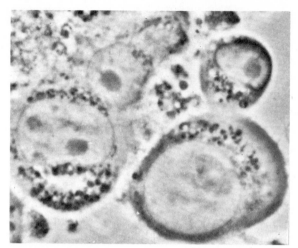

Figure 43. Phase microscopy appearance of peripheral blood 96 hours after in vitro incubation with phytohemagglutinin. Note that at least three of the transformed cells assume now the appearance of large macrophages with a large number of lipoidal droplets in the general region of the Golgi area. (×1,200) *(Courtesy of Dr. Farid Khouri; unpublished data)*

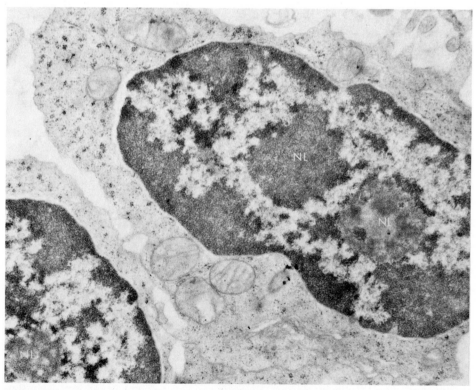

Figure 44. Electron micrograph of peripheral blood preparation 24 hours after in vitro incubation with phytohemagglutinin. Note nucleus with two nucleoli (NL), Golgi region (G), centriole (C), and clusters of polyribosomes (PR). (×24,750) *(Courtesy of Dr. Hiromitsu Okano; unpublished data)*

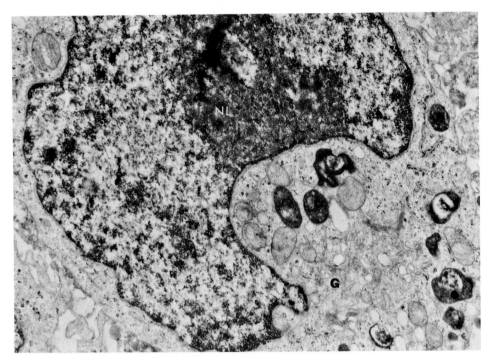

Figure 45. Electron microscopic appearance of peripheral blood 96 hours after in vitro incubation with phytohagglutinin. Note large, irregularly indented nucleus with prominent nucleolus (NL), a prominent Golgi area (G), and many irregularly shaped lysosomal complexes (arrows). The overall appearance is that of a macrophage with residual bodies. (×16,800) *(Courtesy of Dr. Hiromitsu Okano; unpublished data)*

Acute lymphoid (lymphoblastic) leukemia is characterized by cells measuring over 12 μm in diameter, with a high nucleocytoplasmic ratio, a slightly basophilic cytoplasm that is devoid of granules, and with a nucleus having a delicate and regularly distributed chromatin, and usually bearing a single nucleolus. *Macrolymphoblastic, microlymphoblastic,* and *immunoblastic* varieties have been described. Lymphoblastic leukemias may not be easily distinguished, or perhaps cannot be differentiated from *undifferentiated* or *stem cell leukemias.*

Ultrastructurally, the vast majority of acute lymphoblastic leukemias are recognized because of the same cytological features cited above. Most important, electron microscopy should exclude the presence of any cytoplasmic granules that may be interpreted as primary or specific granules of granulocytic precursor cells. Thus, the cells of acute lymphoid leukemias will show a euchromatin type of nuclear configuration, with one or more nucleoli, and a cytoplasm devoid of any form of significant differentiation (Figs. 49 and 50). Leukemic lymphocytes lack any significant peroxidase or esterase activity.

The cells of *chronic lymphoid leukemia* resemble strikingly the small lymphocytes of the peripheral blood (Fig. 51). A small percentage of lymphoblasts or prolymphocytes may be seen in chronic lymphoid leukemia. A *lymphosarcoma cell* variant has been described in association with massive lymphomatous involvement of lymph nodes (96).

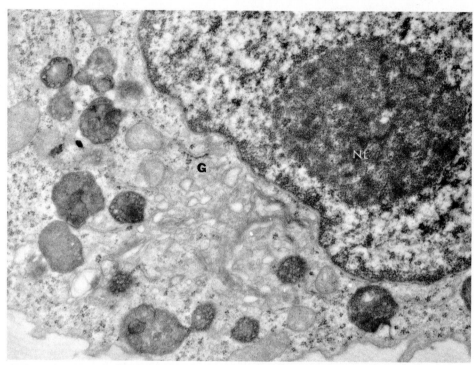

Figure 46. Electron microscopic appearance of peripheral blood 96 hours after in vitro incubation with phytohemagglutinin. The overall appearance is similar to that described above and suggests transformation of lymphocytes into cells with the appearance of macrophages. (×26,000) *(Courtesy of Dr. Hiromitsu Okano; unpublished data)*

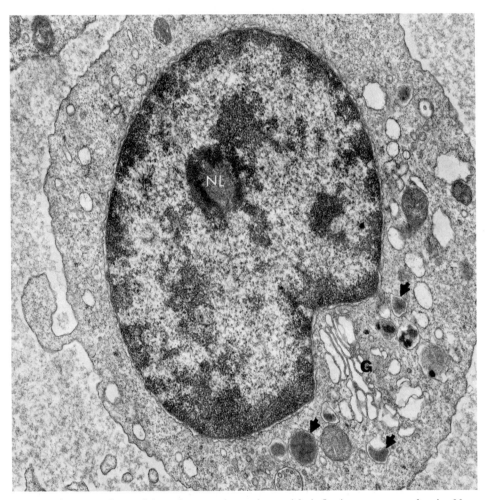

Figure 47. Transformed lymphocytes in patient with infectious mononucleosis. Note prominent nucleolus (NL), prominent Golgi region (G), and scattered lysosomal complexes (arrows). (×20,240)

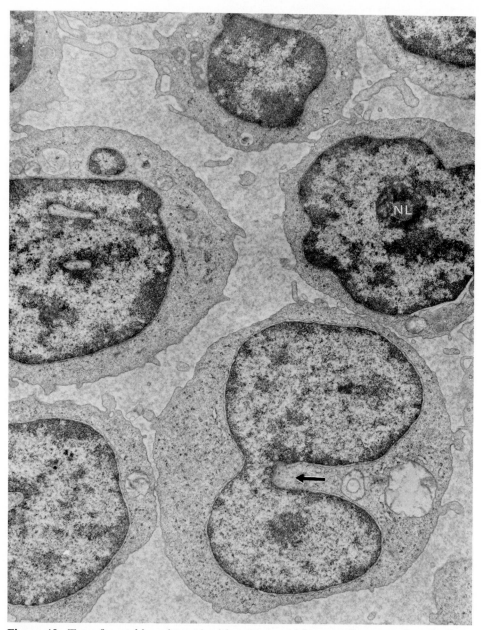

Figure 48. Transformed lymphocytes with monocytoid features from a patient during a bout of influenza. Note prominent nucleolus (NL) and deep invagination of a nucleus (arrow). These cells were interpreted under the light microscope as being monocytes. (×9,500)

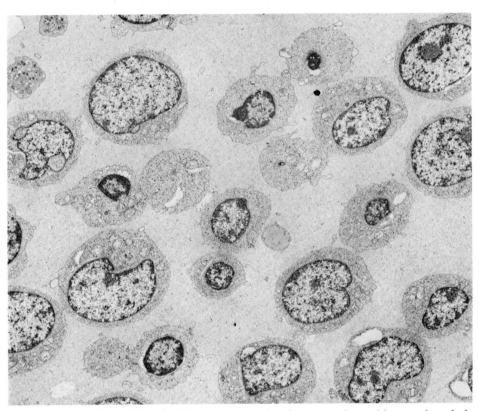

Figure 49. Peripheral blood buffy coat preparation from a patient with acute lymphoblastic leukemia. (×2,875)

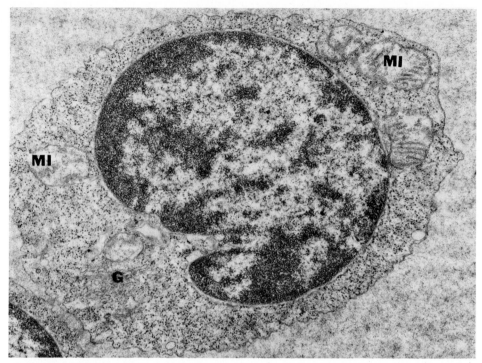

Figure 50. Higher magnification of lymphoid cell from same patient as above. Note total lack of granulocytic differentiation. This by exclusion suggests the diagnosis of acute lymphoblastic leukemia. (×10,000)

There are two special forms of leukemias that are akin to chronic lymphocytic leukemia, although they seldom achieve a high level of lymphocytosis. These are Sézary's syndrome and hairy cell leukemia.

Sézary syndrome may represent several biological processes in which an exfoliative erythroderma is associated with either a true T cell leukemia or a reactive "leukemoid" process (97). In either situation, the Sézary cells can be distinguished by light microscopy because of two distinct features that set them apart from other atypical benign or malignant "atypical" lymphocytes: *a)* The nucleus has a folded, notched, or convoluted appearance that has been described as cerebriform. *b)* The cytoplasm displays vacuolated or clear areas which contain a PAS-positive material. This PAS-positive staining is variable and is removed by diastase digestion, suggesting the presence of glycogen. The cytoplasm is also rich in β-glucuronidase, but not peroxidase or naphthol AS-D esterase activity (98).

The peripheral white cell count is variable in Sézary's syndrome; the highest counts seldom exceed 70,000/mm³. A high proportion of leukocytes is constituted by Sézary cells. Both large and small cell variants have been described. Bone marrow and lymph nodes may be infiltrated by Sézary cells, particularly when these were present in large numbers in the blood (98). Sézary cells have been evaluated by immunological criteria, including surface antigenicity, membrane receptors, response to mitogens, and reactivity in mixed leukocyte cul-

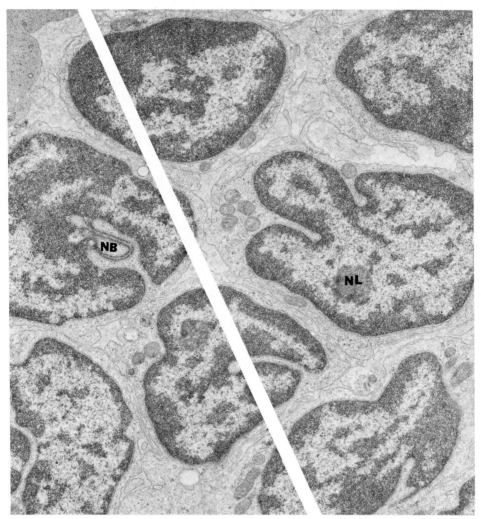

Figure 51. Peripheral blood preparation from a patient with chronic lymphocytic leukemia. Note that one of the lymphocytes has a prominent nucleolus (NL), that there is considerable indentation in the nucleus, and that there is a nucleolar bleb (NB). (×13,050)

tures. In the vast majority of cases, Sézary cells exhibited typical T cell properties (98,99).

Several electron microscopic studies of Sézary cells have been reported (99,100,101). The classical large Sézary cell demonstrates a deeply and intricately indented or cerebriform nucleus (Figs. 52 and 53). The small cell variant shows less prominent nuclear irregularities than the classical large Sézary cell. The demonstration of clusters of glycogen granules is not always possible. This is due either to prior therapy with glucosteroid compounds or to *en bloc* staining with uranyl acetate, which tends to dissolve glycogen. The finding of fibrils and a variety of cytoplasmic and nuclear inclusions probably indicates degenerative changes that are nonspecific.

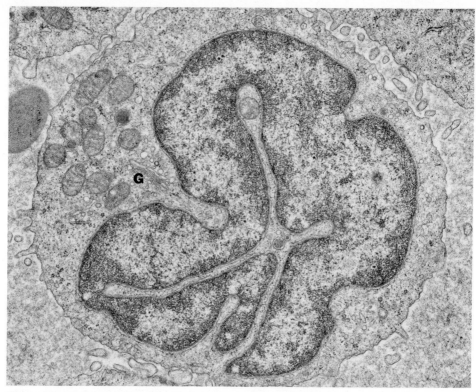

Figure 52. Peripheral blood buffy coat preparation from a patient with Sézary syndrome. Note the marked convolution or cerebriform pattern of nucleus. (×20,240)

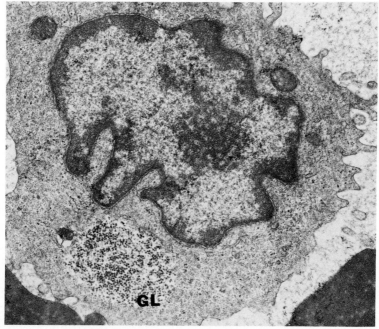

Figure 53. Peripheral blood preparation from another patient with Sézary syndrome. In addition to slight indentation and invaginations of the nucleus, there is a pool made of fine granules of glycogen (GL). (×11,250)

For a detailed discussion of clinical and morphological aspects of the Sézary syndrome, the reader is referred to the proceedings of a symposium held at the Mayo Graduate School of Medicine in 1974 (102). It is still debated whether Sézary syndrome should be considered a premycotic variant of mycosis fungoides (103), or whether it is a "T cell erythroderma" resulting from a delayed hypersensitivity reaction and rarely transforming into a truly malignant condition (97). Regardless of the nature of the relationship between the two conditions, the lymphoid infiltrate of Darier-Pautrier "microabscesses" of mycosis fungoides bears a striking resemblance to Sézary cells (104,105).

Hairy cell leukemia is another example of lymphoid leukemia with circulating lymphocytes that can be distinguished morphologically and cytochemically from the cells of the vast majority of cases of chronic lymphoid (lymphocytic) leukemia.

Though first reported by Bouroncle, Wiseman, and Doan (106) in 1958 under the title of leukemic reticuloendotheliosis, hairy cell leukemia is being increasingly recognized because of its distinct clinical features, and because of our ability to confirm suspected cases by electron microscopic examination of peripheral blood buffy coat preparations, as well as bone marrow and spleen samples.

Clinically, hairy cell leukemia affects mainly adult males and is manifested principally by pancytopenia and splenomegaly. The latter is often massive, whereas lymphadenopathy is rather minimal or absent. Unlike the situation in Sézary syndrome, skin involvement is rare (107).

Attempts at bone marrow aspiration are characteristically unsuccessful. Sections of marrow biopsy samples disclose a moderate to marked increase in cellularity due to a round cell infiltrate which appears loosely packed. The reticulum stain shows a diffuse increase of fine reticulin meshwork.

Freshly prepared, Romanovsky-stained peripheral blood buffy coat preparations and, if successfully aspirated, bone marrow smears may disclose under optimal conditions round lymphoid cells with fuzzy or even hairy cytoplasmic processes. In practice, only the most experienced hematologists are confident about recognizing hairy cells under average light microscopy conditions; hence, the desirability of electron microscopic confirmation in every suspected case. The cytochemical demonstration of tartrate-resistant acid phosphatase activity is also an excellent means of confirming the diagnosis of hairy cell leukemia (8).

Hairy cell leukemia has distinctive ultrastructural features (108,109a). Classically, elaborate and delicate peripheral cytoplasmic ruffles (villous processes) are seen (Figs. 54 and 55). Unfortunately, not all cases are equally "hairy." There is, therefore, need for considerable caution in making this diagnosis. It should be noted that hairy lymphoid cells may be observed occasionally in the peripheral blood and marrow in the absence of hairy cell leukemia. Monocytoid forms of hairy cells may coexist with classical hairy cells.

Classical hairy cells are best seen in buffy coat preparations. In the spleen, the hairy lymphoid cells are packed together in the red pulp, and their hairy processes are closely intertwined (Fig. 56). In the marrow, hairy cells resemble those of the blood; but if closely packed, their processes tend to intertwine. In addition to their characteristic villous processes, hairy cells may display elaborate intracytoplasmic lamellar ribosomal complexes (108). The significance of these is unknown. We have encountered these complexes in only one out of seven cases

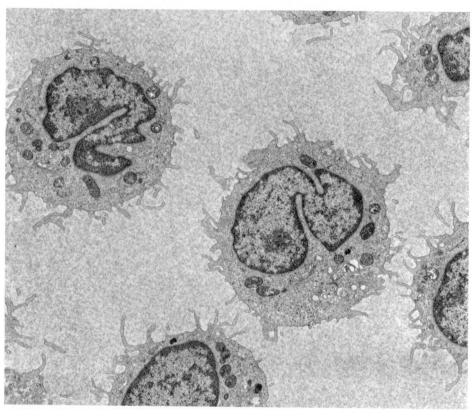

Figure 54. Periperhal blood preparation from a patient with hairy cell leukemia. Note the peripheral filiform processes of the cytoplasm. (×6,256)

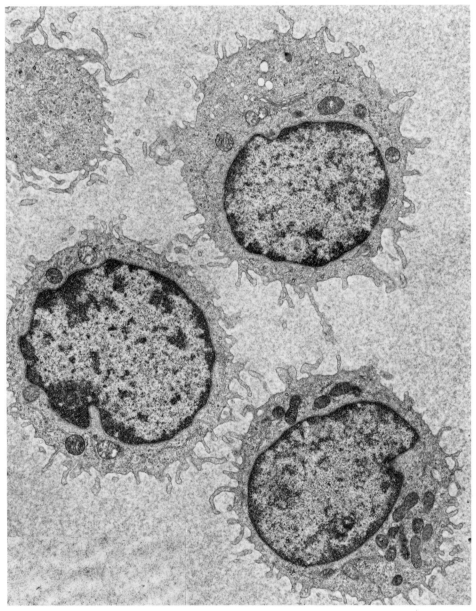

Figure 55. Peripheral blood preparation from a patient with hairy cell leukemia. At higher magnification. (×7,750)

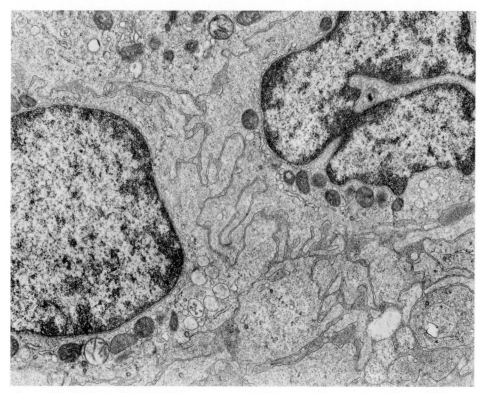

Figure 56. Section of spleen from a patient with hairy cell leukemia. In lieu of obvious peripheral hairy processes, there is considerable interdigitation of cell processes of adjacent lymphoid cells. (×13,064)

of hairy cell leukemia that we had the opportunity of studying during the past two years. The presence of these complexes in lymphoid cells with relatively short villous processes, plus the typical clinical presentation with splenomegaly, pancytopenia, and unsuccessful marrow taps, encouraged us in making a diagnosis of hairy cell leukemia (Fig. 57).

The derivation and nature of hairy cells have not been entirely clarified (109a). There is considerable immunological evidence that they are probably of B cell origin (110). Their monocytoid features and their tartrate-resistant acid phosphatase activity set them apart, however, from both T and B cells. It is clear that hairy cell leukemia is a distinct entity with biological characteristics that are strikingly different from those of chronic lymphoid leukemia. This distinction is not purely of academic interest. Splenectomy is an acceptable mode of palliation in hairy cell leukemia. It has no particular merit in chronic lymphoid leukemia. Hence, the necessity of an accurate diagnosis, preferably backed by electron microscopy.

Malignant Lymphomas

It is obvious that the diagnosis of malignant lymphomas should not be based on electron microscopy alone. What electron microscopy can contribute to the diagnosis of malignant lymphomas is a greater appreciation of cytological detail,

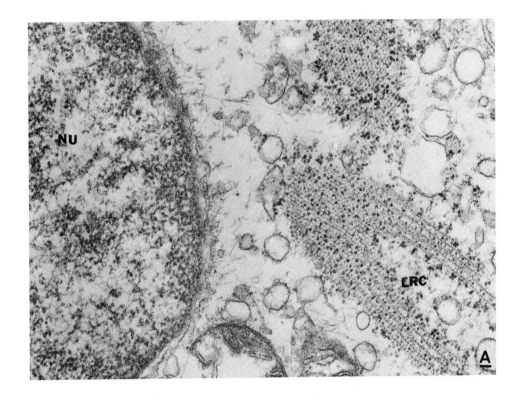

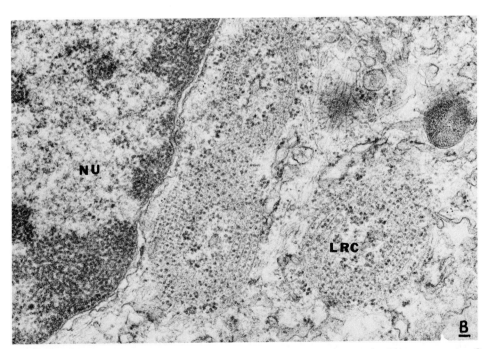

Figure 57. Peripheral blood from a patient with hairy cell leukemia. Note presence of intricate lamellar ribosomal complexes (LRC). These are characteristic but not specific for hairy cell leukemia. (×55,100)

particularly in certain special types, such as mycosis fungoides, Waldenström's macroglobulinemia, lymphoplasmacytic lymphosarcoma, immunoblastic lymphosarcoma, heavy-chain diseases, and reticulum cell sarcoma. Electron microscopy can also help in the differentiation between anaplastic or poorly differentiated carcinoma, neuroblastoma, Ewing's sarcoma, granulocytic sarcoma, other malignant "round cell" tumors, and reticulum cell sarcoma ("histiocytic" or large cell lymphomes).

The classification of malignant lymphomas has undergone several changes during the past decade, and more are expected as functional markers and cytochemical and ultrastructural studies become more widely applied. In practice, most pathologists still use Rappoport's classification (111). There have been more recent attempts to introduce immunological and functional classifications (93,112). It appears, however, that the vast majority of malignant lymphomas are B cell lymphomas, with the exception of mycosis fungoides and rare forms of other T cell lymphomas. Unclassifiable and M cell lymphomas have also been reported (93). In the long run, it would appear that clinical relevance and pragmatic considerations will prevail in the diagnosis of lymphomas, and that a detailed and sophisticated classification may prove to be both cumbersome and economically out of reach. We have favored a traditional classification akin to that proposed by the WHO in 1976. This classification can be conveniently used alongside the Rappaport classification (see Table 4).

Nodular (follicular) lymphosarcoma is considered to be composed of B cells of follicular center cell origin (112). Except for its follicular or nodular architecture, this tumor has essentially the same cell constituents as those of diffuse lymphosarcoma of lymphoblastic origin or of reticulum cell sarcoma. Cleaved and

Table 4. Types of Lymphoid Tumors After the 1976 WHO Classification (95)

A. Lymphosarcomas
   1. Nodular *(follicular)* [a] lymphosarcoma
   2. Diffuse lymphosarcoma
      a) Lymphocytic
      b) Lymphoplasmacytic
      c) Prolymphocytic [b]
      d) Lymphoblastic
      e) Immunoblastic
      f) Burkitt's tumor
B. Mycosis fungoides
C. Plasmacytoma [c]
D. Reticulosarcoma *(reticulum cell sarcoma)* [a]
E. Unclassified malignant lymphomas
F. Hodgkin's disease
   1. With lymphocyte predominance
   2. With nodular sclerosis
   3. With mixed cellularity
   4. With lymphocyte depletion

[a] These terms have been added to the original WHO classification.
[b] This term has not been used by us.
[c] Plasmacytoma is discussed in relation to plasma cell myeloma.

noncleaved nuclei of various sizes have been described. These correspond to related cells of follicular centers (113) Desmosome-associated dendritic reticulum cells were found to be characteristic of both reactive lymphoid follicles and of nodular (follicular) lymphosarcoma (90a, 90b).

Diffuse lymphocytic lymphosarcoma is composed predominantly of small lymphocytes that are often called mature or well differentiated. These appear ultrastructurally identical to the small lymphocytes of chronic lymphocytic leukemia and to the majority of normal small lymphocytes (Fig. 58). Frequently, diffuse lymphocytic lymphosarcomas are associated with an overt chronic lymphocytic leukemia.

Lymphoplasmacytic lymphosarcoma or lymphosarcoma with plasmacytic differentiation has long been recognized as a form of malignant lymphoma associated with a monoclonal gammopathy and constituted by cells exhibiting an intense cytoplasmic pyroninophilia or obvious plasmacytic differentiation (114). The common association between this form of lymphoma and IgM monoclonal gammopathy suggests an intimate relationship with Waldenström's macroglobulinemia. The electron microscopic appearance is that of lymphocytic or lymphoblastic lymphosarcoma, less often a reticulum cell sarcoma, with intermediate or transitional cells showing a well-developed rough endoplasmic reticulum approaching the level of development found in plasma cells (Fig. 59). Typical plasma cells are also found amidst transitional cells and "plain" lymphocytes and lymphoblasts. Heavy-chain diseases of the IgG and IgA classes are also frequently associated with an underlying lymphoreticular growth with a varying degree of plasmacytic differentiation.

Lymphoblastic lymphosarcoma is composed of cells intermediate in size between those of lymphocytic lymphosarcoma and reticulum cell sarcoma, and resembling those of acute lymphoid leukemia (Fig. 60). The nucleus of these cells tends to have at least one prominent nucleolus. The cytoplasm fails to show any appreciable plasmacytic differentiation. The presence of cleaved nuclei has suggested an origin from follicular center cells (113). A less common variety with convoluted nuclei, usually seen in the mediastinum of children, is thought to be of T cell origin (95).

Immunoblastic lymphosarcoma is a more recently described variety of malignant lymphoma composed of large basophilic (or pyroninophilic cells) with plasmacytoid features (115). Unlike Burkitt's tumor, it lacks macrophages. The scarcity of reticulin fibers and obvious plasmacytoid features helps to differentiate this tumor from reticulum cell sarcoma. There is also often a background of autoimmune disease or of an immunoblastic lymphadenopathy (116) evolving into an immunoblastic lymphosarcoma.

Burkitt's tumor is another example of B cell lymphosarcoma composed of basophilic or pyroninophilic cells usually smaller than those of immunoblastic lymphosarcoma or reticulum cell sarcoma. Ultrastructurally, Burkitt cells have the appearance of transitional cells, with a rather well-developed rough endoplasmic reticulum (Fig. 61). Macrophages are scattered among the neoplastic lymphoid cells, giving a "starry-sky" effect.

Mycosis fungoides has already been discussed in relation to Sézary syndrome. The intraepidermal aggregates of neoplastic lymphoid cells, known as Darier-Pautrier abscesses, resemble Sézary cells ultrastructurally. The same type of cells,

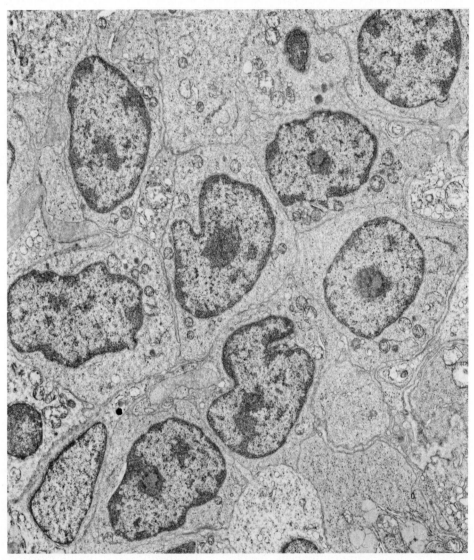

Figure 58. Field of lymph node with lymphocytic lymphosarcoma. The major component is a small lymphocyte, some of which shows a prominent nucleolus. (×5,370)

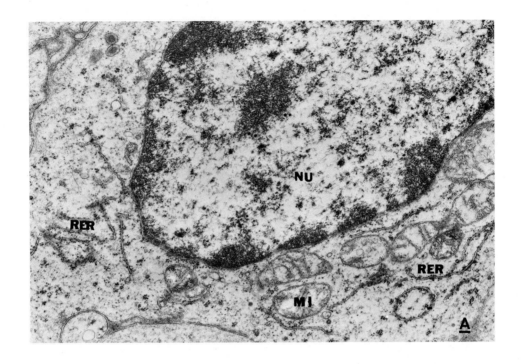

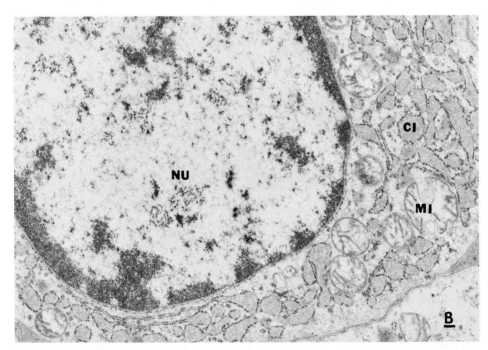

Figure 59. Cells from a patient with lymphoplasmacytic lymphosarcoma associated with an IgM monoclonal gammopathy. Note in *A* the beginning of formation of rough endoplasmic reticulum (RER). In *B*, there is considerable plasmacytic differentiation with slight dilation of the cisternae (CI). (×20,590)

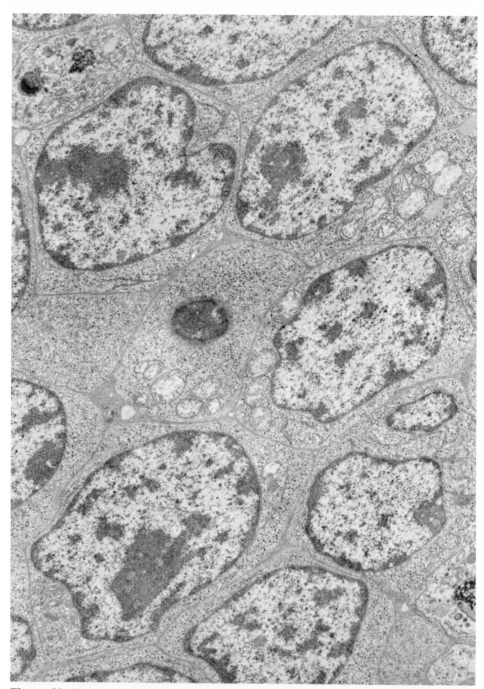

Figure 60. Lymph node from a patient with lymphoblastic lymphosarcoma. The constituent cells are largely medium-sized lymphoblasts with prominent nucleoli. The cytoplasm is rich in polyribosomes, but there is no appreciable plasmacytic differentiation. (×7,700)

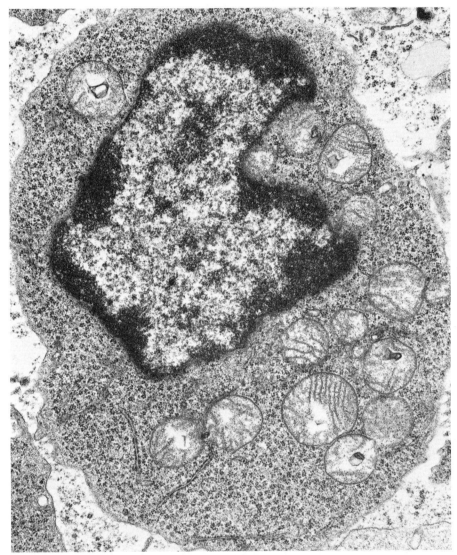

Figure 61. Representative section of a lymphoid cell from a patient with so-called Burkitt's lymphoma, American type. Polyribosomes are abundant in the cytoplasm, and there is formation of segments of rough endoplasmic reticulum. (×14,000) *(From Azar and Potter (eds), ref. 126)*

characterized primarily by their cerebriform nucleus, is observed in dermal infiltrates and in metastatic loci.

Reticulum cell sarcoma, also known in the United States as "histiocytic" malignant lymphoma, is composed of larger lymphoreticular cells that seem to lack any histochemical evidence of histiocytic differentiation (117). The cells in this tumor do not demonstrate as a rule any appreciable basophilia or pyroninophilia, although an uncommon plasmacytic variety of reticulum cell sarcoma has been described (118). This tumor is usually rich in intercellular reticulin fibers, which tend to surround individual tumor cells; hence, the origin of the term reticulum cell sarcoma. It is doubted, however, whether this reticulin or young collagen meshwork is either specific to reticulum cell sarcomas or derived from the so-called reticulum cells. The nucleus of anaplastic reticulum cells may be cleaved or uncleaved. The chromatin tends to be rather evenly dispersed (euchromatin pattern), and there is often at least one distinct nucleolar complex. Electron microscopy fails to disclose elements of histiocytic differentiation, such as the presence of lysosomal granules or phagolysosomes. Likewise, there are no obvious plasmacytic features (Fig. 62). At this time, because of lack of data on histochemical or immunochemical markers in this type of tumor, it is best to consider reticulum sarcomas as undifferentiated or unclassified lymphoreticular growths awaiting a more precise identification. This is not to negate the possibility that some genuine histiocytic lymphomas may exist, and that other reticulum cell sarcomas are B cell lymphomas of follicular center cell origin. Large cell neoplasms, previously classified as "histiocytic" lymphomas, were found to be constituted by cells with ultrastructural and cytochemical characteristics of transformed lymphocytes presumably of follicular center origin (118a).

Hodgkin's disease of all four major subtypes (see Table 4) is characterized by the presence of Reed-Sternberg cells and a related mononucleated variant that is similar to anaplastic reticulum cells. Typical Reed-Sternberg cells are rarely found in the nodular sclerosis subtype. The variant is usually rich in "lacunar" cells that have a rather abundant and pale cytoplasm.

There are several ultrastructural studies of the Reed-Sternberg cell (119,120,121,122). These have not revealed any unusual or pathognomic features. In our own observations, this cell appears as a large and poorly differentiated lymphoreticular cell, which is our own arbitrary definition of reticulum cell. Small vesicles and membrane-bound granules may be observed in variable numbers and principally in the Golgi region. There is, however, insufficient morphologic evidence to consider these large cells as histiocytes or macrophages. Free ribosomal granules, polyribosomes, and segments of rough endoplasmic reticulum are scattered throughout the cytoplasm. The nucleus of the Reed-Sternberg cell is markedly segmented, with a narrow bridge separating the two (or more) nearly symmetrically divided nuclear lobes (Figs. 63 and 64). The nucleolar complexes are markedly developed. This tendency to lobulation, segmentation, or deep indentations (cleaving?) appears also in the mononucleated anaplastic reticulum cells that are readily observed in Hodgkin's disease of the mixed cellularity or lymphocyte depletion subtypes (Figs. 65 and 66). The overall appearance of Reed-Sternberg cells and Hodgkin's mononucleated anaplastic

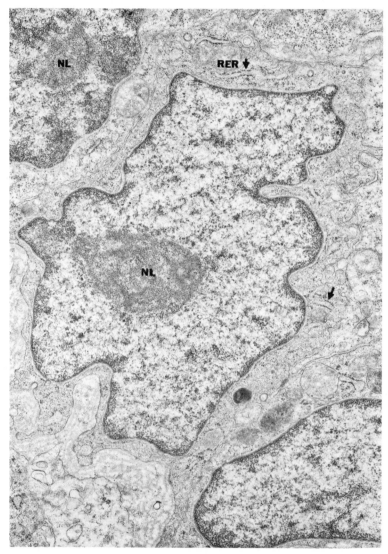

Figure 62. Field from a patient with reticulum cell sarcoma showing a representative cell with prominent nucleolus with dispersed chromatin and a prominent nucleolus (NL). Arrows point to segments of well-developed rough endoplasmic reticulum (RER). (×16,750)

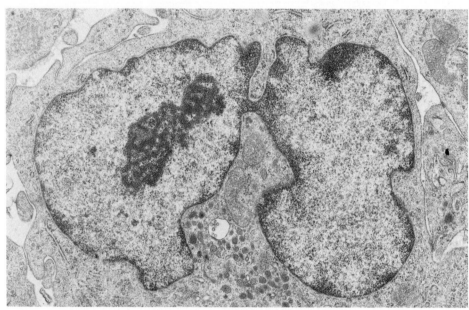

Figure 63. Electron micrograph of a Reed-Sternberg cell showing nuclear bridge connecting two large nuclear lobes. (×12,000) *(From Azar, ref. 122)*

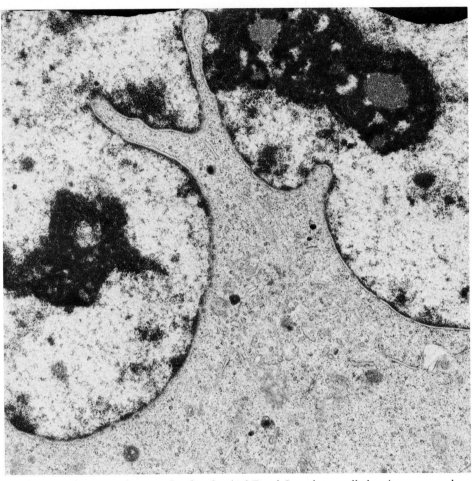

Figure 64. Electron micrograph of a classical Reed-Sternberg cell showing two nuclear lobes, each containing a prominent nucleolar area. (×12,000) *(From Azar, ref. 122)*

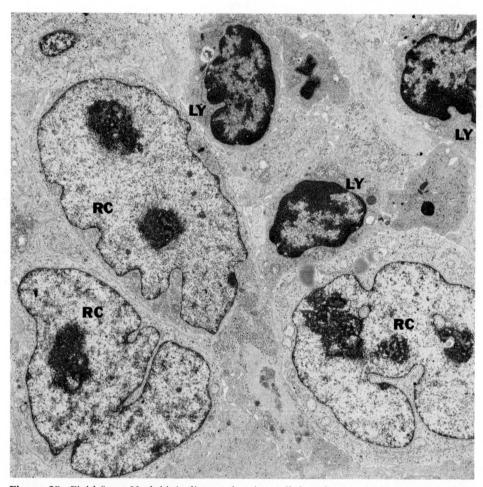

Figure 65. Field from Hodgkin's disease showing cellular pleocytosis. Note lymphocytes (LY) and anaplastic reticulum cells with varying degrees of nuclear indentations and prominent nucleoli labeled RC. (×12,000)

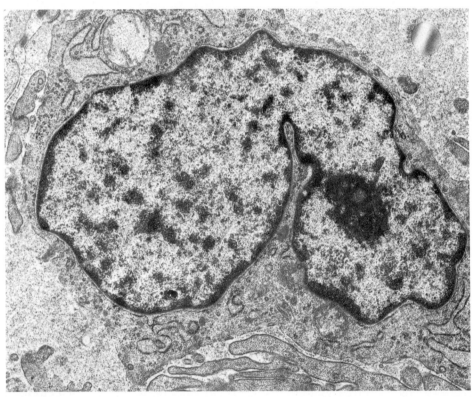

Figure 66. Anaplastic reticulum cell from field of Hodgkin's disease showing marked indentation within a nucleus with dispersed chromatin and prominent nucleolus. (×12,000)

reticulum cells is strongly reminiscent of the mitogen-stimulated blastoid forms of lymphocytes, as well as of the cleaved large follicular center cells. Although the pathogenesis of Hodgkin's disease is not understood, there is considerable evidence that Reed-Sternberg cells represent an end stage of transformed lymphocytes (121,122). There is also some early evidence that they may be of B cell origin (123). It should be emphasized that Reed-Sternberg cells may be occasionally observed in benign conditions, as well as in lymphoreticular growths other than Hodgkin's disease (124).

Occasionally, one observes unusually large intracytoplasmic inclusions in Reed-Sternberg cells (125). These have been suspected to be of viral origin. A recent effort, in the case of a child with Hodgkin's disease, to "carve out" from paraffin sections such intracytoplasmic inclusions and study them under the electron microscope has yielded inconclusive results.

PLASMA CELLS AND RELATED Ig-PRODUCING CELLS

Plasma cells constitute the most differentiated forms of a series of B lymphoid cells that comprises lymphocytes of all sizes and germinal center cells. Whereas it is not possible on an individual cell basis, and without the help of special markers, to determine whether a given cell is of the B or T variety, this is seldom the case with a plasma cell, or even its immediate precursor, the preplasma cell: The eccentric and usually clumped nucleus, the well-delineated Golgi region, the relatively abundant and basophilic cytoplasm, and, on electron microscopy, the well-developed, lamellar or vesicular form of rough endoplasmic reticulum all point to the role of the plasma cell as a site of protein synthesis for export (Figs. 67, 68, and 69). The plasma cell is indeed the prototype of an evolved B lymphocyte. Plasma cell tumors also maintain, as a rule, a level of cytoplasmic differentiation that sets them apart from other lymphoid growths. Apart from tradition, there is a pragmatic basis for discussing disorders of plasma cells separately from those of lymphocytes.

Reactive Plasmacytosis

It can be successfully argued that the very presence of plasma cells at a given site signifies some form of antigenic stimulation. Since the normal state involves "lines of confrontation" with bacteria and products of cellular breakdown, plasma cells are "normally" found in varying numbers, particularly along mucosal surfaces, in lymph nodes, and in the bone marrow. An unusually large number of plasma cells may be observed either locally, as in a syphilitic chancre or in the wall of a granulating abscess, or throughout the lymphoreticular system, as in a variety of inflammatory processes, such as chronic bacterial infections, certain parasitic infestations, liver cirrhosis, and rheumatoid arthritis. A marked reactive plasmacytosis is usually associated with a polyclonal hypergammaglobulinemia.

At times, it is difficult to distinguish an exuberant plasma cell granuloma from a truly neoplastic plasmacytoma. The presence of macrophages and other inflammatory cells, increased vascularity, and the variability in size and level of

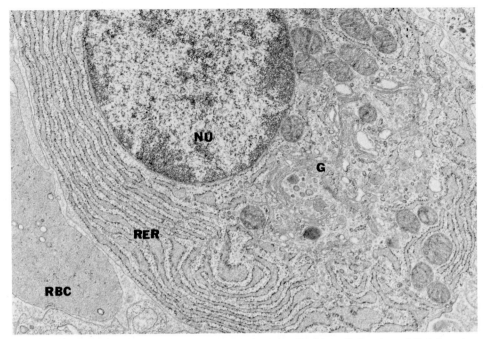

Figure 67. Bone marrow showing a classical plasma cell with well-developed Golgi region (G) and an abundance of lamellar type of rough endoplasmic reticulum (RER). (×12,750)

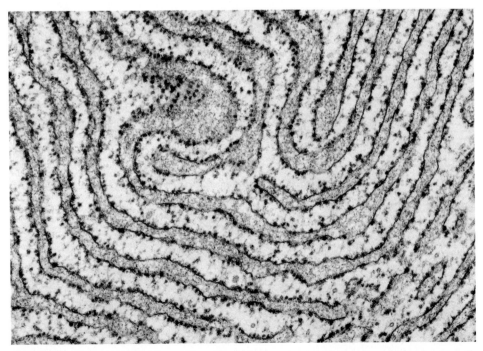

Figure 68. High magnification of rough endoplasmic reticulum of reactive plasma cell. (×72,500)

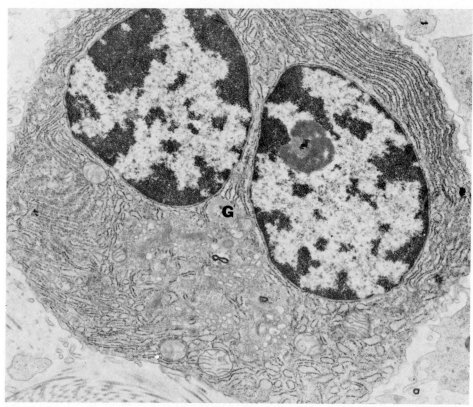

Figure 69. Reactive binucleated plasma cell in syphilitic chancre. Note the Golgi region (G) and the abundance of rough endoplasmic reticulum. (×14,000)

development of plasma cells favor a reactive process rather than a plasmacytoma. Ideally, one should be able by means of immunocytochemical studies to determine whether or not a given plasmacytic population is polyclonal (i.e., constituted by cells of different immunoglobulin classes and types) and, therefore, reactive or inflammatory in nature, or monoclonal (i.e., cells of a restricted immunoglobulin class and type) as to indicate a neoplastic process. It is possible that many examples of extraskeletal "plasmacytomas" reported in the literature are either examples of exuberant plasma cell granulomas or examples of disseminated multiple myeloma (126).

Reactive bone marrow plasmacytosis also may be difficult to differentiate from early plasma cell myeloma or premyeloma. In this situation, the detection of a serum or urine monoclonal immunoglobulin or immunoglubulin fragment indicates a myeloma, whereas polyclonal hypergammaglobulinemia is indicative of reactive plasmacytosis. It is true that myeloma cells have certain distinctive ultrastructural features (see below), but the diagnosis of the average case of myeloma should not depend on the results of electron microscopy.

Plasma Cell Myeloma

In general, myeloma cells, regardless of their degree of differentiation, bear a striking resemblance to their parent plasma cells in their ultrastructural features (Figs. 70, 71, and 72). The general cytologic characteristics of myeloma cells, as contrasted with those of reactive plasma cells, have been described in detail elsewhere (126,127,128) and may be summarized as follows:

1. Myeloma cells are usually larger and more uniform in size and shape than plasma cells in a field of inflammation.
2. The nucleus of the myeloma cells is comparatively larger, and its chromatin more dispersed, thus contrasting with the "cartwheel" nuclear pattern of reactive plasma cells. A prominent nucleolus or nucleoli can be readily seen in myeloma cells.
3. The Golgi region is often more evident in reactive plasma cells, particularly in light microscopic preparations.
4. Russell bodies, though found in good number in occasional cases of myeloma, are frequently observed in fields of reactive plasmacytosis (Figs. 73 and 74). They represent intracytoplasmic storage of immunoglobulin within dilated cisternae.
5. Multinucleation should not be considered a reliable differentiating feature. Binucleated and, less frequently, multinucleated reactive plasma cells are not infrequently observed in fields of chronic inflammation.

Perhaps the most distinctive feature of myeloma cells, as compared with reactive plasma cells, is the striking asynchrony between the level of cytoplasmic differentiation and that of the nucleus. Whereas the nucleus of myeloma cells is considered to be typically "immature" and demonstrates the euchromatin pattern of blastoid or undifferentiated cells, the cytoplasm of these cells often surpasses the level of differentiation of reactive plasma cells by its closely packed, abundant, and delicately folded rough endoplasmic reticulum (Figs. 70, 71, and 72), or by its dilated and empty-looking cisternae (Fig. 75). Even the poorly

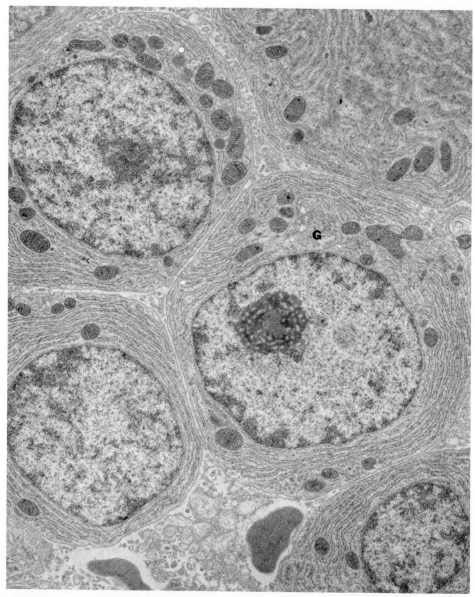

Figure 70. Bone marrow with field of IgG producing multiple myeloma. Note the markedly immature nucleus with dispersed chromatin and prominent nucleolus, as contrasted with the highly developed cytoplasm rich in lamellar type of endoplasmic reticulum. The Golgi region is marked G. (×8,600)

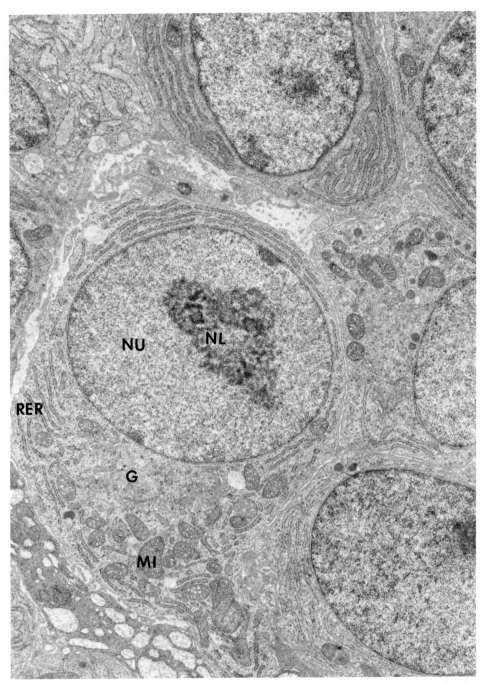

Figure 71. Field in bone marrow of an IgA-secreting multiple myeloma. Here, too, the markedly immature nucleus of myeloma cells is contrasted with the highly developed cytoplasm rich in rough endoplasmic reticulum (RER). (×7,600)

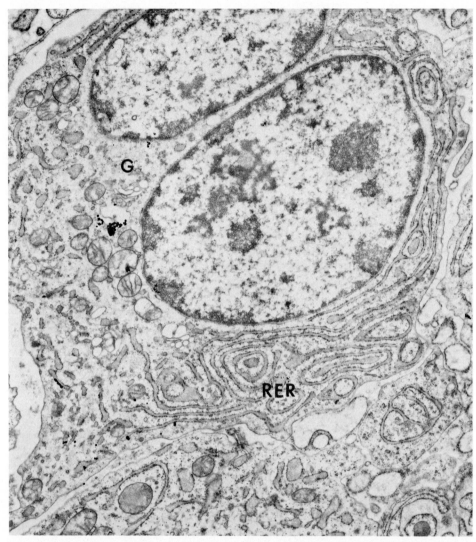

Figure 72. Electron micrograph of a differentiated myeloma cell showing two nuclei or a bilobed nucleus. Note again the immature nucleus with dispersed chromatin, as contrasted with the highly differentiated cytoplasm with abundant rough endoplasmic reticulum (RER). (×15,000) *(From Azar and Potter (eds), ref. 126)*

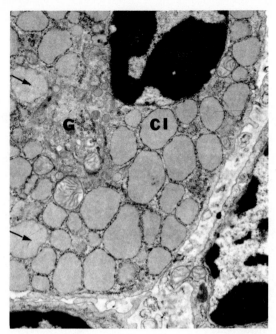

Figure 73. Plasma cell with marked dilatation of cisternae (CI) filled with material of moderate electron density. Arrows point to deposits with a crystalloid appearance. The contents of dilated cisternae approach the size and density of the component of small Russell bodies. (×12,000) *(From Azar and Potter (eds), ref. 126)*

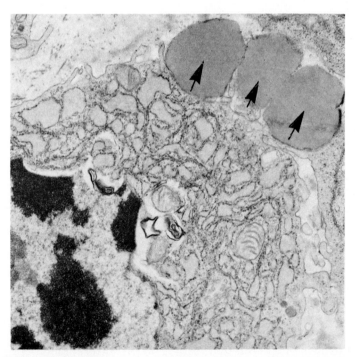

Figure 74. Plasma cell showing moderate dilation of cisternae filled with material of light electron density. The three Russell bodies (arrows) probably belong to a neighboring plasma cell. (×12,000) *(From Azar and Potter (eds), ref. 126)*

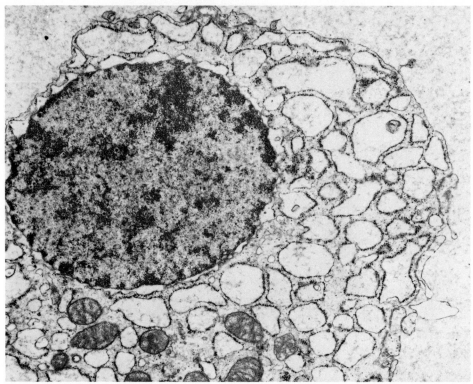

Figure 75. Example of myeloma cell with dilated and empty-looking cisternae. (×15,000) *(From Azar and Potter (eds), ref. 126)*

differentiated form of plasma cell myeloma, the plasmacytic reticulum cell sarcoma (118), shows a moderately evolved rough endoplasmic reticulum.

Both intranuclear and intracytoplasmic inclusions have been observed in myeloma cells. The former probably represent intracytoplasmic inclusions, akin to Russell bodies, invaginating within the nucleus. Similar inclusions have been observed in Waldenström's macroglobulinemia. These tend to be PAS-positive (129) and are often called Dutcher bodies. In addition to Russell bodies, the cytoplasm of myeloma cells may contain crystalloid inclusions (Fig. 76), fat droplets, calcific bodies, and complex phagosomes or phagolysosomes. The latter are particularly prominent in cases of myeloma treated with chemotherapeutic agents (Fig. 77).

There have been several attempts to correlate the morphological features of myeloma cells with the category of monoclonal protein they produce. These correlations have been generally inconsistent; however, the classical myeloma remains that associated with monoclonal IgG. There appears to be some degree of correlation between IgA myeloma and the finding of flaming cells or thesaurocytes (130,131). These exhibit a marked distension of their endoplasmic cisternae. IgD myelomas tend to be rather poorly differentiated. The rare cases of IgE myeloma, now five (132), appear well differentiated. It is argued whether IgM myelomas truly exist; the present tendency is to regard most of them as variants of Waldenström's macroglobulinemia.

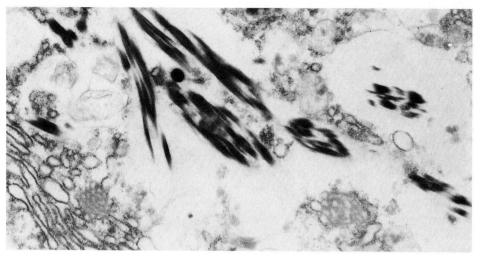

Figure 76. Portion of myeloma cell containing crystalloid formations within the cytoplasm. (×28,800) *(From Azar and Potter (eds), ref. 126)*

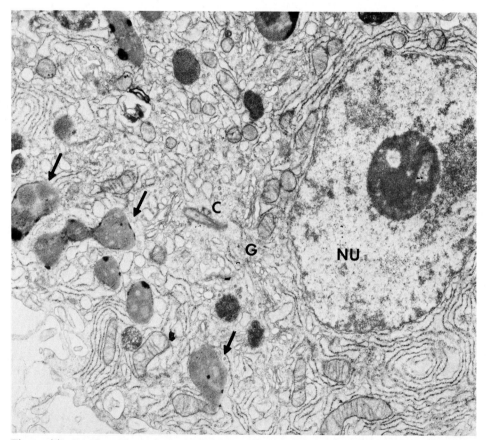

Figure 77. Portion of myeloma cell from patient treated with melphalan. The nucleus (NU) and nucleolus are relatively intact, as opposed to accumulation within the cytoplasm of dense phagolysosomal bodies (arrows). Also note centriole (C) and Golgi region (G). (×11,500). *(From Azar and Potter, (eds), ref. 126)*

135

"Nonsecretory" forms of myeloma represent approximately 1% of all myeloma patients. These have classical radiological and cytological evidence of myeloma, but repeated examinations of their serum and urine fail to reveal a monoclonal immunoglobulin or a Bence Jones protein (133,134). Electron microscopic studies done on such cases reveal either a well-developed rough endoplasmic reticulum (Fig. 78), or, as observed in a more recent case, flaming cells with dilated and empty-looking cisternae lined by an endoplasmic reticulum devoid of ribosomal granules (Figs. 79 and 80). In another case thought to be an advanced osteolytic form of "nonsecretory" myeloma, electron microscopic examination of the marrow yielded granulocytic precursors indicative of a granulocytic sarcoma (see above). Because of the uneasiness in the diagnosis of "nonsecretory" myeloma, it is recommended that the work-up of such cases include electron microscopy.

Waldenström's Macroglobulinemia

Waldenström's macroglobulinemia is characterized by the presence of a serum monoclonal IgM associated with a leukemic or a leukemia-like picture in which IgM-producing lymphocytes, plasma cells, and intermediate cells infiltrate lymph nodes, the marrow, the spleen, the liver, and, to a lesser degree, other viscera. This condition merges imperceptibly with lymphoplasmacytic lymphosarcoma (see above), just as chronic lymphoid leukemia frequently coexists with lymphocytic lymphosarcoma.

Although the emphasis in previous descriptions has been on the finding of lymphocytoid plasma cells (or plasmacytoid lymphocytes) in Waldenström's macroglobulinemia (135), we have observed cases with a predominance of mature-appearing plasma cells, including forms with Russell bodies (Fig. 81). Intranuclear inclusions (Dutcher bodies) are frequently observed in lymphocytoid or plasmacytic cells in this condition (129). Intracytoplasmic crystalloid inclusions may also be noted, as in myeloma cells. See also under lymphoplasmacytic lymphosarcoma (Fig. 59).

Heavy-Chain Diseases

Heavy-chain diseases belong to three classes:

1. IgG H-chain disease (Franklin's disease, which is usually associated with a generalized lymphoplasmacytic proliferation (136,137) involving principally the marrow, lymph nodes, and the speen, without causing any osteolysis. Lymphocytes are intermixed with plasmacytic cells in various stages of development (Figs. 82 and 83).
2. IgA H-chain disease is associated with malabsorption syndrome and a malignant lymphoma primarily involving the small intestine or mesenteric lymph nodes (Mediterranean lymphoma). As in IgG H-chain disease, there is a lymphoylasmacytic infiltrate that may become generalized and involve the marrow (138a).
3. IgM H-chain disease has been observed in association with chronic lymphocytic leukemia. Plasma cells with vacuolated cytoplasm have been found in the bone marrow in addition to lymphocytes (139).

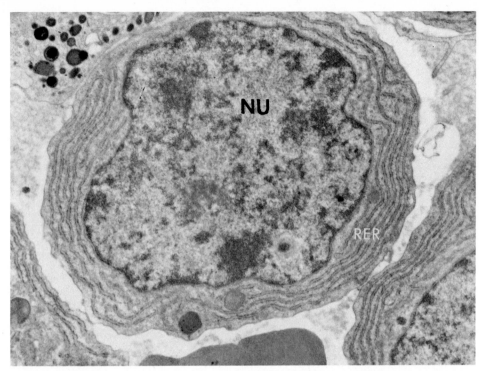

Figure 78. Myeloma cells from a patient with "nonsecretory" myeloma. Note the presence of abundant lamellar rough endoplasmic reticulum (RER) (×16,530)

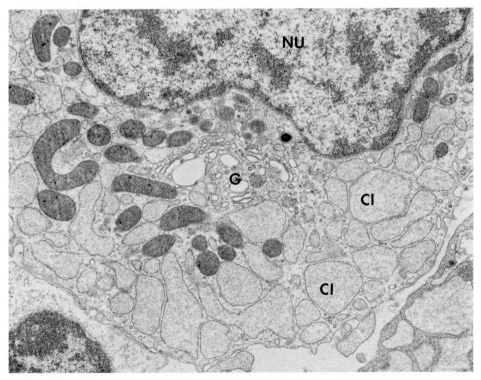

Figure 79. Portion of "nonsecretory" myeloma showing marked distention of cisternae (CI). Also note a well-developed Golgi region (G). (×12,500)

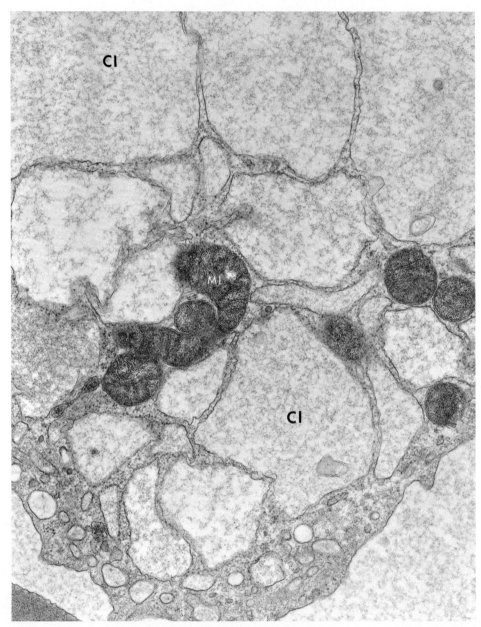

Figure 80. Same patient as above. Portion of "nonsecretory" myeloma cells showing marked dilatation of cisternae (CI). The cisternal lining is made of smooth rather than rough endoplasmic reticulum. (×33,500)

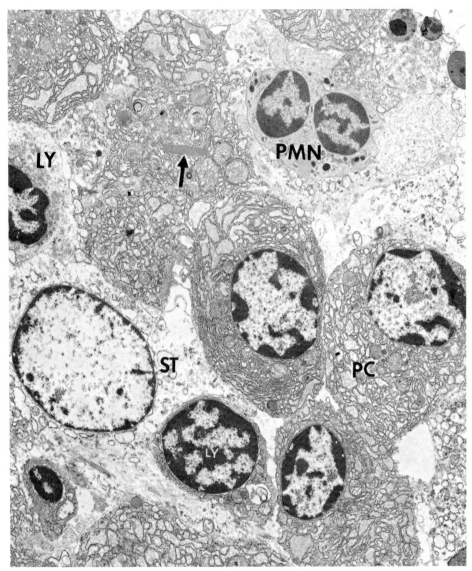

Figure 81. Electron micrograph of bone marrow field from a patient with Waldenström's macroglobulinemia and Raynaud's phenomenon. The predominant cell type is a differentiated plasmacytic cell (PC). Lymphocytes (LY) are also present, as well as a stromal cell (ST) and a neutrophil (PMN). Arrow points to a crystalloid formation within cytoplasm of a plasmacytic cell. (×5,100) *(From Azar and Potter (ds), ref. 126)*

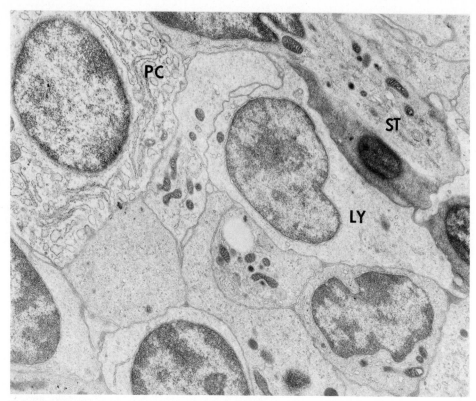

Figure 82. Bone marrow field from a patient with IgG heavy-chain diesease. Note predominance of lymphoid cells (LY) and presence of a rather differentiated plasmacytic cell (PC). A stromal cell is marked ST. (×10,488) *(Courtesy of Dr. Nariosang M. Kandawalla; unpublished data)*

Amyloidosis

Amyloidosis is mentioned here only because a common form of amyloidosis, the primary type, is considered to be constituted by extracellular protein deposits of immunoglobulin derivation (140,141). Amyloid is recognized ultrastructurally as fibrils, haphazardly dispersed or arranged in bundles, and each measuring from 50 Å to 120 Å in diameter and several microns in length. Amyloid fibrils have a tendency to beading, suggesting a helical arrangement (Fig. 84). The primary type of amyloidosis is frequently associated with one form or another of monoclonal gammopathy, including multiple myeloma and Waldenström's macroglobulinemia.

The parenchymal or secondary type of amyloidosis is of nonimmunoglobulin nature and is almost always associated with an underlying chronic inflammatory condition. Various tumor-associated amyloids have been described. All these varieties are characterized by Congo red stainability with greenish birefringence on polarization, and by their fibrillar ultrastructural appearance.

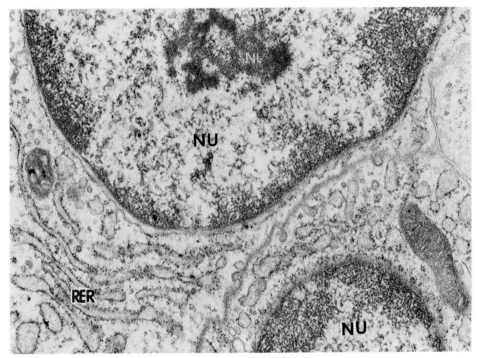

Figure 83. Same patient as above showing two neighboring plasmacytic cells with well-developed rough endoplasmic reticulum (RER). (×27,600) *(Courtesy of Dr. Nariosang M. Kandawalla; unpublished data)*

MACROPHAGES AND MONOCYTES

Phagocytic System

The reticuloendothelial or phagocytic system may be defined as the collectivity of fixed and mobile cells that are primarily concerned with ingestion of particulate or colloidal substances. Three main components are recognized: *a)* blood monocytes, of marrow origin, which are potential macrophages; *b)* macrophages; and *c)* fixed histiocytes.

Macrophages and *histiocytes* are characterized by the presence in their cytoplasm of phagolysosomes or of residual bodies in addition to primary lysosomes. Phagocytosed particles or organisms and lipid droplets may be also recognized (Fig. 85). Histiocytes tend to be elongated, even elliptical, and often resemble fibroblasts with the added characteristics of a phagocytic cell. They are also enmeshed within a network of collagen.

As to tumors or tumor-like proliferations of macrophages, monocytes, and related reticuloendothelial cells, these are classified in Table 5.

Monocytes

Monocytes are difficult to distinguish by light microscopy, and even by ultrastructural study, from neutrophilic metamyelocytes and nonsegmented (band) neutrophils. Although it has been demonstrated under experimental conditions

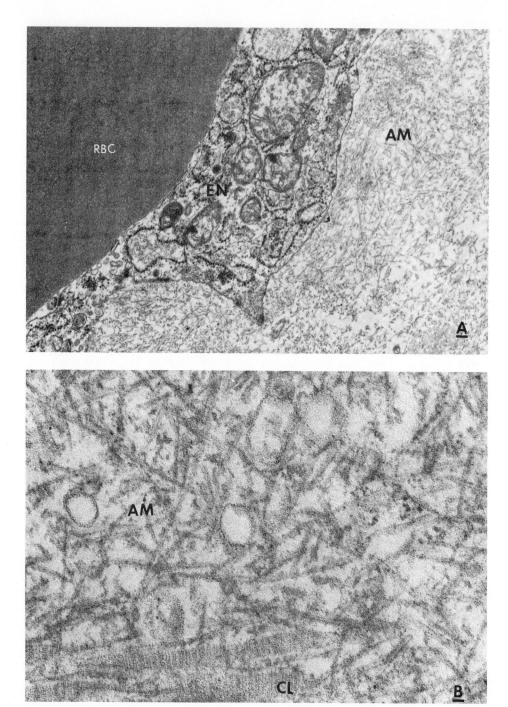

Figure 84. Electron micrograph of field of spleen of patient with λ light-chain disease and generalized amyloidosis. A shows amyloid deposits (AM) under the endothelial lining (EN). The vascular lumen contains a red blood cell (RBC). (×29,810). B shows a higher magnification of the amyloid fibrils. These seem to run in pairs and are slightly beaded or twisted. Compare the size of amyloid fibers to those of two collagen fibers (CL). (×154,470)

142

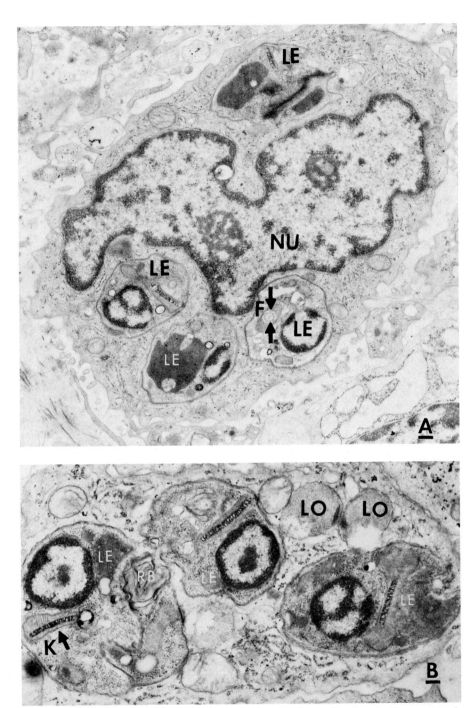

Figure 85. Macrophage of dermis phagocytosing Leishman bodies. In A, note nucleus (NU) of the macrophage and four Leishman bodies (LE). These lie within phagocytic vacuoles. The two arrows point to a cross section of a flagellum (F). (×9,000) B represents portion of a macrophage containing three Leishman bodies (LE). In two instances a lysosomal granule (LO) is adherent and seemingly bursting into its respective phagocytic vacuole. One Leishman organism shows a kinetoplast (K) at arrow and a residual body (RB). (×18,000)

Table 5. Neoplasms and Tumor-Like Conditions of the Macrophage System

Tumor-like conditions [a]
  Xanthomatosis and xanthelesma
  Histiocytosis X
    Eosinophilic granuloma
    Hand-Schüller-Christian disease
    Letterer-Siwe disease

Benign neoplasms
  Xanthoma, fibrous histiocytoma, and variants

Malignant neoplasms
  Malignant fibrous histiocytoma (fibrohistiocytic sarcoma)
  Malignant xanthogranuloma
  Malignant histiocytosis (malignant reticuloendotheliosis, histiocytic medullary reticulosis)
  Histiomonocytic leukemia (Schilling type)
  Monomyelocytic leukemia (Naegeli type)

[a] The genetic lipoidoses or lipoid storage diseases have been excluded from this tabulation.

that monocytes originate from marrow precursors, it is difficult to recognize monocytic precursors in the human bone marrow. By electron microscopy, monocytes have certain features that tend to distinguish them from both granulocytes and stimulated lymphocytes. In addition to the monocyte's characteristically indented nucleus, the cytoplasm contains a varying number of primary lysosomal granules. These tend to be heterogeneous in size and density, but the small, dense, round or ovoid "specific" granules of neutrophils are either absent or in minority (Fig. 86). There is also an abundance of vesicles and agranular endoplasmic reticulum. Microfibrils and microtubules may be readily observed (142,143). It must be emphasized that none of these cytoplasmic features is so characteristic as to distinguish a monocyte unequivocally from a degranulated or stimulated metamyelocyte. Monocytes still need to be characterized by means of selective enzymatic reactions.

Epithelioid Cells and Multinucleated Giant Cells

In one form of transformation of macrophages, their cytoplasm becomes progressively larger, lysosomes are lost or extruded, mitrochondria increase in number, thus giving rise to *epithelioid cells* (Fig. 87). Clusters of epithelioid cells may in turn fuse together by a process of breakdown of contiguous cell membranes to form *multinucleated giant cells* (144,145). The asteroid inclusions of giant cells of histicytic or epithelioid origins appear to be criss-crossing bundles of collagen (Fig. 88) that either are trapped in the process of fusion of neighboring cells or are actually formed in the center of these giant cells (144).

The Lipoidoses

The lipoidoses include principally the genetic lipoidoses or lipoid storage diseases resulting in an accumulation of specific lipoidal substances within macrophages. An enzymatic defect either can be demonstrated or is strongly im-

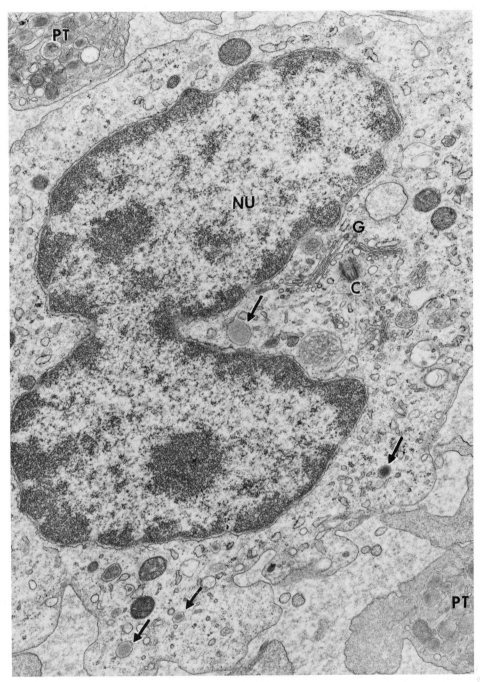

Figure 86. Peripheral blood buffy coat preparation showing a cell conforming to the appearance of a monocyte. Note the bilobed, deeply indented nucleus (NU). The Golgi region (G) is well developed. Note centriole (C). The granulation in the cytoplasm is sparse, and the granules are of different sizes and densities (see arrows). (×6,250)

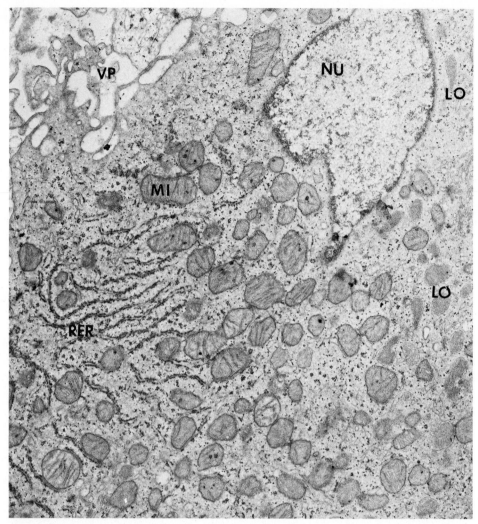

Figure 87. Portion of an epithelioid cell in a lymph node of a patient with sarcoidosis. Note the irregularly shaped nuclear contour with dispersed chromatin. The cytoplasm contains numerous mitochondria (MI). In one area of the cytoplasm is an accumulation of rough endoplasmic reticulum (RER). Primary lysosomal granules (LO) are seen at the right. Also note the intricate peripheral villous processes (VP). (×28,800)

plied. Excessive storage of lipoidal substances may also occur as a result of prolonged and excessive breakdown of cell membranes, usually of erythrocytic and leukocytic origin, thus causing a relative enzymatic insufficiency.

The commonest lipoidosis we have encountered is the chronic adult and non-neurotropic type of Gaucher's disease. The rather insidious onset of this illness in a young adult, male or female, of non-Jewish background can cause a diagnostic problem that can be easily solved by electron microscopy. The classical Gaucher cell is seen to contain membrane-lined, closely packed, and roughly fusiform saccules (Figs. 89 and 90). These are filled with 300- to 400-Å-wide

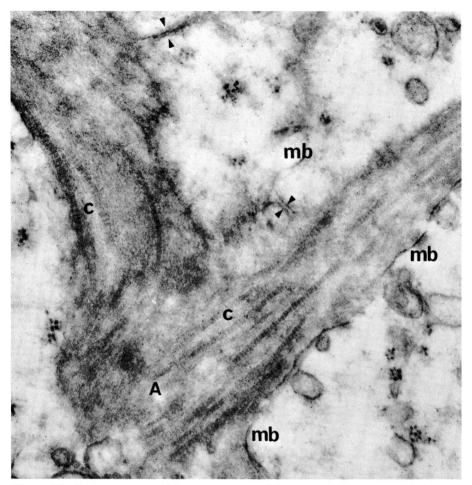

Figure 88. Portion of asteroid body of patient with sarcoidosis. The arms of the asteroid seem to be largely composed of collagen bundles (C). Remnants of neighboring cytoplasmic membranes (mb) of coalescing epithelioid cells are shown between arrows. (×41,000) *(From Azar and Lunardelli, ref. 144)*

microtubules that are thought to be constituted in large part by the stored glucosyl ceramide (146).

Gaucher-type cells, similar but not identical to the classical cells of the genetic type of Gaucher disease, can be observed as a secondary or acquired pheomenon in such conditions as chronic myeloid leukemia (147), thalassemia major (148), and Hodgkin's disease (149). In chronic myeloid leukemia, the excessive breakdown of dead leukemic cells, particularly in the marrow, seems to be the likely source of the stored lipoid substances. The deposits found in the cells of this acquired form are, however, considerably finer than those of classical Gaucher cells, and true microtubules are not observed (Figs. 91 and 92). It has also been suggested that the stored lipoid substances found in this secondary form are not related to the glucosyl cerumide that accumulates in classical Gaucher cells (147).

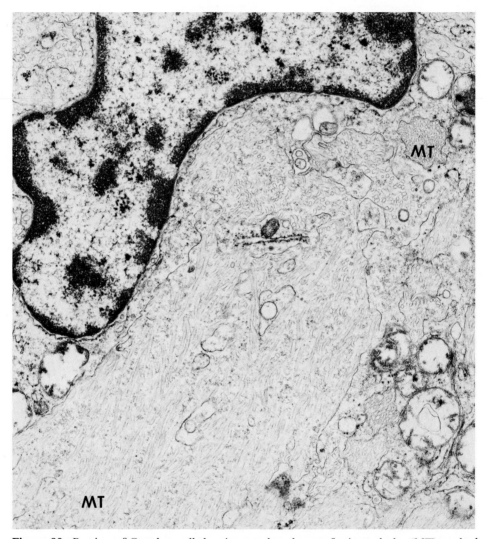

Figure 89. Portion of Gaucher cell showing an abundance of microtubules (MT) packed within its cytoplasm. (×20,000)

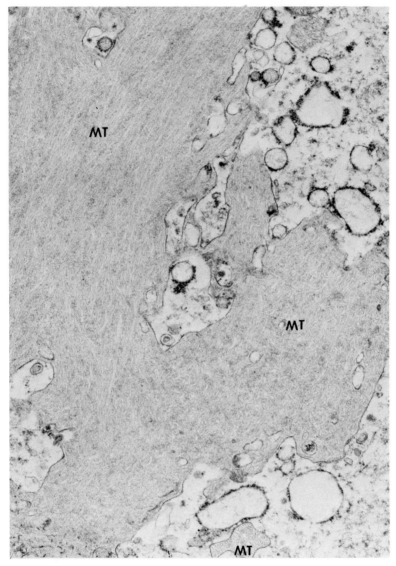

Figure 90. Higher magnification of a portion of Gaucher cell showing detail of the microtubules. (×33,800)

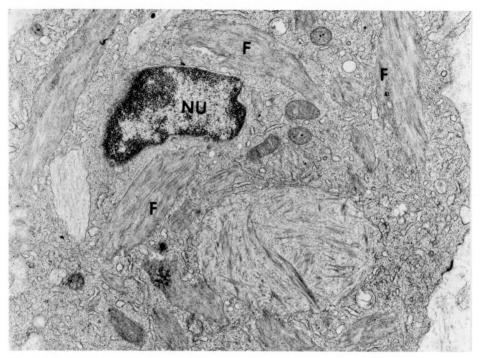

Figure 91. Storage cell from a case of chronic granulocytic leukemia treated for over a two-year period. There is a superficial resemblance to classical Gaucher cells. There are fascicles containing fibrils (F). (×20,240) *(From Azar, ref. 5)*

In the Gaucher-type cells found in chronic thalassemia, the deposits bear a striking resemblance to those of classical Gaucher cells (Figs. 93 and 94). In this situation, the source of the stored lipid was the relatively large amount of destroyed red cells, and the spleen contained an increased amount of monohexyl cerumide (148).

Other forms of storage cells, namely those of Niemann-Pick disease (150), "sea blue" histiocytes (151), exhibit distinctive accumulations or inclusions of stored material within macrophages.

Histiocytosis X

The distinguishing feature of the tumor-like condition of histiocytosis X, and more specifically the form known as Hand-Schüller-Christian disease, is the presence within histiocytes of elongated, "caterpillar-like," tubular cytoplasmic inclusions (152). These are identical to those observed in Langerhans cells of the epidermis (153). Similar inclusions have been described in the histiocytes of Letterer-Siwe disease (154), and of eosinophilic granuloma (155).

Fibrous Histiocytoma

Since the diagnosis of benign and malignant forms of fibrous histiocytoma is at times difficult to reach on the basis of light microscopic criteria, one would assume that electron microscopic studies could facilitate such a task. Indeed, a

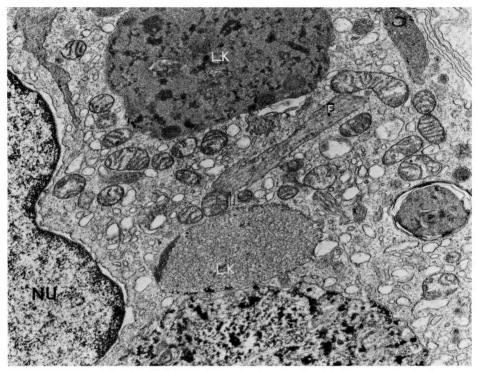

Figure 92. Macrophage in marrow of same patient as above. There are large lysosomal complexes probably representing products of digestion of dead or dying leukocytes (LK). In addition, there is formation of fascicles containing fibrillar deposits (F). (×20,240)

classical tumor of histiocytes discloses its cytological identity because of features denoting its phagocytic functions: presence of primary lysosomes, phagolysosomes, lipoid droplets, and many pinocytic vacuoles. In practical terms, however, connective tissue tumors of all types also have a major fibroblastic or a nondescript mesenchymal component that is seldom suspected on light microscopic examination. To complicate matters, histiocytes and fibroblasts seem to represent distinct poles of a continuum that led Arthur Purdy Stout to write about "facultative" fibroblasts in a number of soft tissue tumors (156). So far, the study of connective tissue tumors, with the exception of occasional "lucky strikes", has not been rewarding, at least in our hands.

Histiomonocytic Leukemia

"Pure" or truly histiomonocytic leukemias, or Schilling-type leukemias, are considerably rarer than the monocytoid or monomyelocytic, Naegeli-type leukemias. The latter have been briefly discussed above in relation to acute leukemias and acute myeloid leukemias. Monomyelocytic leukemias may develop in patients treated for multiple myeloma, and are often associated with high levels of serum and urine lysozyme or muramidase (44). In these cases, degenerating myeloma cells persist in the marrow amidst fields of

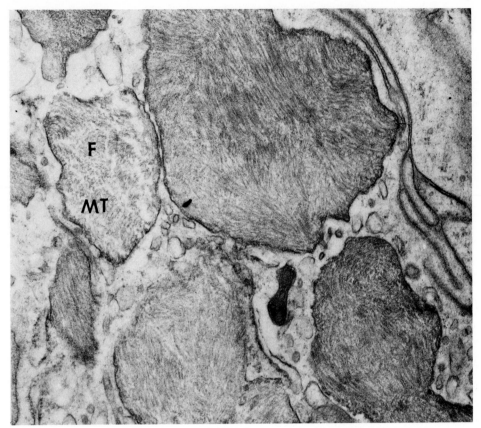

Figure 93. Portion of Gaucher-like cell from a patient with thalassemia. The fascicular formations (F) contain closely packed microtubules (MT). (×37,250)

"monomyelocytic cells" (Fig. 95). A representative lysozyme-producing leukemic cell is depicted in Figure 96.

There have been several attempts at defining the ultrastructure of "pure" monocytic or histiomonocytic leukemias (157,158,159). In addition to their naphthol AS-D acetate esterase positivity, the hallmark of monoblasts appears to be their marked nuclear irregularity, with nuclear folds, pockets, and bridges. Their cytoplasmic granules are scattered and dense and belong to the primary or azurophilic type. Leukemic monocytes and histiocytes can be frequently recognized under the scanning electron microscope by the presence of large, well-developed, broad-based ruffled membranes or prominent ridge-like profiles. These features resemble those of normal monocytes (160).

CONCLUSION

Electron microscopic studies of hematological specimens, such as buffy coat preparations of peripheral blood, bone marrow aspirates, lymph nodes, and spleen, have become, in selected situations, a natural extension of routine light microscopic examinations and have greatly enriched the latter.

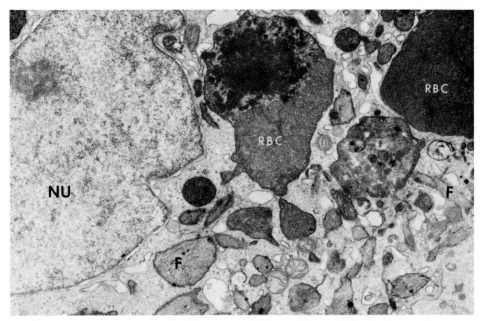

Figure 94. Macrophage in spleen of patient with thalassemia major showing phagocytesed red blood cells (RBC). In addition, there is beginning of fascicle formation (F). These cells contain closely packed microtubules similar to those seen in Figure 93 and in Gaucher's disease. (×15,000)

Two main objectives can be met in ultrastructural studies of hematopoietic tissues: *a)* the solution of diagnostic problems that are important from a clinical management standpoint and cannot be resolved by routine light microscopic studies; and *b)* the development of a keener appreciation of cellular detail and interest in ultrastructural studies that are needed to supplement light microscopic observations and will contribute to a better understanding of both cytogenesis and pathogenesis.

In our electron microscopy unit, which is part of a general diagnostic laboratory, ultrastructural studies of hematopoietic tissues, particularly blood buffy coat and marrow samples, have helped us reach a more precise diagnosis in a variety of serious hematological conditions that include the acute "blast cell" leukemias, hairy cell leukemia, Sézary syndrome, granulocytic sarcoma, "nonsecretory" myeloma, and other unusual forms of monoclonal gammopathy, primary and secondary Gaucher's disease, and sideroblastic anemias of the classical (intramitochondrial) as well as the microsomal variety.

In all diagnostic electron microscopic work-up, the examiner is again alerted about the necessity of interpreting his ultrastructural observations in the light of routine histological and cytological findings and pertinent clinical data.

It should also be obvious that diagnostic electron microscopy has some serious limitations, notably those inherent in the sampling of tissues or cells, and those affecting all morphological disciplines. The enthusiast will need to be reminded that electron microscopy cannot resolve all diagnostic problems and that certain

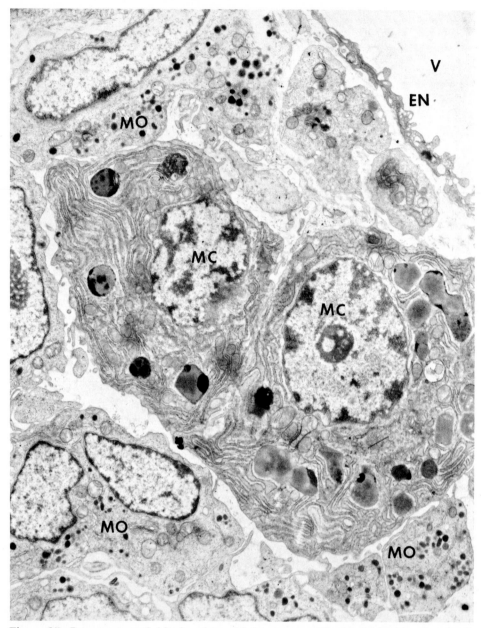

Figure 95. Bone marrow field of patient with lysozyme-producing leukemia whose multiple myeloma was treated with melphalan. The two myeloma cells (MC in center of field show marked cytoplasmic alterations, with formation of irregularly shaped dense granules probably of lysosomal origin. There are also scattered monocytoid cells (MO) in the remainder of the field. In the right upper quadrant is a vascular lumen (V) lined by endothelium (EN). (×9,000) *(From Azar and Potter (eds), ref. 126)*

154

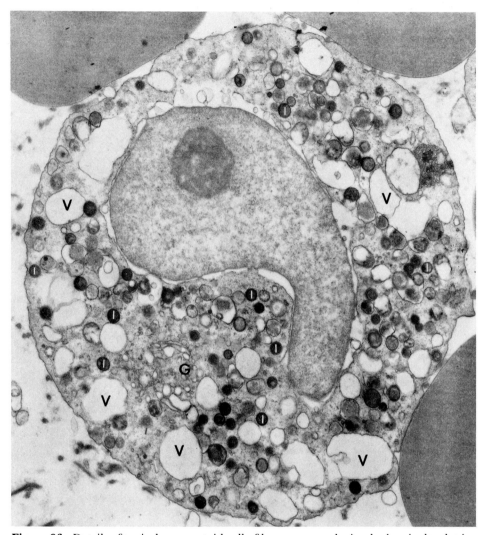

Figure 96. Details of typical monocytoid cell of lysozyme-producing leukemia developing in a patient treated for multiple myeloma. The Golgi region (G) is well developed. Throughout the cytoplasm are numerous granules, most of which are of the primary or azurophilic type. There is, however, considerable heterogeneity in size and density. There are also numerous vesicles (V). (×13,125) *(From Azar and Potter (eds), ref. 126)*

questions will remain unanswerable no matter what method or combination of methods is applied to them.

ACKNOWLEDGMENTS

The author is greatly indebted to Mr. George Kasnic, Jr., supervisor of the Electron Microscopy Unit at Tampa VA Hospital for his technical assistance, and to Mrs. Jo Ellen McKibben for typing this manuscript.

REFERENCES

1. Bessis M: *Living Blood Cells and Their Ultrastructure* (trans Weed RI): New York, Springer-Verlag, 1973.

2. Custer PR: *An Atlas of the Blood and Bone Marrow,* ed. 2. Philadelphia, WB Saunders, 1974.

3. Tanaka Y, Goodman JR: *Electron Microscopy of Human Blood cells.* New York, Harper & Row, 1972.

4. Ruzicka F: *Electronenmicroskopische Hämatologie.* New York, Springer-Verlag, 1976.

5. Azar HA: Application de la microscopie électronique aux études hématologiques. *Vie Méd* (Canada français), 5:1197, 1976.

6. Johari O (ed): *Proceedings 10th Annual SEM Symposium.* Chicago, IIT Research Institute, 1977.

7. Bessis M, Brecher G (eds): *Unclassifiable Leukemias.* New York, Springer-Verlag, 1975.

8. Yam LT, Li CY, Lam KW: Tartrate-resistant acid phosphatase iso-enzyme in the reticulum cells of leukemic reticuloendotheliosis. *N Engl J Med* 284:357, 1971.

9. Hayhoe GFJ, Quaglino D, Doll R: *The Cytology and Cytochemistry of Acute Leukemias. A Study of 140 Cases.* London, M.R.C. Special Report Series, No. 304, Her Majesty's Stationery Office, 1964.

10. Maloney WC, McPherson K, Fliegelman L: Esterase activity in leukocytes demonstrated by the use of naphthol AS-D chloracetate substrate. *J Histochem Cytochem* 8:200, 1960.

11. Taylor FM III, Azar HA: Ultrastructural study of "ringed" sideroblasts. *Fed Proc* 36:1089, 1977.

12. Stetson CA Jr: The state of hemoglobin in sickled erythrocytes. *J Exp Med* 123:341, 1966.

13. Bessis M, Alagille D, Breton-Gorius J: Particularités des érythroblastes et des érythrocytes dans la maladie de Colley. Etude au microscope électronique. *Rev Hematol* 13:538, 1958.

14. Quaglino D, Hayhoe GFJ: Periodic-acid-Schiff positivity in erythroblasts with special reference to Di Guglielmo's disease. *Br J Haematol* 6:26, 1960.

15. Cohn ZA, Hirsch JG: The isolation and properties of the specific cytoplasmic granules of rabbit polymorphonuclear leukocytes. *J Exp Med* 112:983, 1960.

16. Bainton DF, Farquhar MG: Origin of granules in polymorphonuclear leukocytes: two types derived from opposite faces of the Golgi complex in developing granulocytes. *J Cell Biol* 28:277, 1966.

17. Clein GP, Flemans RJ: Involvement of the erythroid series in blastic crises of chronic myeloid leukemia. *Br J Haematol* 12:754, 1966.

18. Baikie AG, Court-Brown WM, Buckton KE, et al: A possible specific chromosome abnormality in human chronic myeloid leukemia. *Nature* 188:1165, 1960.

19. Maniatis AK, Amsel S, Mitus WJ, et al: Chromosome pattern of bone marrow fibroblasts in patients with chronic granulocytic leukemia. *Nature* 222:1278, 1969.

20. Bainton DF, Farquhar MG: Differences in enzyme content of azurophil and specific granules of polymorphonuclear leukocytes. I. Histochemical staining of bone marrow smears. *J Cell Biol* 39:286, 1968.

21. Bainton DF, Farquhar MG: Differences in enzyme content of azurophil and specific granules of polymorphonuclear leukocytes. II. Cytochemistry and biochemistry of bone marrow cells. *J Cell Biol* 39:299, 1968.

22. Watanabe I, Donahue S, Hoggatt N: Method for electron microscopic studies of circulating human leukocytes and observations on their fine structure. *J Ultrastruct Res* 20:366, 1967.

23. Scott RE, Horn RG: Ultrastructural aspects of neutrophil granulocyte development in humans. *Lab Invest* 23:202, 1970.

24. Spitznagel JK: Sorting out lysosomes and other cytoplasmic granules from polymorphs of rabbits and humans: A search for antibacterial factors, in Williams RC Jr, Fudenberg HH (eds): *Phagocytic Mechanisms in Health and disease.* New York, Intercontinental Medical Book, 1972, p 83.

25. Breton-Gorius J: Structures périodiques dans les granulations éosinophiles et neutrophiles des leucocytes polynucléaires du sang de l'homme. *Nouv Rev Fr Hematol* 6:195, 1966.

26. Davidson WM: Inherited variations in leukocytes. *Semin Hematol* 5:255, 1968.

27. Jordan SW, Larsen WE: Ultrastructural studies of the May-Hegglin anomaly. *Blood* 25:921, 1965.

28. Lusher JM, Schneider J, Mizukami I, et al: The May-Hegglin anomaly: Platelet function, ultrastructure and chromosome studies. *Blood* 32:950, 1968.

29. Reilley WA, Lindsay S: Gargoylism (lipochondrodystrophy): A review of clinical observations in eighteen cases. *Am J Dis Child* 75:595, 1948.

30. Page AR, Berendes H, Warner J, et al: The Chediak-Higashi syndrome. *Blood* 20:330, 1962.

31. Efrati P, Jonas W: Chediak's anomaly of leukocytes in malignant lymphoma associated with leukemic manifestations. Case report with necropsy. *Blood* 13:1063, 1958.

32. Douglas SD, Blume RS, Wolff SM: Fine structural studies of leukocytes from patients and heterozygotes with Chediak-Higashi syndrome. *Blood* 33:527, 1969.

33. White JG: The Chediak-Higashi syndrome: a possible lysosomal disease. *Blood* 28:143, 1966.

34. Zucker-Franklin D: Electron microscopic studies of human granulocytes: Structural variations related to function. *Semin Hematol* 5:109, 1968.

35. Kjeldsberg DR, Swanson J: Platelet satellitism. *Blood* 43:831, 1974.

36. Bauer HM: In-vitro platelet-neutrophil adherence. *Am J Clin Pathol* 63:824, 1975.

37. Zucker-Franklin D: Electron microscope study of the degranulation of polymorphonuclear leukocytes following treatment with streptolysin. *Am J Pathol* 47:419, 1965.

38. McCall CE, Katayama I, Cotran RS, et al: Lysosomal and ultrastructural changes in human "toxic" neutrophils during bacterial infection. *J Exp Med* 129:267, 1969.

39. Bennett JM, Reed CE: Acute leukemia cytochemical profile: Diagnostic and clinical implications. *Blood Cells* 1:101, 1975.

40. Spang-Thomsen M, Visfeldt J: Animal model of human disease: Malignant tumors. *Am J Pathol* 84:193, 1976.

41. Lozzio BB, Lozzio CB, Machado E: Brief communication: Human myelogenous (Ph[1+])leukemia cell line: Transplantation into athymic mice. *J Natl Cancer Inst* 56:627, 1976.

42. Machado EA, Lozzio BB: Animal model of human disease: Hyposplenia, asplenia, and immunodeficiency. *Am J Pathol* 85:515, 1976.

43. Corcino J, Krauss S, Waxman S, et al: Release of vitamin B_{12} binding protein by human leukocytes in vitro. *J Clin Invest* 49:2250, 1970.

44. Osserman EF, Lawlor DP: Serum and urinary lysozyme (muramidase) in monocytic and monomyelocytic leukemia. *J Exp Med* 124:921, 1966.

45. Catovsky D, Hoffbrand AV, Ikoku NB, et al: Significance of cell differentiation in acute myeloid leukaemia. *Blood Cells* 1:201, 1975.

46. Pruzanski W, Saito SG: The diagnostic value of lysozyme (muramidase) estimation in biological fluids. *Am J Med Sci* 258:405, 1969.

47. Tan HK, Wages B, Gralnick HR: Ultrastructural studies in acute promyelocytic leukemia. *Blood* 39:628, 1972.

48. Breton-Gorius J, Houssay D: Auer bodies in acute promyelocytic leukemia: Demonstration of their fine structure and peroxidase localization. *Lab Invest* 28:135, 1973.

49. Whang-Peng J, Canellos GP, Carbone PP, et al: Clinical implications of cytogenetic variants in chronic myelocytic leukemia (CML). *Blood* 32:755, 1968.

50. Krauss S: The Philadelphia chromosome and leucocyte alkaline phosphatase in chronic myelocytic leukemia and related disorders. *Ann NY Acad Sci* 155:983, 1968.

51. Shaw MT, Bottomley RH, Grozea PN, et al: Heterogeneity of morphological, cytochemical, and cytogenetic features in the blastic phase of chronic granulocytic leukemia. *Cancer* 35:199, 1975.

52. Rudders RA, Kilcoyne RF: Myeloproliferative disorder with lytic osseous lesions and chromosomal anomalies. *Am J Clin Pathol* 61:673, 1974.

53. Carmichael GP Jr, Lee YT: Granulocytic sarcoma simulating "nonsecretory" multiple myeloma. *Hum Pathol* 8:697, 1977.

54. Gilbert HS: The spectrum of myeloproliferative disorders. *Med Clin North Am* 57:355, 1973.

55. Archer GT, Blackwood A: Formation of Charcot-Leyden crystals in human eosinophils and basophils and study of the composition of isolated crystals. *J Exp Med* 122:173, 1965.

56. Litt M: Eosinophils and antigen-antibody reactions. *Ann NY Acad Sci* 116:964, 1964.

57. Zucker-Franklin D, Stollerman GH (eds): Eosinophil function and disorders. *Adv Intern Med* 19:1, 1974.

58. Benvenisti DS, Ultmann JE: Eosinophilic leukemia: Report of five cases and review of literature. *Ann Intern Med* 71:731, 1969.

59. Chusid MJ, Dale DC, West BC, et al: The hypereosinophilic syndrome: Analysis of fourteen cases with review of the literature. *Medicine* 54:1, 1975.

60. Zucker-Franklin D: Electron microscopic study of human basophils. *Blood* 29:878, 1967.

61. Dvorak HF, Dvorak AM: Basophils, mast cells, and cellular immunity in animals and man. *Hum Pathol* 3:454, 1972.

62. Hastie RA: A study of the ultrastructure of human basophil leukocytes. *Lab Invest* 31:223, 1974.

63. Terry RW, Bainton DF, Farquhar MG: Formation and structure of specific granules in basophilic leukocytes of the guinea pig. *Lab Invest* 21:65, 1969.

64. Shelley WB: Methods of observing the basophil leucocyte degranulation response. *Ann NY Acad Sci* 103:427, 1963.

65. Ishizaka K: Human reaginic antibodies *Ann Rev Med* 21:187, 1970.

66. Ishizaka T, Ishizaka K, Gunner S, et al: Histamine release from human leukocytes by anti-γE antibodies. *J Immunol* 105:1459, 1970.

67. Nan RC, Hoagland HC: A myeloproliferative disorder manifested by persistent basophilia, granulocytic leukemia and erythroleukemic phases. *Cancer* 28:662, 1971.

68. Parmley RT, Spicer SS, Wright NJ: The ultrastructural identification of tissue basophils and mast cells in Hodgkin's disease. *Lab Invest* 32:469, 1975.

69. Burgoon CF Jr, Graham JH, McCaffree DL: Mast cell disease. *Arch Dermatol* 98:590, 1968.

70. Mihm MC, Clark WH, Reed RJ, et al: Mast cell infiltration of the skin and the mastocytosis syndrome. *Hum Pathol* 4:231, 1973.

71. van Kammen E: Generalized mastocytosis. *Acta haematol* 52:129, 1974.

72. Efrati P, Klajman A, Spitz H: Mast cell leukemia. Malignant mastocytosis with leukemia-like manifestations. *Blood* 12:869, 1957.

73. Friedman BI, Will JJ, Freiman DG, et al: Tissue mast cell leukemia. *Blood* 13:70, 1958.

74. Baldini MG, Ebbe S (eds): *Platelets. Production, Function, Transfusion and Storage.* New York, Grune & Stratton, 1974.

75. White, JG: Current concepts of platelet structural physiology and pathology. *Hum Pathol* 5:1, 1974.

76. Behnke O, Kristensen BI, Engdahl NL: Electron microscopical observations on actinoid and myosinoid filaments in blood platelets. *J Ultrastruct Res* 37:351, 1971.

77. Smith TP, Dodds WJ, Tartaglia AP: Thrombasthenic-thrombopathic thrombocytopenia with giant, "Swiss-cheese" platelets. *Intern Med* 79:828, 1973.

78. Epstein CJ, Sahud MA, Piel CF, et al: Hereditary macrothrombocytopathia, nephritis and deafness. *Am J Med* 52:299, 1972.

79. Greipp PR, Gralnick HR: Platelet to leukocyte adherence phenomena associated with thrombocytopenia. *Blood* 47:513, 1976.

80. Maldonado JE, Pintado T, Pierre RV: Dysplastic platelets and circulating megakaryocytes in chronic myeloproliferative diseases. I. The platelets: Ultrastructure and peroxidase reaction. *Blood* 43:797, 1974.

81. Maldonado JE: Dysplastic platelets and circulating megakaryocytes in chronic myeloproliferative diseases. II. Ultrastructure of circulating megakaryocytes. *Blood* 43:811, 1974.

82. Maldonado JE, Pintado T: Ultrastructure of the megakaryocytes in refractory anemia and myelomonocytic leukemia, in Baldini MG, Ebbe S (eds): *Platelets. Production, Function, Transfusion and Storage.* New York, Grune & Stratton, 1974, p 105.

83. Hossfeld DK, Tormey D, Ellison RR: Ph^1− positive megakaryoblastic leukemia. *Cancer* 36:576, 1975.

84. Nowell, PC: Phytohemagglutinin: an initiator of mitosis in cultures of normal human leukocytes. *Cancer Res* 20:462, 1960.

85. Naspitz CK, Richter M: The action of phytohemagglutinin in vivo and in vitro, a review. *Prog Allergy* 12:1, 1968.

86. Ford WL, Gowans JL: The traphic of lymphocytes. *Semin Hematol* 6:67, 1969.

87. Cradock CG: Kinetics of the lymphoreticular tissue, with particular emphasis on the lymphatic system. *Semin Hematol* 4:387, 1967.

88. Norman A, Sasaki MS, Ottoman RE, et al: Lymphocyte lifetime in women. *Science* 147:745, 1965.

89. Bentwich Z, Kunkel HG: Specific properties of human B and T lymphocytes and alteration in disease. *Transplant Rev* 16:29, 1973.

90. Good RA: Immunodeficiency in developmental perspective, in the *Harvey Lectures*. New York, Academic Press, 1973.

90a.Mori Y, Lennert K: *Electron Microscopic Atlas of Lymph Node Cytology and Pathology*. New York, Springer, 1969.

90b.Levine GD, Dorfman RF: Nodular lymphoma: An ultrastructural study of its relationship to germinal centers and a correlation of light and electron microscopic findings. *Cancer* 35:148, 1975.

91. Burnet FM: *Cellular Immunology*. Cambridge, University Press, 1969.

92. Burtin P, Buffe D: Synthesis of human immunoglobulins in germinal centers of lymphoid organs. *J Immunol* 98:536, 1967.

93. Hansen JA, Good RA: Malignant disease of the lymphoid system in immunological perspective. *Hum Pathol* 5:567, 1974.

93a.Parker JW, Taylor CR, Pattengale PK, et al: Morphologic and cytochemical comparison of human lymphoblastoid T-cell and B-cell lines: Light and electron microscopy. *J. Natl Cancer Inst* 60:59, 1978.

94. Okano H, Khouri F, Azar HA: *In vitro* transformation of human lymphocytes into macrophages. *Fed Proc* 27:717, 1968.

95. WHO, *Histological and Cytological Typing of Neoplastic Diseases of Haematopoietic and Lymphoid Tissues*, International Histological Classification of Tumours No. 14. Geneva, World Health Organization, 1976.

96. Zacharski LR, Linman JW: Chronic lymphocytic leukemia versus chronic lymphosarcoma cell leukemia. *Am J Med* 47:75, 1969.

97. Winkelman RK: Clinical studies of T-cell erythroderma in the Sézary syndrome. *Mayo Clin Proc* 49:519, 1974.

98. Flandrin G, Brouet J-C: The Sézary cell. Cytologic, cytochemical, and immunologic studies. *Mayo Clin Proc* 49:567, 1974.

99. Zucker-Franklin D: Properties of the Sézary cell. An ultrastructural analysis. *Mayo Clin Proc* 49:575, 1974.

100. Labaze JJ, Moscovic EA, Pham TD, and Azar HA: Histologic and ultrastructural findings in a case of the Sézary syndrome. *J Clin Pathol* 25:312, 1972.

101. Edelson RL, Lutzner MA, Kirpatrick CH, et al: Morphologic and functional properties of the atypical T lymphocytes of the Sézary syndrome. *Mayo Clin Proc* 49:558, 1974.

102. Winkelman RK (ed): Symposium on the Sézary cell. *Mayo Clin Proc* 49:513, 1974.

103. Clendenning WE, Brecher G, Van Scott EJ: Mycosis fungoides. *Arch Dermatol* 89:785, 1964.

104. Brownlee TR, Murad TM: Ultrastructure of mycosis fungoides. *Cancer* 26:686, 1970.

105. Rosas-Uribe A, Variakojis D, Molnar Z, et al: Mycosis fungoides: An ultrastructural study. *Cancer* 34:634, 1974.

106. Bouroncle BA, Wiseman BK, Doan CA: Leukemic reticuloendotheliosis. *Blood* 13:609, 1958.

107. Katayama I, Finkel HE: Leukemic reticuloendotheliosis. A clinicopathologic study with review of the literature. *Am J Med* 57:115, 1974.

108. Daniel MT, Flandrin G: Fine structure of abnormal cells in hairy cell (tricholeukocytic) leukemia, with special reference to their in vitro phagocytic capacity. *Lab Invest* 30:1, 1974.

109. Schnitzer B, Kass L: Hairy-cell leukemia. A clinicopathologic and ultrastructural study. *Am J Clin Pathol* 61:176, 1974.

109a.Braylan RC, Jaffe ES, Triche TJ, et al: Structural and functional properties of the "hairy" cells of leukemic reticuloendotheliosis. *Cancer* 41: 210, 1978.

110. Catovsky D, Pettit JE, Galetto J, et al: The B-lymphocyte nature of the hairy cell of leukaemic reticuloendotheliosis. *Br J Haematol* 26:29, 1974.

111. Rappaport H: *Tumors of the Hematopoietic System. Atlas of Tumor Pathology.* Washington, DC, AFIP, 1966, section 3, fascicle 8.

112. Jaffe, ES, Shevach EM, Frank MM, et al: Nodular lymphoma: evidence for origin from follicular B lymphocytes. *N Engl J Med* 290:813, 1974.

113. Lukes RJ, Collins RD: New observations on follicular lymphoma. *Gann Monogr Cancer Res* 15:209, 1973.

114. Azar HA, Hill WT, Osserman EF: Malignant lymphoma and lymphatic leukemia associated with myeloma-type serum proteins. *Am J Med* 23:239, 1957.

115. Mathé D, Belpomme D, Dantchev D, et al: Immunoblastic lymphosarcoma, a cytological and clinical entity? *Biomedicine* 22:473, 1975.

116. Fisher RI, Jaffe ES, Braylan RC, et al: Immunoblastic lymphadenopathy. Evolution into a malignant lymphoma with plasmacytoid features. *Am J Med* 61:553, 1976.

117. Dorfman RF: Enzyme histochemistry of the cells in Hodgkin's disease and allied disorders. *Nature* 190:925, 1964.

118. Okano H, Azar HA, Osserman EF: Plasmacytic reticulum cell sarcoma. Case report with electron microscopic studies. *Am J Clin Pathol* 46:546, 1966.

118a.Glick AD, Leech JH, Waldron JA, et al: Malignant lymphomas of follicular center origin in man. II. Ultrastructural and cytochemical studies. *J Natl Cancer Inst* 54: 23, 1975.

119. André R, Dreyfus B, Bessis M: La ponction ganglionnaire dans la maladie de Hodgkin, examinée au microscope électronique. *Presse Méd* 63:967, 1955.

120. Frajola WJ, Greider MA, Bouroncle BA: Cytology of the Sternberg-Reed cell as revealed by the electron microscope. *Ann NY Acad Sci* 73:221, 1958.

121. Dorfman RF, Rice DF, Mitchell AD, et al: Ultrastructural studies of Hodgkin's disease. *Natl Cancer Inst Monogr* 36:221, 1973.

122. Azar HA: Significance of the Reed-Sternberg cell. *Hum Pathol* 6:479, 1975.

123. Cossman J, Deegan MJ, Schnitzer B: Complement receptor B lymphocytes in nodular sclerosing Hodgkin's disease. *Cancer* 39:2166, 1977.

124. Strum SB, Park JK, Rappaport H: Observations of cells resembling Sternberg-Reed cells in conditions other than Hodgkin's disease. *Cancer* 7:167, 1970.

125. Sykes JA, Dmochowski L, Shullenberger CC, et al: Tissue culture studies on human leukemia and malignant lymphoma. *Cancer Res* 22:21, 1962.

126. Azar HA, Potter M (eds): *Multiple Myeloma and Related Disorders.* New York, Harper & Row, 1973, vol 1.

127. Maldonado, JE: *The Ultrastructure of the Myeloma Cell.* Thesis submitted to the Faculty of the Graduate School of the University of Minnesota, September 1965.

128. Smetana J, Gyorkey F, Gyorkey P, et al: Ultrastructural studies on human myeloma plasmacytes. *Cancer Res* 33:2300, 1973.

129. Brunning RD, Parkin J: Intranuclear inclusions in plasma cells and lymphocytes from patients with monoclonal gammopathies. *Am J Clin Pathol* 66:10, 1976.

130. Paraskevas F, Heremans J, Waldenström J: Cytology and electrophoretic patterns in γ1A (β2A) myeloma. *Acta Med Scand* 170:575, 1961.

131. Maldonado JE, bayrd ED, Brown Al Jr: The flaming cell in multiple myeloma. A light and electron microscopy study. *Am J Clin Pathol* 44:605, 1965.

132. Rogers JS II, Spahr J, Judge DM, et al: IgE myeloma with osteoblastic lesions *Blood* 49:295, 1977.

133. Azar HA, Zaino EC, Pham TD, et al: "Nonsecretory" plasma cell myeloma. Observations on seven cases with electron microscopic studies. *Am J Clin Pathol* 58:618, 1972.

134. Stavem P, Froland SS, Haugen HF, et al: Nonsecretory myelomatosis without intracellular immunoglobulin. Immunofluorescent and ultramicroscopic studies. *Scand J Haematol* 17:89, 1976.

135. Zucker-Franklin D: Structural features associated with the paraproteinemias. *Semin Hematol* 1:165, 1964.

136. Franklin EC, Lowenstein J, Bigelow B, et al: Heavy chain disease. A new disorder of serum gammaglobulins. Report of the first case. *Am J Med* 37:332, 1964.

137. Frangione B, Franklin EC: Heavy chain diseases: clinical features and molecular significance of the disordered immunoglobulin structure. *Semin Hematol* 10:53, 1973.

138. Seligmann M: Alpha chain disease. Immunoglobulin abnormalities, pathogenesis and current concepts. *Br J Cancer* 31 (suppl. 2): 356, 1975.

138a. Nassar VH, Salem PA, Shahid MJ, et al: "Mediterranean abdominal lymphoma" or immuno-proliferative small intestinal disease. Part II. Pathological aspects. *Cancer* 41:1340, 1978.

139. Franklin EC, Frangione B, Sheldon C: Heavy chain diseases. *Ann NY Acad Sci* 190:457, 1971.

140. Glenner GG, Terry WD, Isersky V: Amyloidosis: Its nature and pathogenesis. *Semin Hematol* 10:65, 1973.

141. Isobe T, Osserman EF: Patterns of amyloidosis and their associations with plasma-cell dyscrasia, monoclonal immunoglobulins and Bence Jones proteins, *N Engl J Med* 290:473, 1974.

142. van Furth R: Origin and kinetics of monocytes and macrophages. *Semin Hematol* 7:125, 1970.

143. Fedorko ME, Hirsch JG: Structure of monocytes and macrophages. *Semin Hematol* 7:109, 1970.

144. Azar HA, Lunardelli C: Collagen nature of asteroid bodies of giant cells in sarcoidosis. Am J Pathol 57:81, 1969.

145. Black MM, Epstein WL: Formation of multinucleate giant cells in organized epithelioid cell granulomas. *Am J Pathol* 74:263, 1974.

146. Lee RE, Ellis LD: The storage cells of chronic myelogenous leukemia. *Lab Invest* 24:261, 1971.

147. Lee RE, Ellis LD: The storage cells by chronic myelogenous leukemia. *Lab Invest* 24:261, 1971.

148. Zaino EC, Rossi MB, Pham TD, Azar HA: Gaucher cells in thalassemia. *Blood* 38:457, 1971.

149. Sharer LR, Barondess JA, Silver RT, et al: Association of Hodgkin disease and Gaucher disease. *Arch Pathol* 98:376, 1974.

150. Lynn R, Terry RD: Lipid histochemistry and electron microscopy in adult Niemann-Pick disease. *Am J Med* 37:987, 1964.

151. Golde DW, Schneider EL, Bainton DF, et al: Pathogenesis of one variant of sea-blue histiocytosis. *Lab Invest* 33:371, 1975.

152. Basset F, Nezelof C, Turiaf J: Présence en microscopie électronique de structures filamenteuses originales dans les lésions pulmonaires et osseuses de l'histiocytose X. Etat actuel de la question. *Bull Soc Med Hosp Paris* 117:413, 1966.

153. Sagebiel RW, Reed TH: Serial reconstruction of the characteristic granule of the Langerhans cell. *J Cell Biol* 36:201, 1965.

154. Cancilla PA, Lahey ME, Carnes WH: Cutaneous lesions of Letterer-Siwe disease. Electron microscopic study. *Cancer* 20:1986, 1967.

155. Morales AR, Fine G, Honr RC Jr, et al: Langerhans cells in a localized lesion of the eosinophilic granuloma type. *Lab Invest* 20:412, 1969.

156. Stout AP: Fibrous tumors of the soft tissues. *Minnesota Med* 43:455, 1960.

157. Freeman AI, Journey LJ: Ultrastructural studies on monocytic leukaemia. *Br J Haematol* 20:225, 1971.

158. Schumacher HR, Szekely IE, Park SA: Monoblast of acute monoblastic leukemia. *Cancer* 31:209, 1973.

159. Shaw MT, Nordquist RE: "Pure" monocytic or histiomonocytic leukemia: a revised concept. *Cancer* 35:208, 1975.

160. Polliack A, McKenzie S, Gee T, et al: A scanning electron microscopic study of 34 cases of acute granulocytic, myelomonocytic, monoblastic and histiocytic leukemia. *Am J Med* 59:308, 1975.

4

The Role of Electron Microscopy in Ophthalmic Pathology

Ramon L. Font, M.D.

Assistant Chief, Department of Ophthalmic Pathology
Armed Forces Institute of Pathology
Washington, D.C.
Clinical Professor of Pathology and Ophthalmology
Georgetown University Medical Center
Washington, D.C.

Frederick A. Jakobiec, M.D., D.Sc.

Instructor in Clinical Ophthalmology
College of Physicians and Surgeons, Columbia University
New York, New York
Director, Algernon B. Reese Laboratory of Ophthalmic Pathology
Edward S. Harkness Eye Institute
New York, New York.

During the last two decades, electron microscopy has become an increasingly popular method for the study and understanding of a broad variety of diagnostic and research problems encountered by both general pathologists and ophthalmic pathologists. Anatomists and experimentalists prefer optimal methods of fixation of tissues processed for electron microscopy. The use of electron microscopy should not be excluded, however, if the tissues have been fixed in buffered formalin or routinely embedded in paraffin for light microscopy. Quite often, general and ophthalmic pathologists develop a special interest in performing electron microscopic studies after the specimen has been routinely processed for light microscopy. In such cases one can still obtain valuable information about rare tumors that would otherwise require decades using a prospective study. This chapter will illustrate the use of three different types of tissue specimens: a) ideally fixed tissues processed initially for electron microscopy, b) formalin-fixed wet tissues, and c) tissues originally embedded in paraffin.

The opinions or assertions contained herein are the private views of the
authors and are not to be construed as official or as reflecting the views of
the Department of the Army or the Department of Defense.

Electron microscopy can be applied to ophthalmic pathology for a broad variety of non-neoplastic conditions, as well as for neoplastic lesions involving the intraocular structure and the ocular adnexa.

The following outline summarizes the material in the order it will be covered in this chapter:

I.   Non-neoplastic conditions
    A. Metabolic
        1. Hurler's disease
        2. Fabry's disease
    B. Infectious
        1. Herpetic keratitis
        2. Subacute sclerosing panencephalitis (SSPE)
        3. Toxoplasmosis
        4. Sporotrichosis
    C. Metaplastic
        1. Anterior subcapsular cataracts
II.   Neoplastic lesions involving the intraocular structures
    A. Nonpigmented tumors of primitive medullary epithelium and sensory retina
        1. Retinoblastoma
        2. Pure medulloepithelioma
        3. Teratoid medulloepithelioma with rhabdomyoblastic differentiation
        4. Angiomatosis retinae (hemangioblastoma)
    B. Pigmented tumors of uvea
        1. Spindle cell melanoma of iris
        2. Magnocellular nevus (melanocytoma)
        3. Adenoma of retinal pigment epithelium
        4. Melanoma cells versus melanophages
        5. Balloon cells in uveal melanomas
        6. Lipofuscin-containing cells ("orange pigment") overchoroidal melanomas
    C. Myogenous tumors of uvea
        1. Rhabdomyosarcoma of iris
        2. Mesectodermal leiomyoma of ciliary body
III. Neoplastic lesions involving the orbit
    A. Mesenchymal
        1. Fibrosarcoma
        2. Leiomyosarcoma
        3. Chondrosarcoma
        4. Fibrous histiocytoma
        5. Hemangiopericytoma
    B. Neurogenic
        1. Glioma of optic nerve
        2. Meningioma
        3. Schwannoma
    C. Tumors showing diagnostic granules
        1. Carcinoid
        2. Alveolar soft part sarcoma
        3. Granular cell tumor

NON-NEOPLASTIC CONDITIONS

Metabolic

Figure 1 illustrates a corneal keratocyte from a patient with mucopolysaccharidosis, type I-S. The specimen was properly fixed for electron microscropy. The inset (at left) shows the markedly swollen keratocytes containing numerous cytoplasmic vacuoles. By electron microscopy, the vacuoles are surrounded by a single membrane and contain fibrillogranular material. In some of the vacuoles, electron-dense concentric membranous structures are observed (arrow, inset at right). Histochemically, these vacuoles contain abundant acid mucopolysaccharides that are resistant to hyaluronidase digestion. This group of genetic mucopolysaccharidoses, which comprises seven subgroups, has been well characterized clinically as well as biochemically as having specific enzyme deficiencies with excretion of mucopolysaccharides in the urine (1,2).

Fabry's disease is a systemic disease of glycolipid metabolism with abnormal accumulations of lipid deposits in practically all organs of the body. The affected male as well as the female carriers frequently exhibit the characteristic whorl-like corneal opacities. Figure 2 illustrates the limbal epithelium near the peripheral cornea, disclosing numerous membranous cytoplasmic bodies located mainly between the nuclei of the basal cell layer (displaced apically) and the epithelial basement membrane (see also insert A, Fig. 2). Histochemical studies of frozen

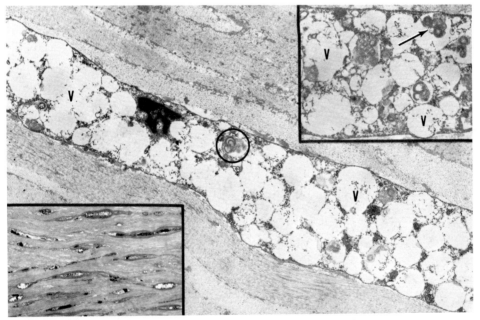

Figure 1. Stromal keratocyte from a patient with mucopolysacharidosis (MPS-type I-S) showing numerous cytoplasmic membrane-bound vacuoles (V) containing fibrillogranular material and occasional concentric membranous structures (circle). (×11,400, AFIP Neg. 77-4536). Inset, left: numerous swollen keratocytes with vacuolated cytoplasm. (Toluidine blue, ×290) *Inset, right:* the vacuoles (V), some of which contain membranous bodies (arrow). (×15,500)

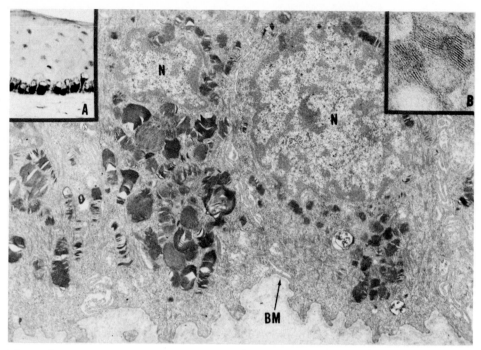

Figure 2. Lamellar bodies are present within the basal cells of the limbal epithelium. N = nuclei of basal epithelial cells; BM = basement membrane. (x7,000, AFIP Neg. 71-1652-6) *Inset A:* a similar region of limbal epithelium. The densely stained deposits involve most of the basal cells, but only a few of the next layer of cells. (Paraphenylenediamine, ×275) *Inset B:* fine lamellar material (like a fingerprint), at high magnification. (×80,000)

sections of choroid and retina show the vacuolated smooth muscle cells of the small and medium-sized arterioles to contain granules that are birefringent under polarized light (Fig. 3, top) and stain positively with the oil red 0 technique (Fig. 3, bottom), as well as with Sudan black B (inset, Fig. 4). Electron microscopy demonstrates the presence of dense, laminated cytoplasmic inclusions within the smooth muscle cells of the media of the choroidal arterioles (Fig. 4). Similar laminated inclusions are also observed in endothelial cells and pericytes of all the vasculature of the eye, including the choriocapillaris. Detailed histochemical and electron microscopic studies of an eye from a patient with Fabry's disease have been reported (3). Weingeist and Blodi also studied the eye of a female carrier by light and electron microscopy (4).

Infectious

Figure 5 (top) depicts a keratectomy specimen showing a large area of ulceration with impending perforation and formation of a descemetocele. At the edge of the ulcer, numerous multinucleated syncytial giant cells are observed (Fig. 5, bottom). Electron microscopy of the paraffin-embedded, formalin-fixed tissue demonstrates the presence of mature viral particles that morphologically are typical of herpes simplex virus (Fig. 6). This case has been reported in detail

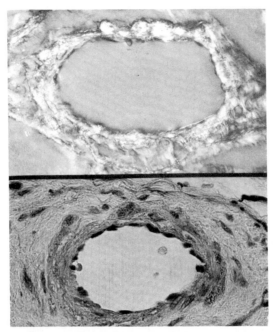

Figure 3. *Top*: under polarized light, this choroidal arteriole shows birefringent lipid in the vacuolated cells of the media. (frozen section, ×300). *Bottom*: oil red 0 stained positively the lipid granules in the smooth muscle cells of the media of an adjacent choroidal vessel. (Oil red 0, ×300; AFIP Neg. 77-4521-2)

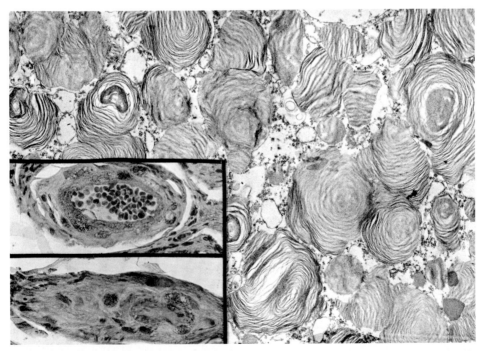

Figure 4. *Inset*: a choroidal arteriole with many black granules within their walls. (Sudan black B, ×300) The electron micrograph shows the membranous cytoplasmic inclusions ("zebra bodies") within the smooth muscle cells of the media. (×12,000, AFIP Neg. 71-1652-4)

167

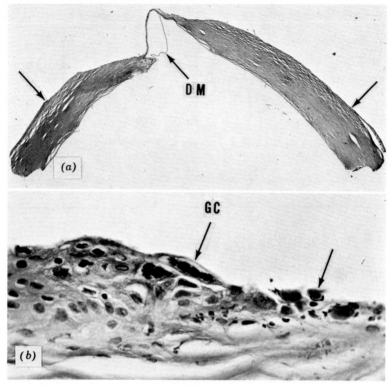

Figure 5. *Top*: keratectomy specimen showing an extensive ulceration of epithelium (between free arrows) with descemetocele that has almost perforated the cornea. DM = Descemet's membrane. (Hematoxylin-eosin, ×14, AFIP Neg. 69-9832). *bottom*: edge of corneal ulcer (left arrow) showing epithelial cells, some of which are forming syncytial giant cells (GC) containing intranuclear reddish-purple inclusions with margination of nuclear chromatin (free arrow). Underlying superficial stroma shows areas of scarring. (Hematoxylin-eosin, ×350, AFIP Neg. 69-9834)

elsewhere (5). Herpes virus particles have also been found in the iris (6) and retina (7) by electron microscopy.

Figure 7 illustrates the postmortem eyes from a young boy who died with a rapidly progressive neurologic syndrome. A pigmentary maculopathy was observed in the right eye (Fig. 7, top inset). Sections through the retina show myriad intranuclear inclusions with margination of the nuclear chromatin (Fig. 7, bottom inset). By electron microscopy, the intranuclear inclusions proved to be myriad interwoven microtubular structures that are characteristic of the paramyxovirus group (Figs. 7 and 8). The term "measles maculopathy" was proposed because of the specific immuno-ultrastructural localization of measles antigen on the viral nucleocapsids of the nuclear and cytoplasmic inclusions (Fig. 9). This case was reported by Font and coworkers (8). Landers and Klintworth (9) first demonstrated by electron microscopy the presence of viral nucleocapsids within the retinal lesions in a child with subacute sclerosing panencephalitis (SSPE).

Whereas light microscopic studies of toxoplasmic retinochoroditis have been

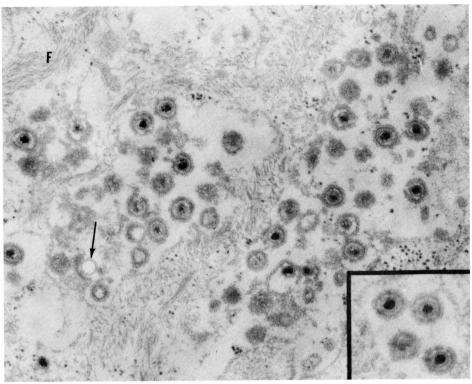

Figure 6. Intracytoplasmic viral particles displaying electron-dense nucleoid surrounded by inner membrane as well as outer membrane or envelope, characteristic of infective mature particle. Empty capsid (arrow) is seen. Cytoplasmic filaments (F) are typical of basal cells of corneal epithelium. (×46,500) *Inset*: several viral particles with thick, fluffy material attached to outer membrane. (×63,000, AFIP Neg. 77-4516-2)

extensively described, only rarely have electron microscopic studies of the offending parasite been performed. Figures 10 and 11 demonstrate the ultrastructural features of viable and necrotic cysts of *Toxoplasma gondii* (10). The studies confirmed the presence of a true wall for the cyst, as well as additional ultrastructural features that are highly characteristic of the parasite (Figs. 12 and 13).

Figure 14 is an example of a granulomatous necrotizing retinochoroiditis caused by *Sporotrichum schenkii*. With the McCallum-Goodpasture stain, the fungus, which was present within the necrotic retina and subretinally, displays a "safety-pin" appearance (top inset, Fig. 14). By electron microscopy, the fungus shows a rather impressive thick capsule and a distinct cell wall (Fig. 14) (11). A peculiar pattern of granular filamentous material can be seen radiating from the cell wall of the organism (Fig. 14, bottom inset). Streeten and coworkers (12) reported a case of sporotrichosis of the orbital margin in an immunosuppressed patient who had an underlying systemic disease. Although some workers have questioned the presence of a capsule in *Sporotrichum,* our electron microscopic observations (11) and those of Streeten and coworkers (12) clearly demonstrate its presence.

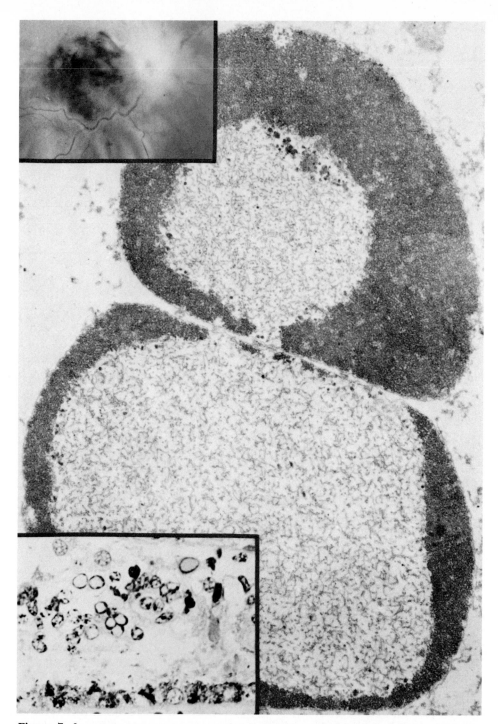

Figure 7. *Inset, top*: gross appearance of the posterior pole of the right eye, which was obtained postmortem. There is an extensive pigmentary maculopathy almost reaching the edges of the optic disc, temporally. *Inset, bottom*: numerous intranuclear inclusions within the retina in the macular region. (Toluidine blue, ×485) By electron microscopy, the intranuclear inclusions in the retinal cells are composed of myriad interwoven microtubular structures that have displaced the nuclear chromatin peripherally. (×27,200, AFIP Neg. 77-4532)

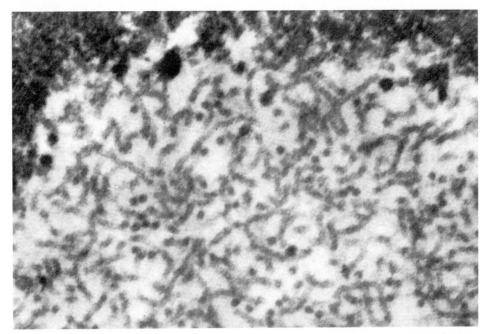

Figure 8. High-power view of the viral nucleocapsids shown in Figure 7. (×97,500, AFIP Neg. 77-4531)

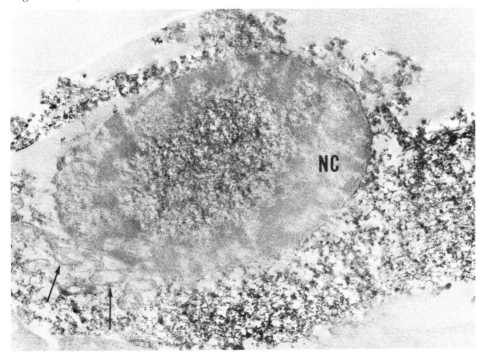

Figure 9. Single retinal cell with specific immunostructural localization of measles antigen on the nucleocapsids of the nuclear and cytoplasic inclusions. Section treated with guinea pig antimeasles (GPAM) antiserum and horseradish peroxidase-conjugated rabbit anti-guinea pig gamma globulin (HRP-RAGP-γG) without additional heavy metal counterstaining. NC = nuclear chromatin; M = mitochondria (×14,400,AFIP Neg. 72-13452-3)

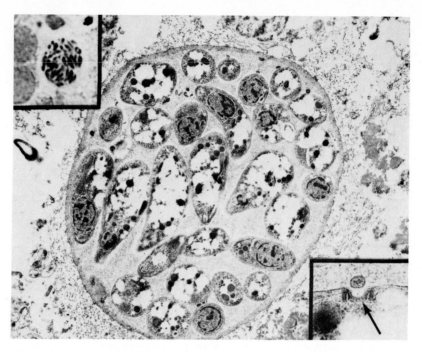

Figure 10. Electron micrograph of a presumably viable cyst of *Toxoplasma gondii* containing numerous elongated or crescent-shaped organisms and a prominent, slightly irregular wall. (×7,200) *Inset, top*: viable cyst in retina. (Toluidine blue, ×750) *Inset, bottom*: micropyle (arrow). (×35,000, AFIP Neg. 76-4145-1)

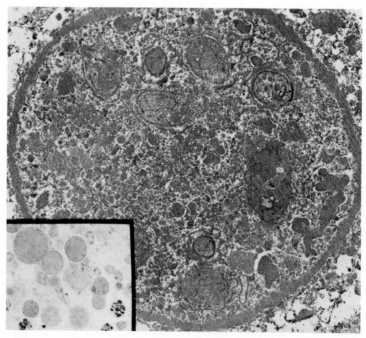

Figure 11. Electron micrograph of a necrotic cyst disclosing poorly defined internal structures and a thick wall that is homogeneously electron-dense and without a discernible double layer. (×7,200) *Inset*: necrotic cysts. (Toluidine blue, ×395, AFIP Neg. 76-4145-2)

172

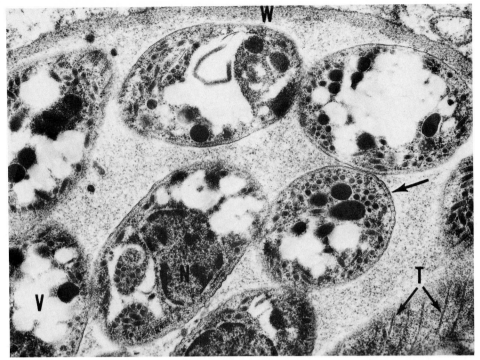

Figure 12. Viable cyst of *Toxoplasma gondii* containing numerous oval organisms, each surrounded by a double membrane (pellicle) (arrow). W = wall of the cyst disclosing thick inner granular layer and a thin, more electron-dense outer layer; T = toxonemes; N = nucleus; V = vacuole. (×21,000, AFIP Neg. 76-4145-4)

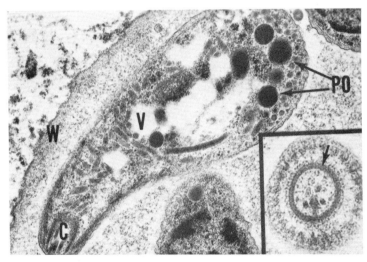

Figure 13. Elongated organism disclosing a pointed anterior end containing conoid (C) and a broad, rounded posterior end. Electron-dense round structures at posterior end are segments of paired organelles (PO). Anterior end contains micronemes (toxonemes). V = lucent vacuoles; W = wall of cyst. (×27,000) *Inset*: cross section of anterior end of organism displaying conoid ring (arrow) that is surrounded by 22 toxonemes. (×45,000, AFIP Neg. 76-4145-3)

173

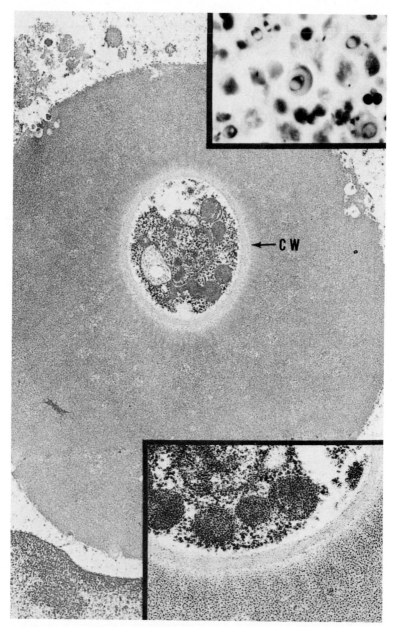

Figure 14. Electron micrograph demonstrates markedly thickened capsule with well-developed cell wall (CW) surrounding protoplasm of fungus. (×27,000) *Inset, top*: well-developed capsule around protoplasm of fungi, which exhibit "safety-pin" appearance. (MacCallum-Goodpasture, ×1,300) *Inset, bottom*: peculiar radiating pattern of granular filamentous material projecting from cell wall of fungus. (×54,000, AFIP Neg. 75-10706-1)

174

Metaplastic

Another example of a non-neoplastic condition is the plaque of an anterior subcapsular cataract (Fig. 15, top). At the edge of the plaque, the lens epithelium is continuous with a single flat layer of epithelial cells having a polarized basement membrane located between the fibrous plaque and the superficial cortex of the lens (Fig. 15, bottom). It has been suggested that the spindle-shaped cells within the plaque (Fig. 16, bottom inset) represent fibroblasts. By electron microscopy, the spindle-shaped cells are almost entirely surrounded by a basement membrane and display interdigitations of their plasmalemmas and well-developed desmosomes (macula adherens) between adjacent cells (Fig. 16, top right inset). In addition, mature collagen fibrils having a periodicity of approximately 60 to 70 nm can be seen within the plaques. These findings are strong evidence in support of the epithelial origin of the spindle-shaped cells, since they have retained their epithelial characteristics (13).

NEOPLASTIC LESIONS INVOLVING INTRAOCULAR STRUCTURES

Nonpigmented Tumors of Primitive Medullary Epithelium and Sensory Retina

Retinoblastoma is the most common malignant intraocular neoplasm of childhood. The origin of retinoblastoma has been a controversial subject for slightly more than a century. Ts'o and coworkers (14–17), in a retrospective study using mostly paraffin-embedded tissues, clearly established that retinoblastomas are in fact neuronal neoplasms and not gliomas of the retina as originally proposed by Virchow. Figure 17 illustrates an unucleated eye containing a retinoblastoma. The pale areas exhibiting photoreceptor differentiation are usually small and often grossly undetectable prior to microscopic examination of the eye (Fig. 17). Within these light-stained areas, Ts'o and coworkers (14–17) first described structures that resemble a bouquet of flowers (fleurettes) (inset, Fig. 17). By electron microscopy, the latter structures disclose a girdle of zonula adherentes simulating the external limiting membrane of the retina, bulbous expansions of cytoplasm packed with mitochondria (analogous to photoreceptor inner segments) (Fig. 18), and stacks of membranous structures (analogous to photoreceptor outer segments). They concluded that the fleurette is a more differentiated structure than the Flexner-Wintersteiner rosette and herefore phylogenetically more closely related to the mature adult retina.

Medulloepithelioma is an embryonic tumor of childhood believed to arise from the primitive medullary epithelium, involving mainly the anterior segment structures of the eye. Figure 19 illustrates the clinical appearance of the eye from a child who had a medulloepithelioma. The diagnosis was established by electron microscopic examination of a cyst obtained from the anterior chamber. Tumor cells are forming lumina and display a girdle of zonula adherentes, the presence of microtubules, and cilia disclosing nine peripheral doublets without a central doublet (inset, Fig. 20). The results of the electron microscopic studies clearly established the diagnosis of medulloepithelioma (18).

Figure 21 illustrates a problem in the cytological interpretation of large ganglioform cells with abundant fibrillated acidophilic cytoplasm (inset, Fig. 21)

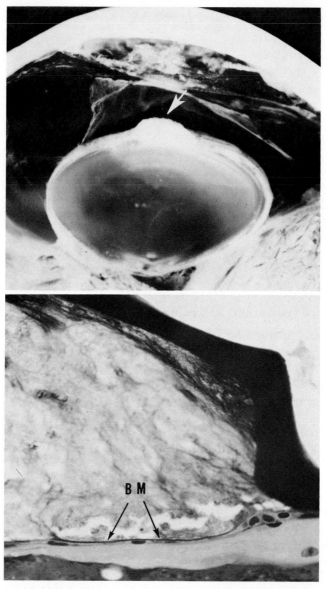

Figure 15. *Top*: gross appearance of the anterior segment of the eye showing a white plaque (arrow) involving mainly the anterior pole of the lens. (AFIP Neg. 71-9827-2) *Bottom*: edge of the anterior subcapsular plaque showing a relatively normal cuboidal epithelium in continuity with a flat layer of lens epithelial cells interposed between the fibrous plaque (above) and the underlying cortex (below). A thin basement membrane (BM) is evident only on the anterior surface of these polarized cells. (PAS, ×350, AFIP Neg. 71-8779)

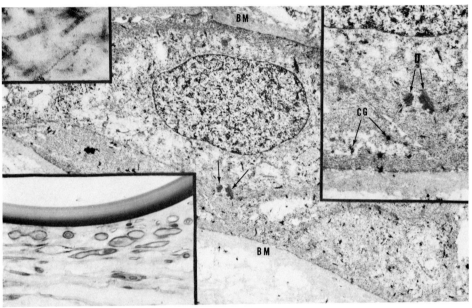

Figure 16. Electron micrograph demonstrating the basement membrane (BM) surrounding the spindle-shaped cells within the subcasular plaques. Desmosomes are present (free arrows). (×12,000, AFIP Neg. 77-4528) *Inset, bottom*: several elongated cells surrounded by basement membranes. (Toluidine blue, ×440) *Inset, top right*: desmosomes (D) and several electron-dense irregular deposits compatible with calcific granules (CG). N = nucleus. (×27,000) *Inset, top left*: collagen fibers with a periodicity of 60 to 70 nm. (×33,600)

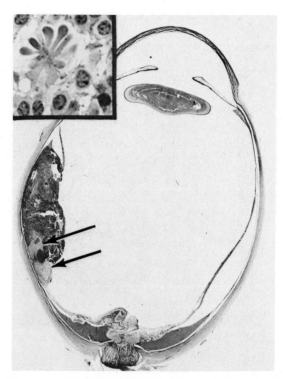

Figure 17. Retinoblastoma with photoreceptor differentiation. A solid tumor occupies less than one-third of the posterior segment of the eye. Arrows show the pale-staining areas displaying photoreceptor differentiation, which are clearly demarcated from the remainder of the tumor. (Hematoxylin-eosin, ×8) *Inset*: flower-like arrangement (fleurette) of the cell processes. (Paraphenylenediamine, ×575, AFIP Neg. 77-4524)

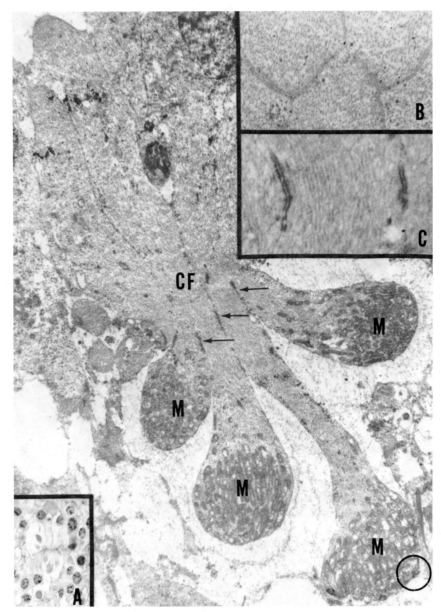

Figure 18. Electron micrograph of photoreceptor cell elements (fleurette) in retinoblastoma. Tumor cells with long conducting fibers (CF) are joined by dense cell attachments (arrows). The cell processes extend beyond the plane of cell attachments for a variable distance. Longitudinally oriented mitochondria (M) in large numbers are packed in the bulbous end of the cytoplasmic processes. A cilium (cut in cross section—circle) is present at one of the cell processes. (×5,300) *Inset A*: similar fleurette by light microscopy. (×400) *Inset B*: a portion of the hexagonal pattern observed in a tangential section of the cell attachments. (×6,700) *Inset C*: cross section of similar cell attachments (terminal bars). (×16,100, AFIP Neg. 69-4552-8)

179

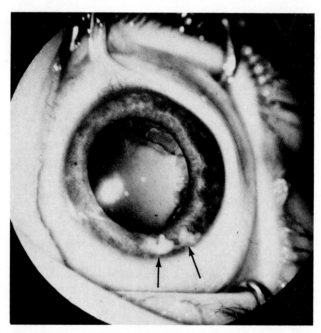

Figure 19. Clinical appearance of the left eye. Two white flocculi are present in the angle, inferiorly (arrows). A mass is seen superior temporally behind the dilated iris. (AFIP Neg. 75-11731)

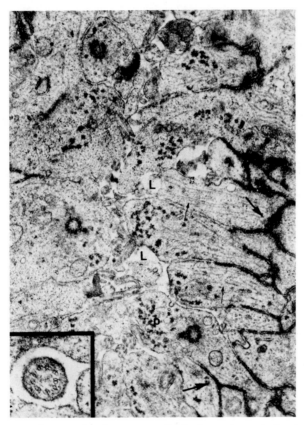

Figure 20. Electron micrograph of a rosette displaying microtubules (small arrows) located in the apical cells near a lumen (L). Polyribosomes (P) are clustered near the lumen. A partial girdle of zonula adherentes (large arrows) is present. Several cilia are cut on cross section in the lumen. (×22,000) *Inset*: cross section of cilium displaying nine peripheral doublets without a central doublet. (×32,000, AFIP Neg. 75-11733)

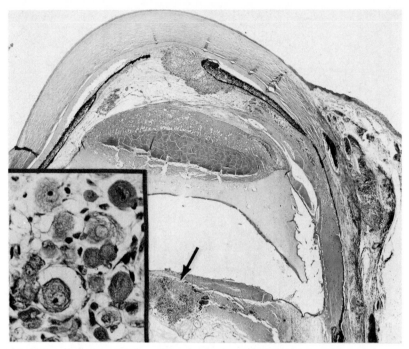

Figure 21. The posterior chamber and pupillary region are filled by a malignant teratoid medulloepithelioma that has infiltrated the ciliary body (right) and extended to the episclera subconjunctivally. (Hematoxylin-eosin, ×12, AFIP Neg. 70-7855) *Inset:* high magnification of the cells within the tumor nodule (arrow) showing large ganglioform cells (actually rhabdomyoblasts) among the smaller undifferentiated cells. (Hematoxylin-eosin, ×305, AFIP Neg. 69-9903)

present within a malignant teratoid medulloepithelioma of the ciliary body (Fig. 21). Based on the morphological features, the cells in question (Fig. 21) were suspected of being ganglion cells. By electron microscopy, however, they were found to be rhabdomyoblasts. The rhabdomyoblasts are characterized by their cytoplasmic content of interlacing bundles of thick (myosin) and thin (actin) myofilaments with formations of I bands and Z lines (Fig. 22). This case, as well as three additional examples of embryonal intraocular medulloepitheliomas with rhabdomyosarcomatous differentiation, was reported by Zimmerman and co-workers (19). These embryonic tumors may contain other heterologous mesenchymal elements, such as islands of hyaline cartilage, in addition to the myoblasts. Zimmerman has proposed a classification of intraocular medulloepitheliomas (20), emphasizing the marked polymorphism of these tumors arising from the ciliary epithelium (21). Kroll and coworkers, on the basis of electron microscopic observations, proposed a cytological classification of orbital rhabdomyosarcoma (22).

Another interesting problem depicted in Figures 23, 24, and 25, was the study of the "stromal cells" within an angioma (hemangioblastoma) of the retina. The tumor cells frequently show nests of clear cells with abundant honeycombed or vacuolated cytoplasm (Fig. 23, top right). Electron microscopic studies revealed that these vacuolated cells appear to be fibrous astrocytes, some of which display

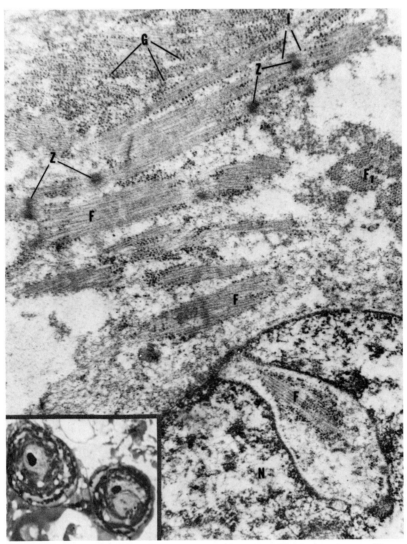

Figure 22. Rhabdomyoblast showing numerous bundles of cytoplasmic myofilaments cut longitudinally (F) and transversely (F1). Numerous glycogen particles (G) are oriented along the myofilaments. I bands (I) and Z lines (Z) are readily recognizable. Myofilaments can be seen in a portion of the cytoplasm that is present in an indentation of the nucleus (N). (×34,000, AFIP Neg. 70-9143-2) *Inset*: large round cells (shown also in inset of Figure 21) present within the medulloepithelioma of the ciliary body. (Toluidine blue, ×750, AFIP Neg. 69-9830)

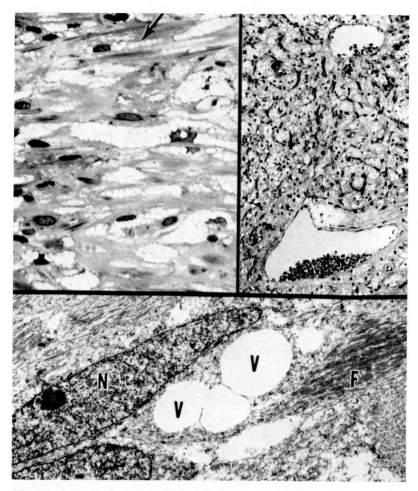

Figure 23. *Top right*: classical picture of retinal henangioblastoma showing a plexiform anastomosing capillary network surrounded by nests of clear vacuolated cells. (Paragon, ×100) *Top left*: glial cells with homogeneous cytoplasm in various stages of lipidization. Arrow points to early vacuolization with a linear array of cytoplasmic vacuoles. (Paragon, ×380) *Bottom*: electron micrograph showing vacuoles (V) in a fibrous astrocyte containing masses of conglutinated glial filaments (F), which are also present in adjacent nonlipidized astrocytes. Note the typical elongated nucleus (N) of the fibrous astrocyte. (×5,400, AFIP Neg. 77-4522-1)

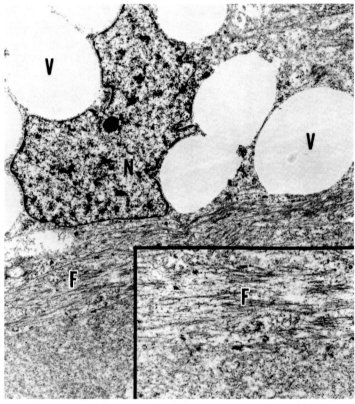

Figure 24. Moderately lipidized stromal cell containing empty vacuoles (V) indenting the nucleus (N). Some filaments (F) are preserved, but others have undergone granular degeneration. (×5,400) *Inset* of area at lower left shows at higher magnification the juxtaposition of intact filaments (F) and granular debris. (×10,800, AFIP Neg. 75-10045-4)

185

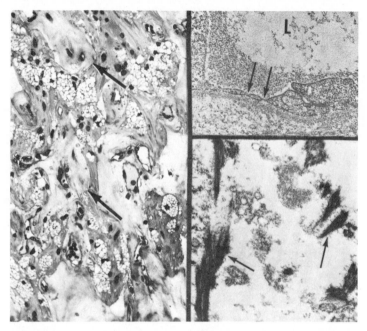

Figure 25. *Left*: extravasated plasmatic exudate (arrows) forms a hyaline mantle around the capillaries. (Paragon, ×185) *Bottom right*: strands of fibrin (arrows) in the perivascular spaces. *Top right*: arrows point to fenestrations bridged by diaphragms in the endothelial cells. L= lumen. (×14,400, AFIP Neg. 77-4522-2)

conglutinated masses of glial filaments resembling early Rosenthal fibers (Fig. 23, bottom). The more fully lipidized cells contain large vacuoles that indent the nuclear membrane (Fig. 24). The glial filaments in some areas disclose areas of granular degeneration (Fig. 24, inset). Electron microscopy of the vessels of the tumor shows endothelial cells with fenestrations (Fig. 25, top right), with perivascular deposition of fibrin (Fig. 25, bottom right) forming a hyaline mantle around some of the vessels in the tumor (Fig. 25, left). Lipid analysis of the tumor by three special techniques revealed the presence of cholesterol stearate, a plasma lipid. Our ultrastructural observations suggested that the lipidized "stromal cells" in the tumor represent fibrous astrocytes that have passively imbibed the lipid which has passed through fenestrated, leaky endothelial cells of the vessels within the tumor (23).

Pigmented Tumors of the Uvea

Another important practical application of electron microscopy involves the study and classification of intraocular melanotic lesions. Figure 26 is from a moderately pigmented tumor of the iris that proved to be a low-grade spindle cell melanoma. By electron microscopy, the spindle-shaped cells (circle, Fig. 26) disclose the presence of immature and mature melanosomes in their cytoplasm (Fig. 27) (24). The spindle-shaped cells typically lack basement membranes. Figure 28, by contrast, reveals the presence of melanin-laden macrophages or

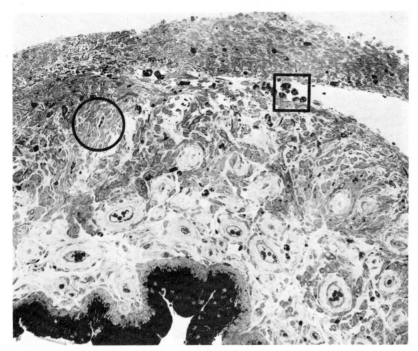

Figure 26. Spindle cell melanoma of iris (low-grade malignant). The tumor cells have formed a placoid, cohesive mass extending along the anterior iridic surface. The circle encloses spindle-shaped melanoma cells (see Figure 27). Scattered melanophages are present within the square. (Paraphenylenediamine, ×140, AFIP Neg. 74-9102)

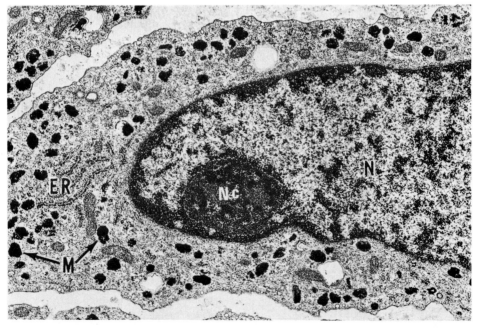

Figure 27. Spindle B type melanoma cell (within circle in Figure 26) showing a moderate number of melanosomes (M) and rather scanty endoplasmic reticulum (ER). The plasmalemma shows minimal interdigitations with adjacent cells. N = nucleus; Nc = nucleolus. (×14,400, AFIP Neg. 74-10013-3)

187

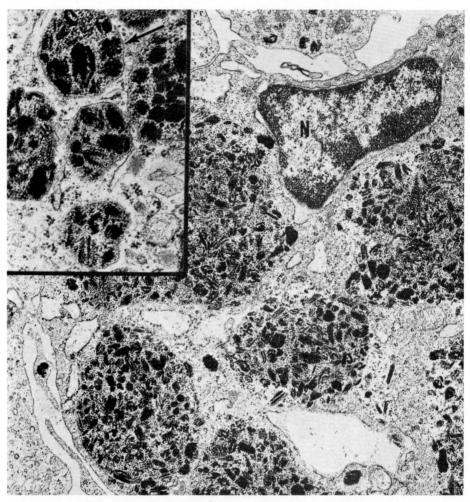

Figure 28. Pigment-laden macrophage (melanophage) (located within square in Figure 26) containing numerous phagocytic vacuoles or phagosomes predominantly filled with immature melanosomes. The nucleus (N) is eccentric. (×18,000, AFIP Neg. 74-10013-4) *Inset*: several melanosomes in phagocytic vacuoles, each surrounded by a single membrane (arrow). (×36,000)

melanophages (square, Fig. 26) containing myriad melanosomes within phagocytic vacuoles, many of which are enclosed by a single membrane (inset, Fig. 28). Another interesting melanocytic lesion is the magnocellular nevus (melanocytoma), that typically contains abundant large mature melanosomes, many of which are giant melanosomes (Fig. 29). This large number of mature melanosomes probably explains the jet black color of these tumors (25).

Other interesting intraocular melanotic lesions that differ from uveal melanomas are the tumors originating from the ciliary and retinal pigment epithelium (26). Typically, the tumors are quite small and jet black (top inset, Fig. 30). By light microscopy, the tumor is composed of light and dark cells. Large cytoplasmic vacuoles, some of which contain a sialomucin, are predominantly observed within the dark cells (bottom, inset, Fig. 30.). The tumor had a flat base implanted on the inner surface of the choroid (Fig. 30). Electron microscopic studies show the presence of basement membranes, fine collagen fibrils, and well-developed desmosomes (Fig. 31), as well as junctional complexes (i.e., zonula occludens and zonula adherens) and microvilli formed by the tumor cells (Fig. 32) (26). Although the normal retinal pigment epithelium possesses no desmosomes or fine collagen fibrils, both were present in this adenoma of the pigment epithelium. The latter findings, however, have been observed in reactive proliferations of the retinal pigment epithelium overlying choroidal tumors (27). A similar tumor of the pigmented ciliary epithelium has been studied by electron micrsocopy by Streeten and McGraw (28).

Comparison of Figures 33 and 34 clearly distinguishes a malignant melanoma cell and a melanophage. Figure 33 (inset) shows epithelioid melanoma cells seeding the chamber angle structures from a malignant melanoma of the anterior choroid, mixed cell type. The tumor cells show evidence of melanin production, with melanogenesis in different stages of evolution (Fig. 33). By contrast, Figure 34 shows a melanophage located in the subretinal space containing numerous melanosomal complexes that indicate origin of the melanosomes from the pigment epithelial cells, in contrast to the melanin granules derived from uveal melanocytes as depicted in Figure 27.

Certain uveal melanomas contain lipid-laden cells (balloon cells, inset, Fig. 35) (24, 29). We believe that these cells represent a peculiar lipoidal degeneration of melanocytic cells and not lipid-laden macrophages. Figure 35 demonstrates that these cells contain numerous intracytoplasmic round to oval vacuoles bounded by a single membrane. In addition, the convolutions of the nuclear membrane, the prominent nucleolus, and the presence of immature and mature melanosomes between the cytoplasmic vacuoles are strong evidence in favor of the theory that they represent a lipoidal degeneration of melanoma cells, rather than lipid-laden macrophages. The cytoplasmic vacuoles in Figure 35 appear empty because the lipid has been dissolved during storage of the tissue in 60% alcohol. By light microscopy, the lipidized melanoma cells show foamy, vacuolated cytoplasm (Fig. 36, top). The lipid can be vividly demonstrated, however, on frozen sections stained with oil red 0, as well as under polarized light (Fig. 36, bottom). Balloon cells have been described in nevi of the skin, (30) conjunctiva (31,32), and choroid (33), as well as in primary (34) and metastatic (35) cutaneous melanomas. Riley reported a 10% incidence of balloon cells in 200 cases of choroidal melanomas (36).

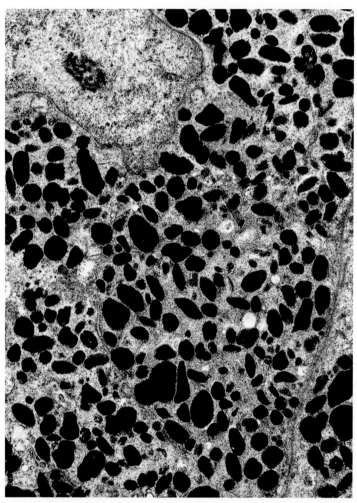

Figure 29. *Left*: melanocytoma of ciliary body. Tumor cells containing numerous mature melanosomes of variable size and shape. (×9,000) *Right*: high-power view of adjacent tumor cell discloses a giant melanosome (GM). The cytoplasm contains fine filaments. (×18,000, AFIP Neg. 77-4517-1 and 2).

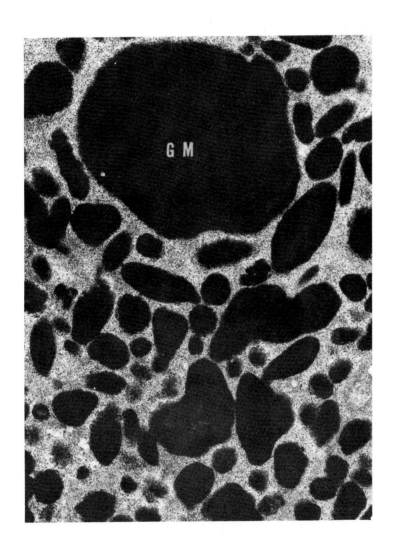

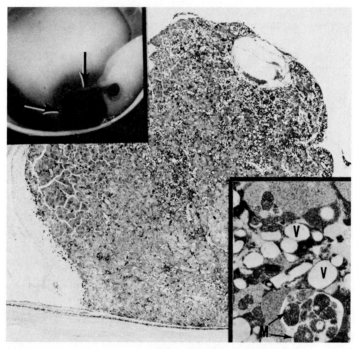

Figure 30. Adenoma of the retinal pigment epithelium. Under low magnification, the pigmented tumor has a flat base toward the sclera (bottom of picture). Cells with vacuolization are more numerous toward the apical portion than toward the base. (Hematoxylin-eosin, ×28, AFIP Neg. 77-4520-2) *Inset top*: a jet black tumor (arrows) located between the equator and the ora serrata. *Inset, bottom*: light and dark cells. Intracellular vacuoles (V) are mainly located in the dark cells. Melanophages (M) are present within a cystoid space. (Paraphenylenediamine, ×245)

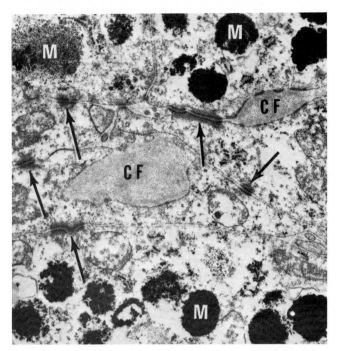

Figure 31. The tumor cells are attached to each other by desmosomes (arrows). Fine collagen filaments are present between the tumor cells (CF). M = melanosomes. (×16,500, AFIP Neg. 71-2004)

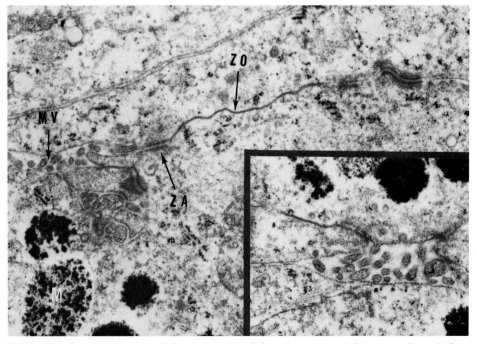

Figure 32. Another tumor cell shows junctional complexes containing a zonula occludens (ZO) and a zonula adherens-like configuration (ZA). Numerous microvilli (MV) are present at places where the apical portions of several cells meet. (×21,000, AFIP Neg. 71-2004-1; *inset,* ×20,000)

193

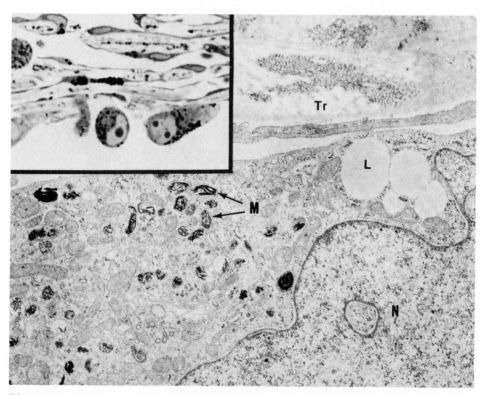

Figure 33. This electron micrograph illustrates that pigment-containing cells in the trabecular meshwork spaces (*inset*) are in fact melanoma cells and not macrophages, a distinction that is not readily made by light microscopy. Note the presence of immature melanosomes (M) in the nonfilamentous cytoplasm (epithelioid cell type). L = lipid vacuoles; Tr = trabecular spaces; N = nucleus. (×16,500; *inset,* paraphenylenediamine, ×750, AFIP Neg. 74-10013-6)

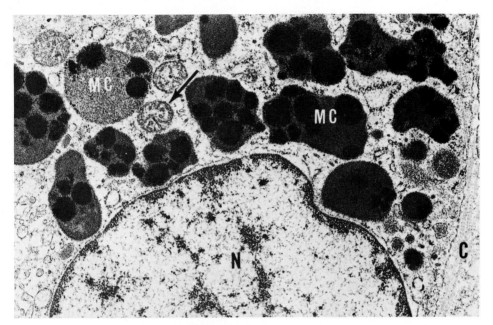

Figure 34. Another pigment-laden macrophage (melanophage) filled with melanosomal complexes (MC). The pigment granules here are indicative of origin from pigment epithelium (compare with uveal melanin granules in macrophage of Figure 28). Arrows point to mitochondria. N = nucleus; C = collagen. (×18,000, AFIP Neg. 74-10013-5)

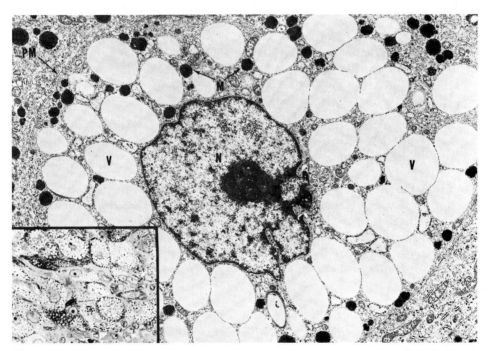

Figure 35. Electron micrograph of "balloon cells" in choroidal melanoma (*inset*) showing numerous large, round to oval empty vacuoles (V). The lipid has been mainly dissolved out, since the specimen was stored in 60% alcohol. The indentations of the nuclear membrane, the prominent nucleolus, and the presence of melanosomes (M) are strong evidence indicating lipoidal degeneration of a melanoma cell rather than lipid engorgement by a macrophage. PM = plasma membrane; N = nucleus. (×10,800, AFIP Neg. 10013-7; *inset,* paraphenylenediamine, ×440, AFIP Neg. 72-12985)

195

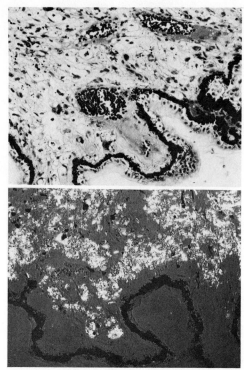

Figure 36. *Top*: lightly pigmented malignant melanoma of ciliary body containing numerous balloon cells. (Hematoxylin-eosin, ×110, AFIP Neg. 74-9103) *Bottom*: adjacent field reveals numerous birefringent crystals within the cytoplasm of the tumor cells. This finding probably represents an in vitro crystallization of the intracytoplasmic lipoidal material. (Polarized light, ×140, AFIP Neg. 74-9106)

Another practical application of electron microscopy was in establishing the nature of a peculiar orange pigment that was observed clinically on the surface of a choroidal melanoma (37). Fluorescein angiography revealed early fluorescence of the tumor and a blockage of background fluorescence in exactly the same areas that corresponded to the location of the yellowish-orange pigment over the surface of the tumor. Microscopic examination disclosed that the orange pigment was contained within proliferated retinal pigment epithelial cells and in macrophages located both within and under the degenerated retina. Histochemical and electron microscopic studies provided conclusive evidence that the pigment was lipofuscin (Fig. 37) (37).

Myogenous Tumors of the Uvea

Another puzzling diagnostic problem was a yellowish-tan tumor of the right iris in a young child (Fig. 38) that was clinically suspected of being a juvenile xanthogranuloma. Histopathologically, the iridectomy specimen reveals two distinct cell types (Fig. 38, inset). The larger cells show ecentric or paracentral nuclei with abundant acidophilic cytoplasm, and the smaller cells display small pyknotic nuclei and a scanty cytoplasm. Two histopathologic diagnoses were

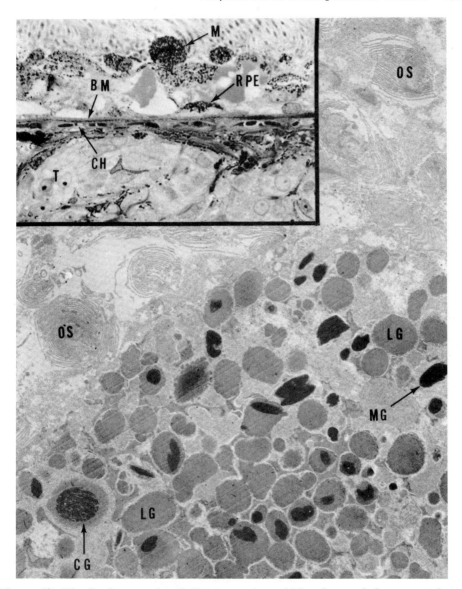

Figure 37. Lipofuscin granules (LG) are present within pigment-laden macrophage. Membranous structures are fragments of outer segments (OS). MG = melanin granule; CG = compound granule. (×11,800) *Inset*: macrophages (M) attached to outer retina. BM = Bruch's membrane; T = tumor cells-melanoma; CH = choriocapillaris; RPE = retinal pigment epithelium. (Toluidine blue, ×530, AFIP Neg. 72-12745)

considered: a reticulohistiocytic lesion and a tumor of striated muscle origin (rhabdomyoma versus rhabdomyosarcoma). After several recurrences of the tumor, which were initially controlled by radiation therapy, the eye was enucleated. Electron microscopy of the formalin-fixed recurring iridic tumor, which had infiltrated the ciliary body, showed that the large globoid tumor cells possessed the characteristic features of rhabdomyoblasts (Fig. 39). Details of this unusual case have been reported by Naumann and coworkers (38). The first

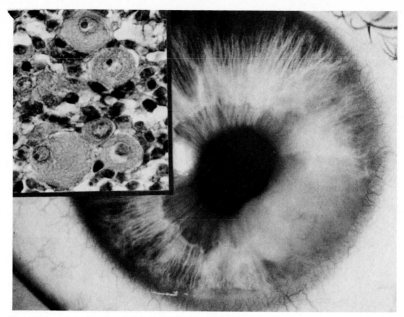

Figure 38. A grayish mass, which has led to irregularity of the pupil, is located between the 2- and 6-o'clock positions of the iris in the right eye. *Inset* demonstrates that the iris tumor is composed of two cell types. The larger cells show round contours with abundant fibrillated eosinophilic cytoplasm and eccentric or paracentral nuclei containing prominent nucleoli. The smaller cells, which show small pyknotic nuclei and scanty cytoplasm, represent undifferentiated mesenchymal cells. (Hematoxylin-eosin, ×160, AFIP Neg. 77-4518-1)

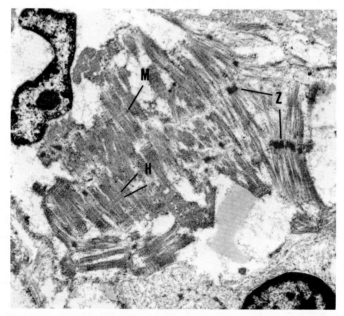

Figure 39. Electron micrograph of formalin-fixed tissue from the recurring tumor involving the ciliary body unequivocally demonstrates that the large cells show the ultrastructural features of rhabdomyoblasts. These include the presence of bundles of myofilaments with the formation of I bands (I) with Z lines (Z), as well as light H bands (H) contining M lines (M). (×10,500, AFIP Neg. 77-4518-2)

198

example of a primary rhabdomyosarcoma of the iris, confirmed by electron microscopic studies, was reported by Woyke and Chwirot (39).

Another unusual myogenous tumor of the uvea is depicted in Figure 40, top. A circumscribed tumor of the ciliary body, initially thought to represent a malignant melanoma, led to enucleation of the eye. By light microscopy, the tumor cells exhibit slightly large pleomorphic nuclei with a fine fibrillary background suggestive of a neuropil (Fig. 40, bottom). The latter findings suggested a benign neurogenic tumor, probably a neurilemoma or a neurofibroma. By electron microscopy, the tumor cells show "hybrid" features of both neuroglial and smooth muscle cells. The portion of the cytoplasm adjacent to the nucleus (perikaryon) discloses stacks of rough-surfaced endoplasmic reticulum and clusters of mitochondria suggesting a neuroglial tumor. The cell processes, however, exhibit bundles of myofilaments with fusiform densities and occasional formation of a fibrillogranular body (Fig. 41). Some cells are more fully differentiated and possess, in addition to abundant myofilaments with fusiform densities, long segments of basement membrane (Fig. 42). Recent experimental embryological studies in lower animals suggest that the smooth muscle of the ciliary body is of neural crest origin (40). The neural crest cells that contribute to the connective tissue of the eye and orbit, as well as to the formation of bone, cartilage, and smooth muscle in the regions of the head and neck, have been called mesectoderm. We proposed that the unusual appearance of this tumor of the ciliary body is a reflection of their origin from the mesectodermal smooth muscle of the ciliary body (40).

NEOPLASTIC LESIONS INVOLVING THE ORBIT

Mesenchymal

Figure 43 represents an electron micrograph of a spindle cell fibrosarcomatous neoplasm that originated from the paranasal sinuses and secondarily invaded the orbit. The tumor cells disclose microtubules, as well as abundant rough-surfaced endoplasmic reticulum with a prominent dichotomous branching pattern. The cells lack basement membrane and possess fine actin-like filaments located under the plasmalemma (Fig. 43). These ultrastructural features are characteristic of fibroblasts, and the tumor was diagnosed as a fibrosarcoma (41).

Figure 44 illustrates the problems in differential diagnosis of a spindle cell sarcomatous tumor (inset, Fig. 44) that displays moderate nuclear pleomorphism, areas of necrosis, and increased mitotic activity. Electron microscopy of the formalin-fixed tissue reveals that the cells were packed with myofilaments containing scattered fusiform densities (Fig. 44), some of which appear to be inserted on the plasmalemma. In other areas, segments of basement membrane and numerous micropinocytotic vesicles can be found. All these five criteria are highly characteristic of smooth muscle cells, and the tumor was classified as a leiomyosarcoma of the orbit. Jakobiec and coworkers (42) compared the clinical behavior and histopathological features of two orbital leiomyomas with two orbital leiomyosarcomas.

Figure 45 is an example of a chondrosarcoma of the paranasal sinuses with

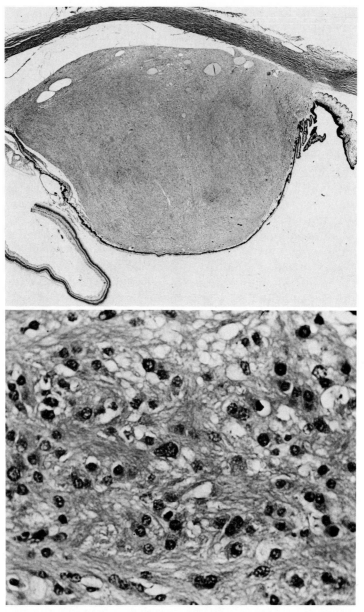

Figure 40. *Top*: an ovoid well-circumscribed tumor of the ciliary body extends from the ciliary processes anteriorly (right) to the pars plana posteriorly (left), elevating the intact ciliary epithelium. (Hematoxylin-eosin, ×9, AFIP Neg. 76-447) *Bottom*: high magnification of the tumor displays a "neural" fibrillated appearance and somewhat hyperchromatic, slightly pleomorphic nuclei. (Hematoxylin-eosin, ×300, AfIP Neg. 76-456)

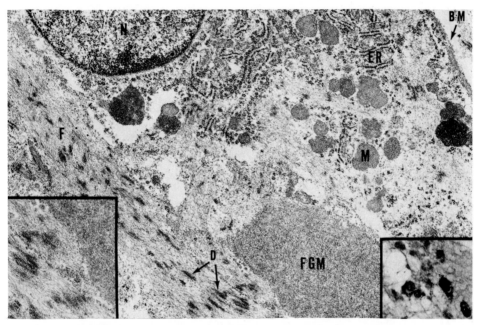

Figure 41. Mesectodermal leiomyoma. Tumor cell exhibiting "hybrid" features of neuroglial and smooth muscle differentiation. On the right, numerous profiles of rough-surfaced endoplasmic reticulum (ER) and degenerated mitochondria (M) are seen. Thin filaments (F) with fusiforn densities (D) are concentrated on the left and merged with a globoid mass composed of fibrillogranular material (FGM). BM = basement membrane. (×18,000, AFIP Neg. 76-6363-2) *Inset, left*: blending of thin filaments into the globoid mass is highlighted. (×30,000) *Inset, right*: cells with PAS-positive globules extending toward the cell processes. (PAS ×485)

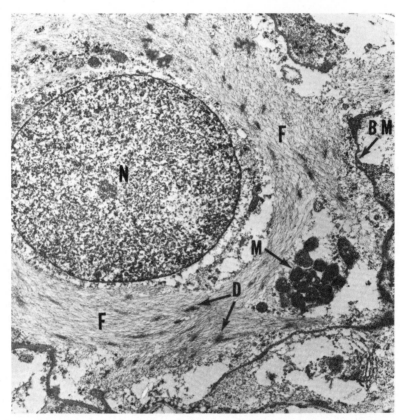

Figure 42. Mesectodermal leiomyoma. Well-differentiated tumor cell displaying a vortex of cytoplasmic filaments (F) exhibiting scattered fusiform densities (D). Except for a few clusters of mitochondria (M), no other cellular organelles are present. A long segment of basement membrane (BM) invests the plasmalemma. N = nucleus. (×9,000, AFIP Neg. 76-6363-3)

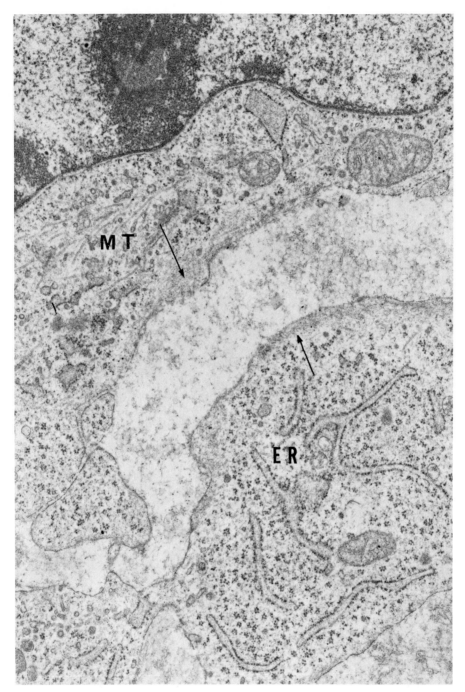

Figure 43. This electron micrograph is from a fibrosarcomatous neoplasm that originated in the left antrum and invaded the floor of the orbit. Tumor cell displays prominent endoplasmic reticulum (ER) forming dichotomous branches. Fine cytoplasmic filaments are compacted beneath the plasmalemma (arrows). Clusters of polyribosomes fill the cytoplasm. Note the absence of basement membrane. MT = microtubules. (×17,000, AFIP Neg. 77-4541)

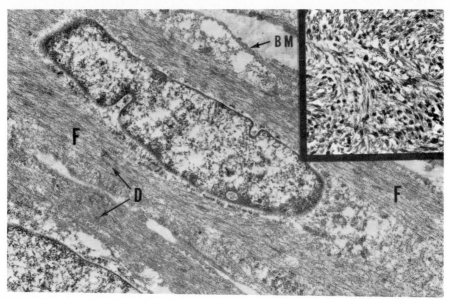

Figure 44. Leiomyosarcoma of orbit. Spindle-shaped cell contining myriad thin filaments (F), as well as fusiform densities (D). Segments of plasmalemma are invested by a basement membrane (BM). (×13,500) *Inset*: interlacing fascicles of tumor cells. (Hematoxylin-eosin, ×180, AFIP Neg. 77-4534)

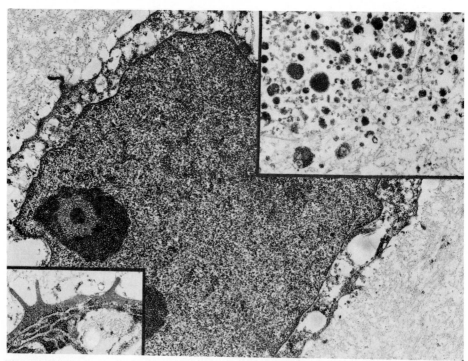

Figure 45. Chondrosarcoma invading the orbit. Tumor cell showing prominent nucleolus and relatively scanty cytoplasm. The extracellular matrix surrounding the lacuna contains numerous fine fibrils. (×18,000, AFIP Neg. 77-4539) *Inset, bottom*: scalloping on the cell surface and dilated cisternae of endoplasmic reticulum. *Inset, top*: extralacunar matrix containing vesicles and granules (glycosaminoglycan granules) enmeshed in fine fibrils. (×21,000)

204

orbital invasion. The figure depicts a tumor cell within the lacuna surrounded by fine fibrils. The bottom inset displays the typical ruffled border of the cell and dilated cisterna of the endoplasmic reticulum. The top inset discloses the extra-lacunar matrix containing abundant glycosaminoglycan granules enmeshed in fine fibrils. Erlandson and Huvos studied by light and electron microscopy eight examples of chondrosarcomas of bone (43).

In contrast to the rare orbital fibrosarcoma, fibrous histiocytoma is the most common primary mesenchymal tumor of the orbit in adults. This fact has only recently been appreciated after a review of the collection of orbital tumors on file at the Armed Forces Institute of pathology (unpublished data). Ultrastructurally, the cells of fibrous histiocytoma are usually closely applied to one another and display myriad interdigitating cell processes (Fig. 46) (44). Abundant lysosomal bodies and variable lipid are contained within the cytoplasm (Fig. 47, top); the rough-surfaced endoplasmic reticulum may be quite prominent (Fig. 47, bottom). There is no basement membrane formation, but the tumor cells may form poorly developed desmosomes. The foregoing characteristics, coupled with the recent finding of myofibroblasts (45) in some nonorbital tumors, reinforce our impression that most of these tumors are probably derived from a primitive mesenchymal cell that is related to the fibroblast, which has adopted some of the morphological and/or functional aspects of the histiocyte.

Figure 48 is an example of a hemangiopericytoma of the orbit. The tumor cells are intimately attached to the vessel walls and possess a clear cytoplasm and a relative scarcity of organelles (Fig. 48). Other features of pericytes include the presence of plasmalemmal hemidesmosomes and their being almost totally invested by a basement membrane. The electron microscopic features of soft tissue hemangiopericytomas have been described (46,47).

Neurogenic Tumors

Figure 49 is an example of a glioma of the optic nerve. The tumor cells disclose numerous thin, straight glial filaments and scattered electron-dense bodies that appear to be lysosomes (48). The cells are focally invested by a basal membrane at places where they abut on connective tissue (inset, Fig. 49). Anderson and Spencer studied histochemically and by electron microscopy two gliomas of the optic nerve (49). Both tumors showed abundant extracellular mucopolysaccharides that they believe contributed to a sudden increase of proptosis in one patient.

Figure 50 is an example of a meningioma invading the orbit. The tumor cells exhibit prominent interdigitations of their plasmalemmas and possess abundant fine filaments, haphazardly arranged. Formations of desmosomes with tonofilaments were frequently present (50) (inset, Fig. 50).

Figure 51 is an example of a neurilemoma (schwannoma) of the orbit. The Schwann cells, which were invested by a continous basement membrane, are surrounded by fusiform bundles of 1,000-Å banded basement membrane (29) ("long-spacing collagen") oriented mainly along their plasmalemmas (50). Although banded basement membrane has been found in a variety of other mesenchymal neoplasms, its occurrence is highly characteristic of neurogenic

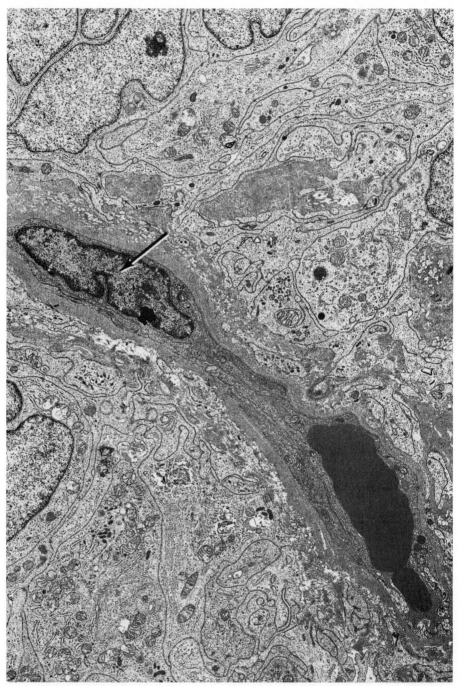

Figure 46. Fibrous histiocytoma. Tumor cells create a tangle of cell processes. Numerous lysosomal dense granules are scattered in the cytoplasmic processes. Arrow points to the nucleus of an endothelial cell, and a compressed erythrocyte is shown toward the bottom right (×4,000, AFIP Neg. 77-4515-1)

206

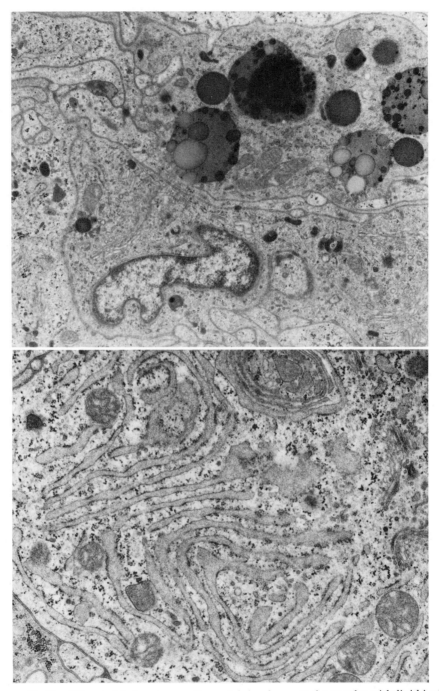

Figure 47. *Top*: Fibrous histiocytoma cells containing lysosomal granules with lipid inclusions. *Bottom*: abundant profiles of rough-surfaced endoplasmic reticulum are present in another tumor cell. (*top,* ×6,000, AFIP Neg. 77-4515-3; *Bottom,* ×7,200, AFIP Neg. 77-4515-2)

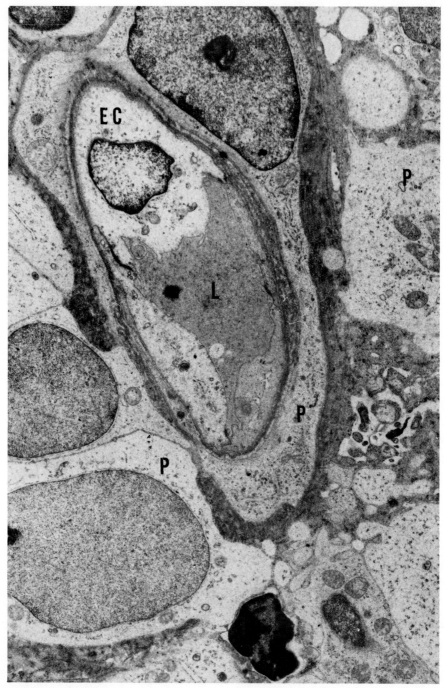

Figure 48. Hemangiopericytoma of orbit. The pericytes (P) are ovoid and show an electron-lucent cytoplasm that contains few cytoplasmic organelles. EC = endothelial cell; L = capillary lumen. (×5,000, AFIP Neg. 77-4542)

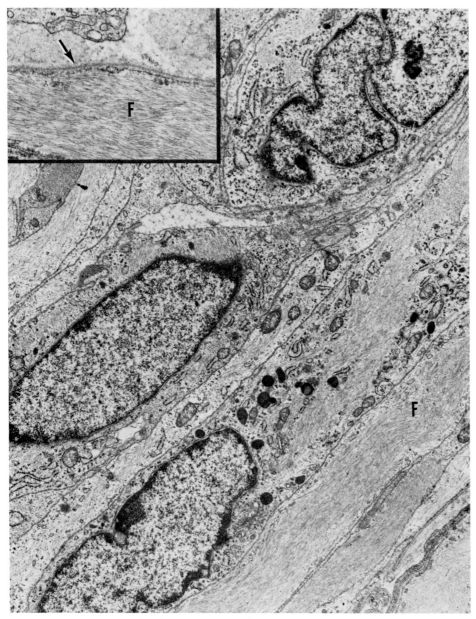

Figure 49. Optic nerve glioma. The fibrous astrocytes contain longitudinally oriented cytoplasmic filaments (F) without fusiform densities. Numerous lysosomal bodies are also present. *Inset*: cytoplasmic filaments (F), at higher magnification, and formation of basement membrane (arrow) where tumor cells abut on connective tissues. (×2,400; *inset*, ×28,000, AFIP Neg. 77-4535)

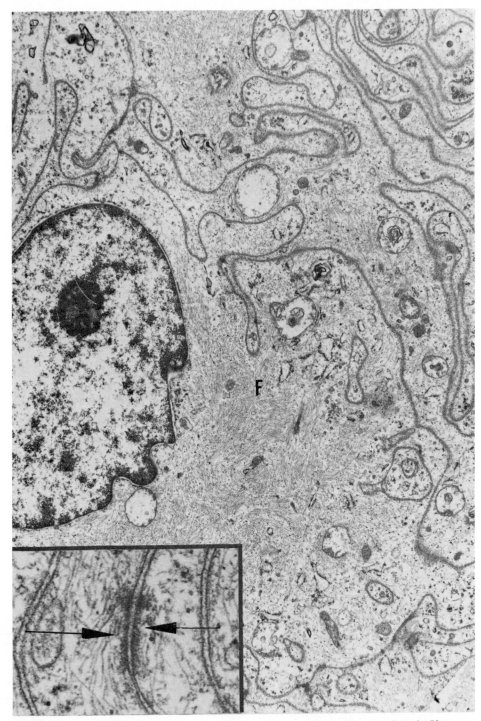

Figure 50. Meningioma of orbit. The tumor cells are filled with cytoplasmic filaments (F) and show prominent interdigitations of their plasmalemmas. *Inset*: well-developed desmosome formation (arrows). (×5,000; *inset*, ×22,000, AFIP Neg. 77-4529)

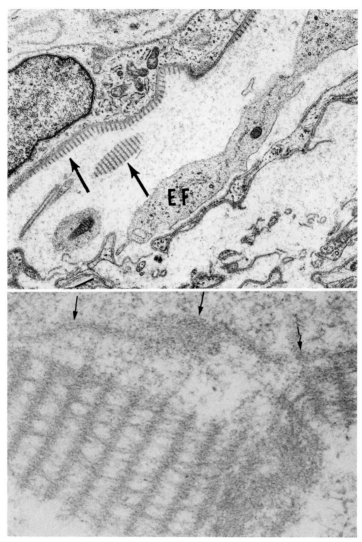

Figure 51. Schwannoma. *Top*: banded basement membrane material ("fibrous long-spacing collagen") (arrows) is present along the plasmalemma of a Schwann cell shown at upper left. An endoneural fibroblast (EF) is not surrounded by basement membrane. *Bottom*: banded basement membrane has a periodicity of 1,000 Å and lies against the plasmalemma of the Schwann cell (arrows). (*top*, ×4,200; *bottom*, ×35,000, AFIP Neg. 77-4519)

tumors, especially neurilemomas and neurofibromas, (51), and some astrocytic neoplasms of the brain.

Tumors Showing Diagnostic Granules

Certain tumors that have the capacity to metastasize to the eye and orbit may disclose characteristic histochemical and electron microscopic findings. Figure 52 represents a metastatic carcinoid tumor to the orbit (24). By light microscopy, the tumor lobules contained light and dark cells. The dark cells are confined predominantly to the periphery of the tumor lobules embracing the light cells more centrally located (bottom inset, Fig. 52). Histochemical stains showed the presence of a positive argentaffin reaction (top inset, Fig. 52). By electron microscopy of the paraffin-embedded tissue, the two cell types could be clearly demonstrated, both containing myriad neurosecretory granules that exhibited variable shape and density (Fig. 52) (24). The granules in the light cells were less numerous and more electron-dense, and varied from round to elongated (rod-shaped) structures (top, Fig. 3), whereas the granules in the dark cells were more densely packed, less electron-dense, and spherical (bottom, Fig. 53). Carcinoid tumors metastatic to the eye and orbit have been previously reported by light microscopy (52–54). Electron microscopic studies of the neurosecretory granules of carcinoid tumors have been described elsewhere (55,56).

Certain poorly differentiated neoplasms may contain characteristic crystalline inclusions that are highly diagnostic and may be of great help in the differential diagnosis from other tumors that may exhibit an alveolar pattern (e.g., alveolar rhabdomyosarcoma versus alveolar soft part sarcoma, versus nochromaffin paraganglioma). A distinctive and highly diagnostic feature in favor of alveolar soft part sarcoma, not seen in other tumors with alveolar pattern, is the presence of PAS-positive, diastase-resistant crystalline inclusions. These crystals vary in size and in shape from rectangular to rhomboidal and from rod-like to spiked needles (Fig. 54) (24). The crystals in some areas are incompletely formed and demonstrate longitudinal striations with a periodicity of 80 Å (left inset, Fig. 54). Shipkey and coworkers found these peculiar crystalline inclusions in 9 out of 13 tumors (57). Their histochemical studies revealed that the crystals were composed of a protein-carbohydrate complex. A recent electron-microscopic study described neurosecretory granules and suggested that these tumors may arise from paraganglia (58).

Figure 55 is an example of a granular cell tumor of the orbit. The top inset in Figure 55 discloses the light-microscopic features of the tumor cells, exhibiting plump fusiform or polyhedral cells totally packed with large spheroidal granules. By electron microscopy, the lysosomal-type granules are of variable size and density and appear to be enclosed by single membranes (bottom inset, Fig. 55). The interstitial or stromal cells of the tumor contain filamentous fusiform structures ("angulated bodies") that are believed to be highly characteristic of these tumors (Fig. 56). These bodies appear to be related to dilated, rough-surfaced endoplasmic reticulum. Morgan recently studied by electron microscopy a granular cell tumor of the orbit (59). From his own observations and a review of the literature, he proposed that the tumor originates from a primitive mesenchymal cell resembling a fibroblast.

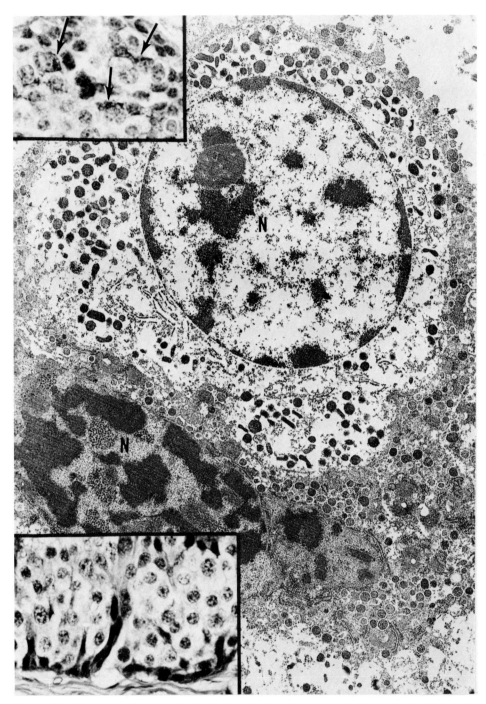

Figure 52. Electron micrograph of a light cell (above) surrounded by a dark cell (below). Both cells contain numerous neurosecretory granules that vary in shape, density, and distribution. N = nucleus. (×14,000, AFIP Neg. 74-10013-9) *Inset, bottom*: both dark and light cells at the periphery of one of the tumor lobules. (Hematoxylin-eosin, ×440, AFIP Neg. 74-9513) *Inset, top*: positive argentaffin reaction (see black cytoplasmic granules indicated by arrows). (Fontana reaction, ×575, AFIP Neg. 74-9511)

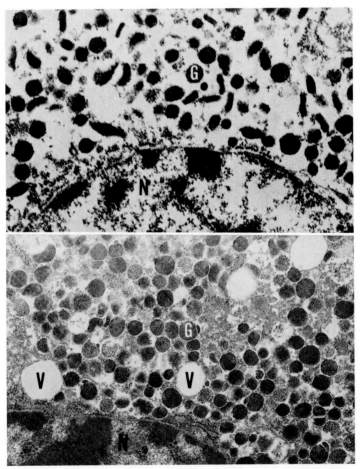

Figure 53. *Top*: light cell containing electron-dense spherical or elongated (rod-shaped) granules (G). The cytoplasm is more translucent, and the granules are widely dispersed throughout. N = nucleus. (×21,000, AFIP Neg. 74-10013-11) *Bottom*: dark cell disclosing more closely packed, less electron-dense spherical granules (G) and scattered cytoplasmic vacuoles (V). N = nucleus. (×21,000, AFIP Neg. 74-10013-10)

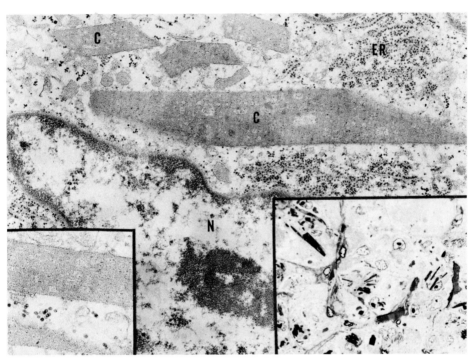

Figure 54. Electron micrograph of alveolar soft part sarcoma showing several rectangular or rhomboidal crystals (C) and abundant rough-surfaced endoplasmic reticulum (ER). N = nucleus. (×30,000, AFIP Neg. 74-10013-1) *Inset, right*: numerous intracytoplasmic crystalline inclusions within tumor cells. (Toluidine blue, ×485, AFIP Neg. 74-9100) *Inset left*: one of the crystals with a periodicity of 80 Å. (×63,000)

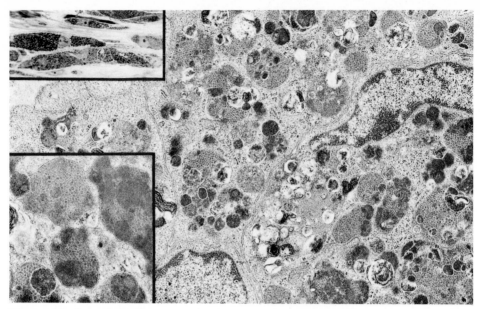

Figure 55. Granular cell tumor of orbit. The cytoplasm of the tumor cells contains granules of variable density, size, and shape, most of which are surrounded by a membrane. (×18,000, AFIP Neg. 77-4533) *Inset, top:* the plump fusiform tumor cells packed with cytoplasmic granules. (Toluidine blue, ×395) *Inset, bottom:* several irregular granules each containing smaller round particles (resembling viral particles) and encased by a single membrane. (×27,000)

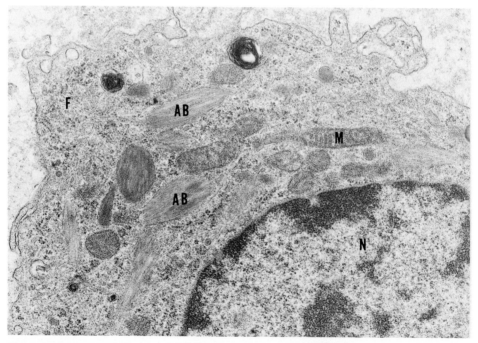

Figure 56. Interstitial (stromal) cell with blunt cytoplasmic processes. Several fusiform filamentous structures (angulated bodies—AB) are present in the cytoplasm. Fine cytoplasmic filaments (F) are also evident. M = mitochondria. (×27,000, AFIP Neg. 77-4534)

216

REFERENCES

1. Goldberg MF, Maumenee AE, McKusick VA: Corneal dystrophies associated with abnormalities of mucopolysaccharide metabolism. *Arch Ophthalmol* 74:516, 1965.

2. Kenyon KR, Quigley HA, Hussels IE, Wyllie RG: The systemic mucopolysaccharidosis: Ultrastructural and histochemical studies of the conjunctiva and skin. *Am J Ophthalmol* 73:811, 1972.

3. Font RL, Fine BS: Ocular pathology in Fabry's disease. *Am J Ophthalmol* 73:419, 1972.

4. Weingeist TA, Blodi FC: Fabry's disease: Ocular findings in a female carrier. *Arch Ophthalmol* 85:169, 1971.

5. Font RL: Chronic ulcerative keratitis caused by herpes simplex virus. Electron microscopic confirmation in paraffin-embedded tissue. *Arch Ophthalmol* 90:382, 1973.

6. Witmer R, Iwamoto T: Electron microscopic observations of herpes-like particles in the iris. *Arch Ophthalmol* 79:331, 1968.

7. Minckler DS, McLean EB, Shaw OM, Hendrickson A: Herpes virus hominis encephalitis and retinitis. *Arch Ophthalmol* 94:89, 1976.

8. Font RL, Jenis EH, Tuck KD: Measles maculopathy associated with subacute sclerosing panencephalitis. Immunofluorescent and immuno-ultrastructural studies. *Arch Pathol* 96:168, 1973.

9. Landers MB, Klintworth GK: Subacute sclerosing panencephalitis (SSPE): A clinicopathologic study of the retinal lesions. *Arch Ophthalmol* 86:156, 1971.

10. Rao NA, Font RL: Toxoplasmic retinochoroiditis. Electron microscopic and immunofluorescence studies of formalin-fixed tissue. *Arch Ophthalmol* 95:273, 1977.

11. Font RL, Jakobiec FA: Granulomatous necrotizing retinochoroiditis caused by *Sporotrichum schenkii*. Report of a case including immunofluorescence and electron microscopical studies. *Arch Ophthalmol* 94:1513, 1976.

12. Streeten B, Robuzzi D, Jones D: Sporotrichosis of the orbital margin. *Am J Ophthalmol* 77:750, 1974.

13. Font RL, Brownstein S: A light and electron microscopic study of anterior subcapsular cataracts. *Am J. Ophthalmol* 78:972, 1974.

14. Ts'o MOM, Fine BS, Zimmerman LE, Vogel MH: Photoreceptor elements in retinoblastoma. A preliminary report. *Arch Ophthalmol* 82:57, 1969.

15. Ts'o MOM, Fine BS, Zimmerman LE: The Flexner-Wintersteiner rosettes in retinoblastoma *Arch Ophthalmol* 88:664, 1969.

16. Ts'o MOM, Zimmerman LE, Fine BS: The nature of retinoblastoma. I. Photoreceptor differentiation: A clinical and histopathologic study. *Am J Ophthalmol* 69:339, 1970.

17. Ts'o MOM, Fine BS, Zimmerman LE: The nature of retinoblastoma. II. Photoreceptor differentiation: An electron microscopic study. *Am J Ophthalmol* 69:350, 1970.

18. Jakobiec FA, Howard GM, Ellsworth RM, Rosen M: Electron microscopic diagnosis of medulloepithelioma. *Am J Ophthalmol* 79:321, 1975.

19. Zimmerman LE, Font RL, Andersen SR: Rhabdomyosarcomatous differentiation in malignant intraocular medulloepitheliomas. *Cancer* 30:817, 1972.

20. Zimmerman LE: Verhoeff's "teratoneuroma." *Am J Ophthalmol* 72:1039, 1971.

21. Zimmerman LE: The remarkable polymorphism of tumors of the ciliary epithelium. *Trans Cong Austral Coll Ophthalmol* 2:114, 1970.

22. Kroll AJ: Fine structural classification of orbital rhabdomyosarcoma. *Invest Ophthalmol* 6:531, 1967.

23. Jakobiec FA, Font RL, Johnson FB: Angiomatosis retinae. An ultrastructural study and lipid analysis. *Cancer* 38:2042, 1976.

24. Font RL: Electron microscopy of tumors of the eye and ocular adnexa, in Reese A: *Tumors of the eye.* New York, Harper & Row, 1976, chap 12.

25. Zimmerman LE: Melanocytes, melanocytic nevi and melanocytomas. *Invest Ophthalmol* 4:11, 1965.

26. Font RL, Zimmerman LE, Fine BS: Adenoma of the retinal pigment epithelium. Histochemical and electron microscopic observations. *Am J Ophthalmol* 73:544, 1972.

27. Wallow IHL, Ts'o MOM: Proliferation of the retinal pigment epithelium over malignant choroidal tumors. A light and electron microscopic study. *Am J Ophthalmol* 73:914, 1972.

28. Streeten BW, McGraw JL: Tumor of the ciliary pigment epithelium. *Am J Ophthalmol* 74:420, 1972.

29. Yanoff M, Fine BS: *Ocular Pathology. A Text and Atlas.* New York, Harper & Row, 1975, p 674 (Fig. 17–73B); p 638.

30. Schrader WA, Helwig EB: Balloon cell nevi. *Cancer* 20:1502, 1967.

31. Jao W, Fretzin DF, Sundaran L, Frydman JE: Balloon cell nevus of the conjunctiva. *Arch Pathol* 96:124, 1973.

32. Pfaffenbach DD, Green WR, Maumenee AE: Balloon cell nevus of conjunctva. *Arch Ophthalmol* 87:192, 1972.

33. Naumann G, Yanoff M, Zimmerman LE: Histogenesis of malignant melanomas of the uvea. I. Histopathologic characteristics of nevi of the choroid and ciliary body. *Arch Ophthalmol* 76:784, 1966.

34. Gardner WA Jr, Vazquez MD: Balloon cell melanoma. *Arch Pathol* 89:470, 1970.

35. Ranchod M: Metastatic melanoma with balloon cell changes. *Cancer* 30:1006, 1972.

36. Riley FC: Balloon cell melanoma of the choroid. *Arch Ophthalmol* 92:131, 1974.

37. Font RL, Zimmerman LE, Armaly MF: The nature of the orange pigment over a choroidal melanoma. *Arch Ophthalmol* 91:359, 1974.

38. Naumann G, Font RL, Zimmerman LE: Electron microscopic verification of primary rhabdomyosarcoma of iris. *Am J Ophthalmol* 74:110, 1972.

39. Woyke S, Chwirot R: Rhabdomyosarcoma of the iris. Report of the first recorded case. *Br J Ophthalmol* 56:60, 1972.

40. Jakobiec FA, Font RL, Tso MOM, Zimmerman LE: Mesectodermal leiomyoma of the ciliary body: A tumor of presumed neural crest origin. *Cancer* 39:2102, 1977.

41. Jakobiec FA, Tannenbaum M: The ultrastructure of orbital fibrosarcoma. *Am J Ophthalmol* 77:899, 1974.

42. Jakobiec FA, Howard GM. Rosen M, Wolff M: Leiomyoma and leiomyosarcoma of the orbit. *Am J Ophthalmol* 80:1028, 1975.

43. Erlandson RA, Huvos AG: Chondrosarcoma: A light and electron microscopic study. *Cancer* 34:1642, 1974.

44. Jakobiec FA, Jones IS: Mesenchymal and fibro-osseous tumors of the orbit, in Duane T (ed): *Clinical Ophthalmology.* New York, Harper & Row, 1976, vol 2, chap 44.

45. Churg A, Kahn L: Myofibroblasts and related cells in malignant fibrous and fibrohistiocytic tumors. *Hum Pathol* 8:205, 1977.

46. Kuhn C, Rosai J: Tumors arising from pericytes: Ultrastructure and organ culture of a case. *Arch Pathol* 88:653, 1969.

47. Battifora H: Hemangiopericytoma: Ultrastructural study of five cases. *Cancer* 31:1418, 1973.

48. Eggers H, Jakobiec FA, Jones IS: Tumors of the optic nerve. *Doc Ophthalmol* 41:43, 1976.

49. Anderson DR, Spencer WH: Ultrastructural and histochemical observations of optic nerve gliomas. *Arch Ophthalmol* 83:324, 1970.

50. Jakobiec FA, Jones IS: Neurogenic tumors of the orbit, in Duane T (ed): *Clinical Ophthalmology.* New York, Harper & Row, 1976, vol 2, chap 41.

51. Fisher ER, Vuzevski VD: Cytogenesis of schwannoma (neurilemoma), neurofibroma, dermatofibroma and dermatofibrosarcoma as revealed by electron microscopy. *Am J Clin Pathol* 49:141, 1968.

52. Font RL, Kaufer G, Winstanley RA: Metastasis of bronchial carcinoid tumor to the eye. *Am J Ophthalmol* 62:723, 1966.

53. Rosenbluth J, Laval J, Weil JV: Metastasis of bronchial adenoma to the eye. *Arch Ophthalmol* 63:47, 1960.

54. Honrubia FM, Davis WH, Moore MK, Elliot JH: Carcinoid syndrome with bilateral orbital metastases. *Am J Ophthalmol* 72:1118, 1971.

55. Bensch KG, Gordon GB, Miller LR: Electron microscopic and biochemical studies on the bronchial carcinoid tumor. *Cancer* 18:592, 1965.

56. Rosai J, Rodriguez HA: Application of electron microscopy to the differential diagnosis of tumors. *Am J Clin Pathol* 50:555, 1968.

57. Shipkey FH, Lieberman PH, Foote FW, Jr, Stewart FW: Ultrastructure of alveolar soft part sarcoma. *Cancer* 17:821, 1964.

58. Unni KK, Soule EH: Alveolar soft part sarcoma. An electron microscopic study. *Proc Mayo Clin* 50:591, 1975.

59. Morgan G: Granular cell myoblastoma of the orbit. *Arch Ophthalmol* 94:2135, 1976.

5
Ultrastructural Pathology of the Human Urinary Bladder

Myron Tannenbaum, M.D., Ph.D.
Departments of Pathology and Urology
College of Physicians and Surgeons, Columbia University
New York, New York

A tissue organ system in man that is readily accessible for serial surgical biopsies and that has been inadequately studied ultrastructurally from a pathobiological or clinical aspect is the transitional epithelium (or urothelium) of the genitourinary system. The urothelium lines a urine conduit that extends from the renal pelvis to the ureters, bladder, prostatic urethra, and periurethral prostatic ducts to the penile urethra or female urethra. The urothelium is perpetually subject to infections, trauma, malformations, and a neoplastic process. The last is the main concern of this chapter.

In many instances, the development of a cystoscopically visible tumor or an in situ neoplastic process produces associated symptoms such as dysuria, hematuria, and pyuria that will bring the affected person to the attention of the urologist. These patients are then cystoscoped, and biopsy material is obtained for the surgical pathologist to evaluate. There is a great unawareness, however, of the conditions under which the biopsy specimens are obtained. The visualization of the urothelium in the bladder by means of a cystoscope is done through a fluid that is instilled into the urinary bladder. The cystoscopic fluid is usually distilled water or a glycine solution with a low osmotic pressure of at least 150 mOsm. Osmotically, this fluid compares very unfavorably to urine, but for varying periods of time it comes into contact with all parts of the urothelium of the genitourinary system.

In retrospect, there are very few ultrastructural descriptions or studies of normal human and neoplastic urothelium (1–4). When such studies are reevaluated or future ones contemplated, the manner in which the urothelium material was or is to be obtained will be of great importance. The following morphological descriptions and the accompanying photomicrographs were obtained from noncystoscoped urothelium unless otherwise stated.

THE NORMAL UROTHELIUM

In some specimens, notably those obtained from surgical specimens of the ureter and renal pelvis, the urothelium is composed of a very tight arrangement of compact layers of cells. Whenever these cells show disruptive interdigitation, some attention must be paid to the manner in which the specimen was collected. These disruptive features are not noted to occur with any frequency in animal bladders or ureters that are isolated and fixed very rapidly (5). The normal urothelium is composed, for the most part, of three different layers of cells that can be described ultrastructurally as comprising a layer of a) *superficial cells,* b) one or more *intermediary* layers of *cells,* and finally c) a third layer of cells called the *basal cells.* The latter are attached to a basement membrane by means of numerous hemidesmosomes or half-desmosomes.

Superficial Cells

The luminal surface of the bladder and other portions of the genitourinary system are lined by superficial cells (Fig. 1). The superficial cells may measure 20 to 30 μ in width and half that in height. The cytoplasm of these cells contains numerous mitochondria, as well as numerous profiles of thick-walled, rounded or fusiform vesicles that vary greatly in size and number (Figs. 2–5) (6–8). In man, the vesicles are not as numerous as they are in the rodent. It is, of course, extremely important to consider what is meant by "normal urothelium." The illustrations (Figs. 1–8, 10–19) that appear in this section show bladder mucosa that was not inflamed or involved by a neoplastic process and that was obtained during a suprapubic prostatectomy procedure. Here the urothelium is almost similar to what has been described in the rodent. When favorable sections are seen, a tripartite junctional complex may be seen in the superficial cells (Figs. 1–4). This consists of a zonula occludens, a zonula adherens, and a macula adherens.

The function of these junctions and their relationship with the vesicles and ridged surface of the normal urothelium with its asymmetrical membranes are not clearly understood. The exact role of the vesicles and asymmetrical membranes may have to do with fluid transport and consequently the systemic hydration of the individual. It has been speculated that the vesicles may represent structures for storage or transport of the plasmalemma of transitional epithelial cells. It has also been suggested that the vesicles may function as part of an excretory system for the excessive fluid that might enter through the surface cells. In the human material illustrated in this chapter, such vesicles can also be seen in the intermediaty layer of cells.

Intermediary Cells

The cells of the intermediary layer are more irregular in shape than surface cells. They tend to be stellate and are separated from their adjacent intermediary cells, surface cells, or basal cells. The degree of separation may depend on how the biopsy material is obtained for study. A cisternal system or extracellular fluid compartment exists in greater volume and area between the cells of this layer than between the surface cells (Figs. 1, 10–14, and 16). The intermediary cells

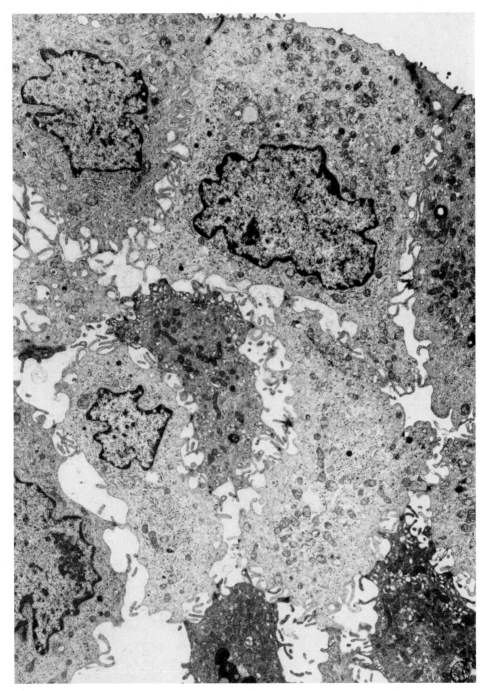

Figure 1. Electron micrograph of normal urothelium with surface cells (top). Two to three layers of intermediary cells are below the surface cells. Intermediary cells are separated by intercellular cisternal spaces. Intermediary cells are light and dark and contain numerous microvilli, intracytoplasmic vesicles, and desmosomes. (×3,625)

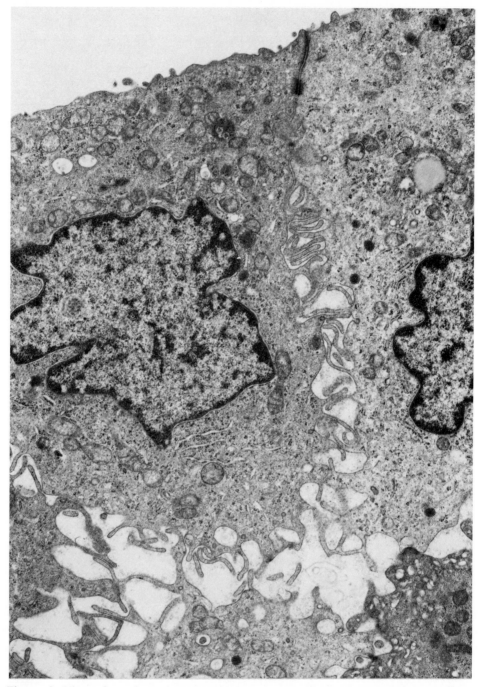

Figure 2. View of previous section at higher power showing surface cells and some intermediary cells. Note tripartite junctional complex in surface cells. (×7,750)

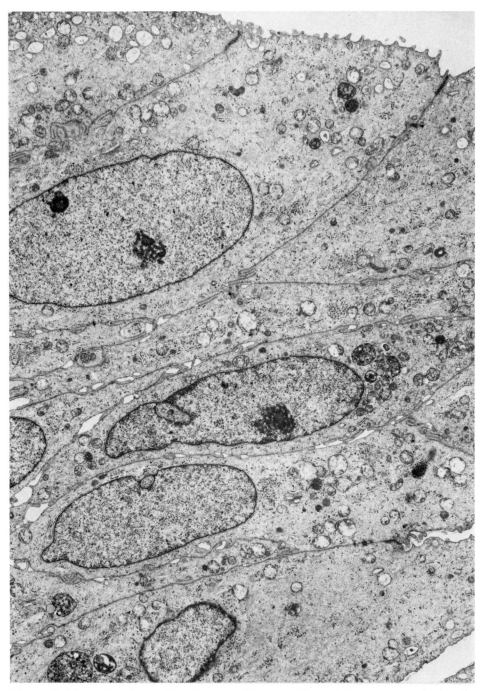

Figure 3. Normal surface cells with microridges, vesicles, telolysosomes, and tripartite junctional complexes. (×3,625)

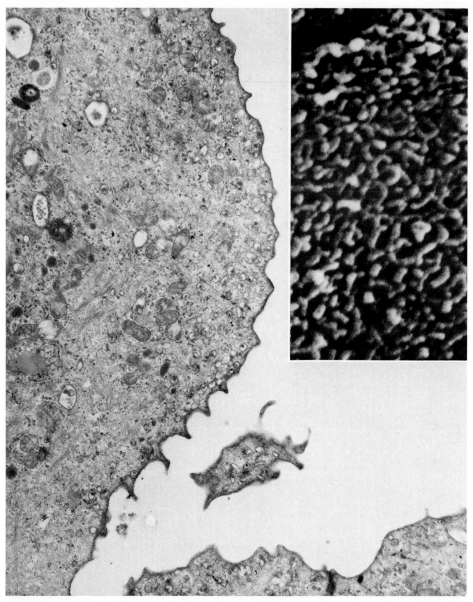

Figure 4. Transmission electron micrograph (TEM) of surface cell with microridges, subluminal vesicles, fibrillar cytoplasm, and mitochondria. (×13,000) *Insert:* Scanning electron micograph (SEM) of surface cell with anastomosing microridges. (×8,000)

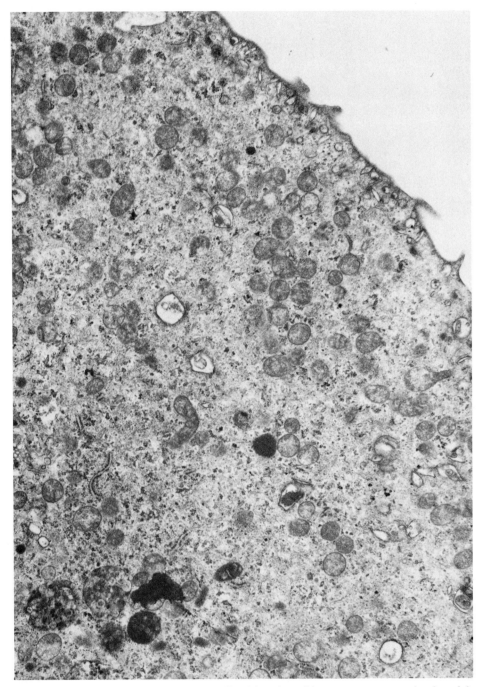

Figure 5. TEM of normal surface cell with microridges, numerous mitochondria, telolysosomes, subluminal vesicles, and Golgi apparatus. (×7,750)

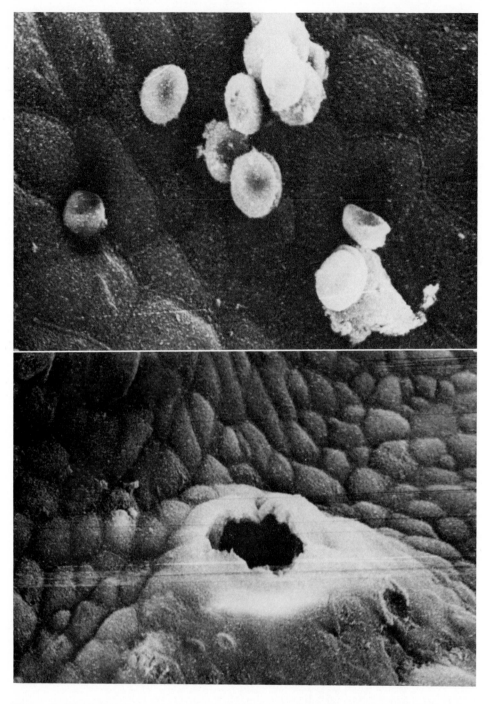

Figure 6. (Top): SEM of non-neoplastic urothelium with several blood cells and inflammatory cells. (×2,000)

Figure 7. (Bottom): SEM of trigone of bladder with pit entrance to cystitis cystica at lower right. (×1,100)

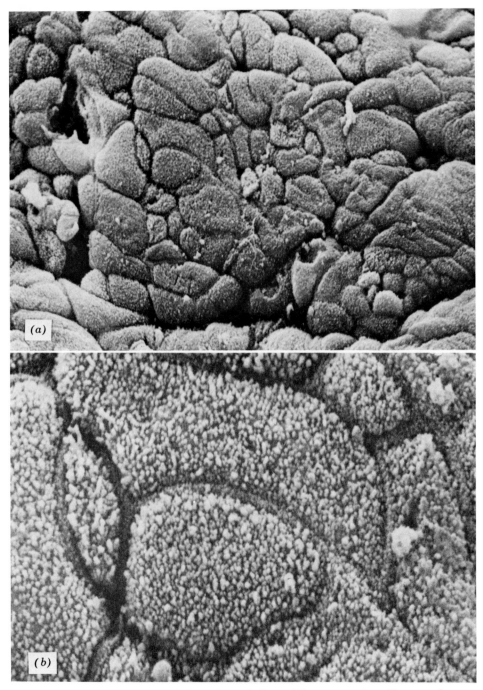

Figure 8. (a) SEM of slightly infamed urothelium. There are microridges, and moats interposed between hexagonal and irregular luminal outlines of surface cells. Many of the surface cells have microvilli. There are some pit entrances. (×900) (b) Higher magnification of (a) (×3,600)

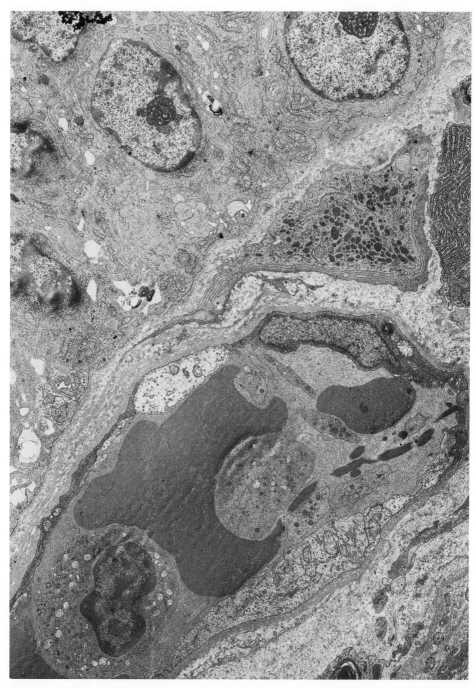

Figure 9. TEM of basal cells covering the lamina propria. Note the basement membrane, capillary, plasma cells, and portions of a polymorphonuclear leukocyte (PMNL). The epithelium is partially damaged by exposure to distilled water. (×6,000)

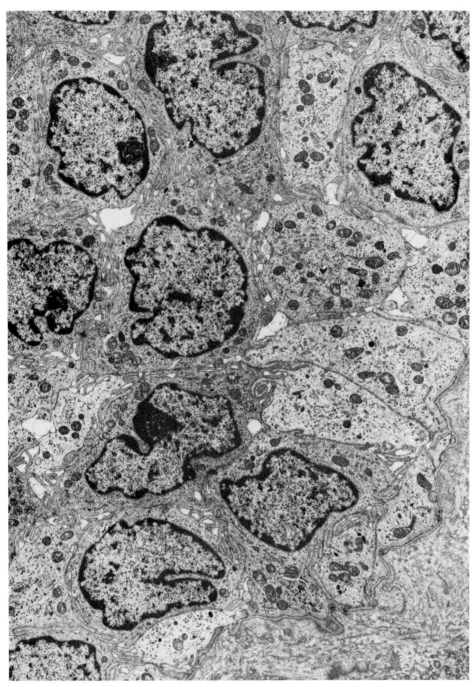

Figure 10. TEM of several layers of intermediary cells above basal cells. The intermediary cells contain telolysosomes, mitochondria, desmosomal cellular interconnections, and a very small intercellular space. Basal cells are attached to the basement membrane by hemidesmosomes. Lamina propria is at lower right. (×6,000)

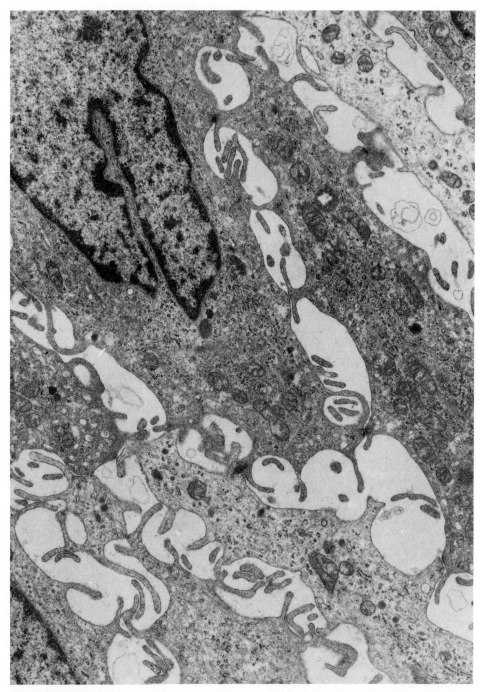

Figure 11. TEM of intermediary cells that are separated by a large intercisternal space. These cells contain microvilli, desmosomes, and Golgi apparatus, as well as numerous clear vesicles. (×13,000)

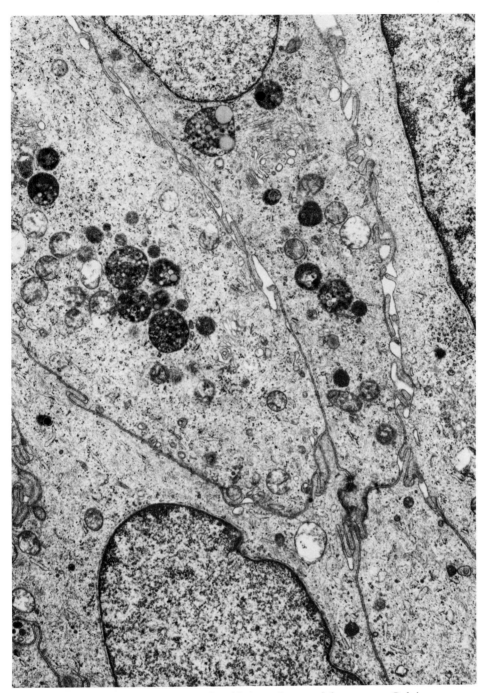

Figure 12. TEM of intermediary cells with prominent telolysosomes, Golgi apparatus, and a very small intercellular cisternal space. Numerous desmosomes are seen. (×13,000)

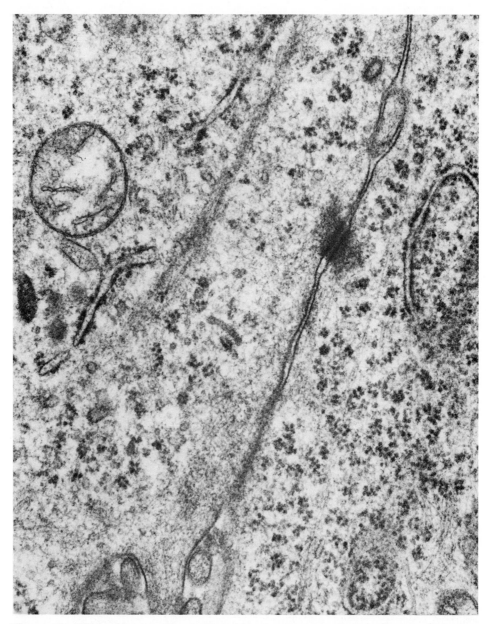

Figure 13. TEM of intermediary cells with prominent desmosomes, intracytoplasmic filaments, and polyribosomes. Note the microvilli in lower portion of picture (×23,000)

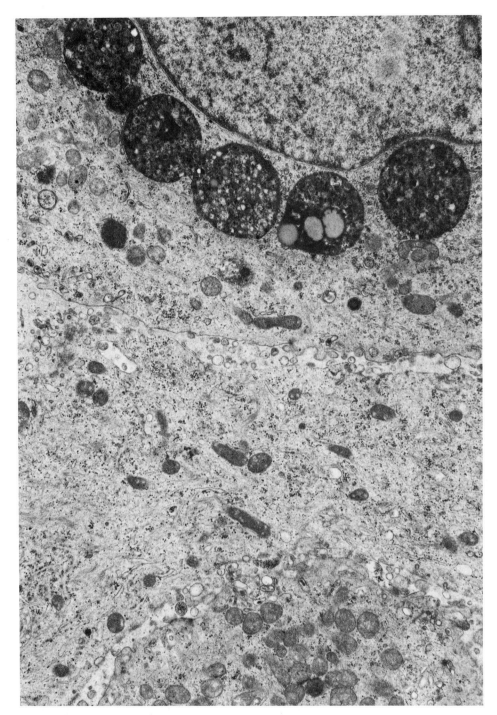

Figure 14. TEM of intermediary cells with prominent telolysosomes. Cells also contain numerous cytoplasmic filaments, mitochondria, Golgi apparatus, and clear vesicles with a limiting membrane. (×13,000)

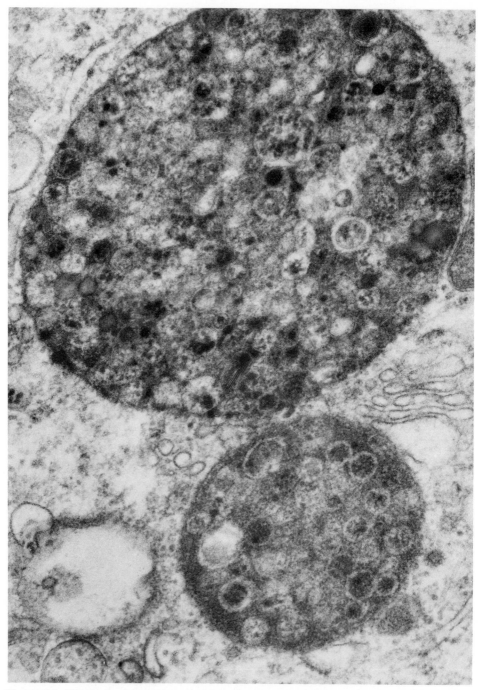

Figure 15. TEM of telolysosomes from intermediary cells, Note the Golgi apparatus at right, as well as the membrane substructure of the telolysosomes. (×60,000)

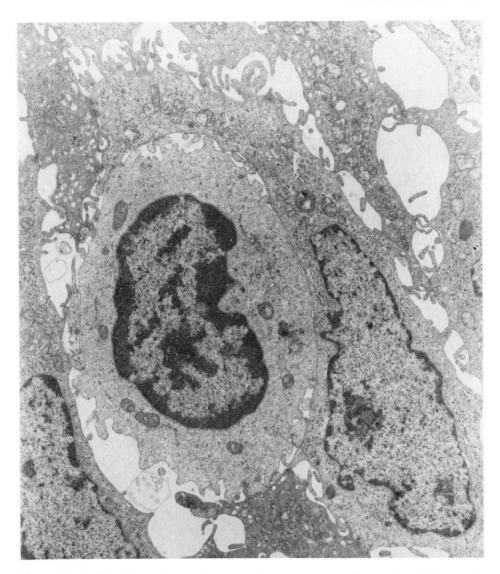

Figure 16. TEM of lymphocyte interposed between intermediary cells which contain prominent vesicles of the dark cell type. (×9,250)

also have numerous microvilli, with desmosomes between some of the long, slender microvilli (Figs. 9, 11, 13, 14 and 16). Pinocytotic vesicles are also noted within these cells (Figs. 11, 12, and 16). The cytoplasm of these cells, as compared to that of the basal cells beneath them, is much more electron-dense (Fig. 10).

Another distinctive feature of the intermediary cells is the abundant number of small and large telolysosomes seen within them, as compared to the few found within the surface luminal cells (Figs. 5, 12, 14, and 15) (2). The telolysosomes probably represent the end stage of autodigestive processes that are involved in the turnover or sequential removal of aging or damaged cell organelles. It has been demonstrated that the surface cells are capable of phagocytosing bacteria when there is an active urinary bladder infection, and perhaps the organelles also help get rid of noxious stimuli.

Telolysosomes are very difficult to find when the urothelium is involved by a neoplastic process. However, if the urothelium is atypical and is obtained from an inflamed but definitely non-neoplastic urothelium, telolysosomes are still seen. Some types of urothelium are constantly inflamed. In the exstrophied bladder or bladder turned inside out, the microridged system (Fig. 4, insert) is replaced by microvilli on the surface of the luminal cells (Figs. 6-8) (9). Telolysosomes are scant in the intermediary cells and luminal cells, but still do not disappear.

When the normal bladder urothelium is compared to the well-differentiated neoplastic process, there does not seem to be any objective alteration in the number of mitochondria, Golgi apparatus, or polyribosomes in the superficial or intermediary layer. However, the telolysosomes and vesicles are markedly decreased in number.

Basal Cells

The basal cells, the cells immediately beneath the intermediary cells, are attached to a basement membrane by means of hemidesmosomes (Figs. 17, 19, and 20). The basement membrane may be either single-stranded or, as in Figures 17 and 18, a lace-like network of single strands. The adjacent cells may be attached to each other by means of desmosomes, and some of the intercellular spaces may open freely on the basement membrane. When there has been inflammation in the lamina propria, the basement membrane is usually multilaced and thickened. This thickening of the basement membrane is readily demonstrated in Figure 19. The basal cells can be either of a dark cell or light cell nature, and there is usually abundant fibrillar material in them, in close proximity to the basement membrane (Fig. 18).

The bladder urothelial surface also abuts on or encompasses certain pits in the bladder mucosa that may be found in the trigone or urinary outlet area of the bladder near the urethra. These pit-like areas are readily demonstrated by scanning electron microscopy (SEM), as seen in Figure 7. The pit-like areas lead into dilated glandular areas, cystitis cystica, or grape-like clusters of transitional epithelium known as Brunn's nests. Non-neoplastic urothelium surrounds these pits, and very often, if there has been chronic inflammation, the microridge surface of the surface cells (Fig. 4, insert) may give way to microvilli (Fig. 8). The non-neoplastic luminal surface cells usually will show a hexagonal or irregular shape by scanning electron microscopy (Fig. 8). These surface cells are separated by a very fine ridge that can be readily seen in Figure 6 and the two micrographs of Figure 8. Transmission electron microscopy of the elevated ridges between the cells is demonstrated in Figures 1, 2, and 3. These elevated ridges and moats are further represented by the tripartite junctional complexes below the surfaces. The elevated ridges that form the border or moat surrounding the cells will partially disappear if the urothelium becomes either preneoplastic (dysplastic) or carcinomatous in situ (Fig. 35). The same phenomenon is seen in many of the well-differentiated and/or poorly-differentiated papillary exophytic lesions of the bladder.

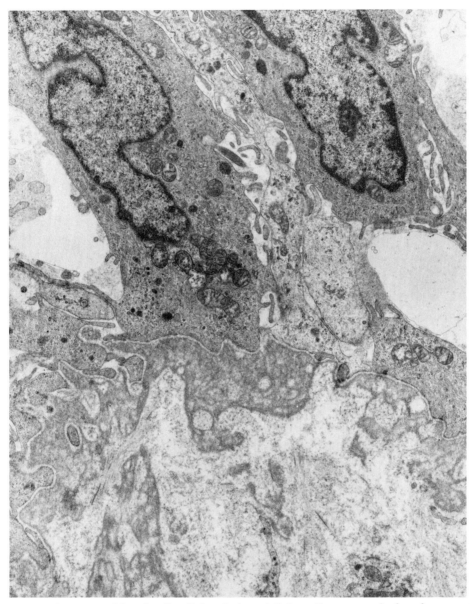

Figure 17. TEM of basal cells of the dark and light cell type attached to basement membrane which is multilayered in several foci. Note attachment to the basement membrane by hemidesmosomes. (×10,750)

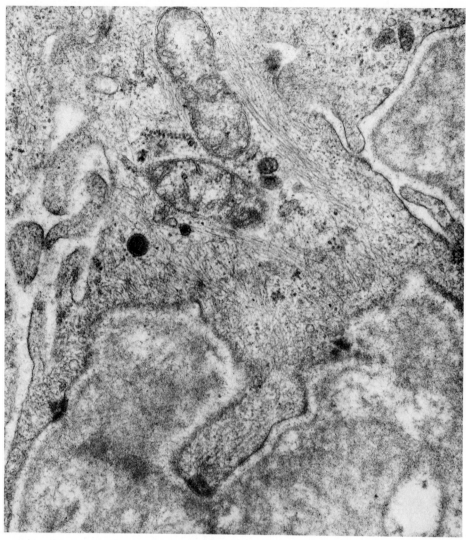

Figure 18. TEM of basal cell attachment to basement membrane by means of hemidesmosomes. Note the extensive cytoplasmic fibrillar network in the basal portion of the cytoplasm of these basal cells. (×40,750)

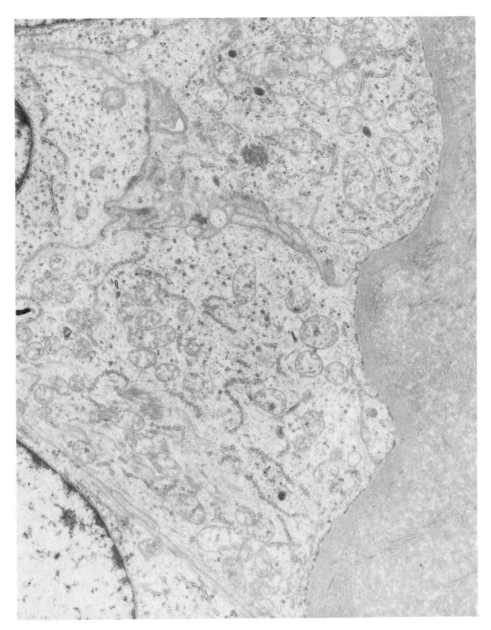

Figure 19. TEM of light type of basal cell as it attaches to a very thickened basement membrane and a lamina propria that is highly collagenized. Note the numerous hemidesmosomes and fibrillar network of these cells. (×11,750)

Changes induced by diagnostic procedures are seen in many biopsy specimens obtained at cystoscopy because the urothelium has been bathed in distilled water or solutions of low osmotic pressure. These fluids cause marked ultrastructural alteration of the cytoplasm of the cells. This can be seen in Figure 9, where there is marked distortion of various mitochondrial units and separation of the intercellular spaces between the basal cells in the normal urothelium. Other changes that have been observed in urinary bladder samples taken at cystoscopy and examined in the author's laboratory are numerous dense bodies in the mitochondria, as well as large cytoplasmic vacuoles that are surrounded by double-layered membranes (Figs. 20 and 21). The nuclei of these cells are in general markedly swollen as compared to those of specimens that are obtained as general surgical pathology specimens.

UROTHELIAL CANCER—CARCINOMA IN SITU

Two and a half decades ago, the phrase "carcinoma in situ" was applied to the urothelium that was cytologically neoplastic and that was interposed between or surrounding overt exophytic papillary tumors of the urinary bladder removed in cystectomy specimens (10,11). A little more than a decade later, by means of urinary cytology, a group of patients were selected out of an even larger group of patients with cancer of the bladder (12). The patients in the selected group had overt large cancer cells in their urine. When they were examined cystoscopically, they were found to be free of visible tumor; however, they were positive by means of urinary cytology. These patients were then followed cystoscopically for a period of 3 months to 7 years before there were visible recurrences of urinary bladder cancer. Bladder mucosa was obtained from cystectomy specimens that were cystoscopically normal but pathologically carcinomatous. This was examined by both scanning and transmission electron microscopy. Bladder mucosa was also examined by similar methodology from patients who had had no previous history of cystoscopically visible lesions, but only positive cytology.

The light microscopy of these lesions is in part demonstrated in Figures 22 and 23. Figure 22a reveals a carcinoma in situ over an inflamed and vascular stroma (see the lower portion of the micrograph). In the upper portion, there is an exuberant urothelial lesion that contains intraepithelial blood vessels. The cytological characteristic of these carcinoma in situ cells is further demonstrated to be pleomorphic and anaplastic when compared to cytologically normal cells found in a cystitis cystica urothelial inclusion cyst (Figs. 23a and b). The nuclei in the carcinoma in situ at upper right in both micrographs are two to three times the nuclear size in the urothelium that is normal, as shown in the lower half of the micrographs. But there is intense vacuolization and fine chromatin dispersion in both the neoplastic and normal urothelial cells due to the fact that this biopsy specimen was taken under less than normal osmotic pressure conditions: more than distilled water but less than urine. There are small nonmuscular blood vessels in the intraurothelial proliferation (Fig. 22b). The nuclei here too are distorted because of the cystoscopy fluid. Transmission electron microscopy of these lesions reveals numerous light and dark cells abutting a lamina propria that contains numerous inflamed cells and blood vessels. However, the distinc-

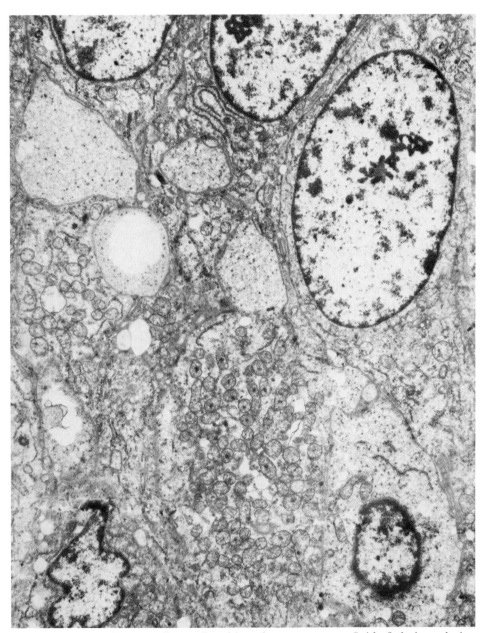

Figure 20. TEM of intermediary cells subjected to a cystoscopy fluid of glycine solution with an osmotic pressure of 150 mOsm. Note the electron-dense bodies in the mitochondria and the areas of cytoplasmic swelling surrounded by double-layered membranes. There is also swelling of the nuclei with some vacuolization of them and the surrounding cytoplasm. (×9,000)

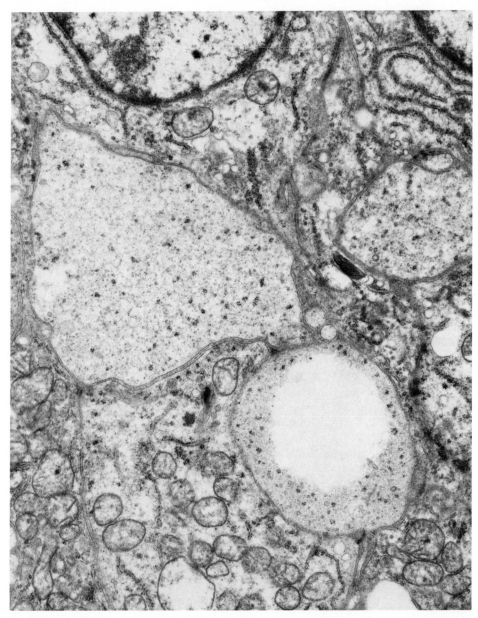

Figure 21. Figure 20 at higher power. (×21,000)

tive feature noted in more than one-half of the approximately 10 cases so far examined was the disappearance of the distinctive basement membrane, as well as telolysosomes seen in the normal urothelium.

Figures 24 through 29 demonstrate, in major part, the disappearance of the normal basement membrane, as well as the partial appearance of electron-dense deposits beneath these cells. These electron-dense deposits are reminiscent ultrastructurally of the immune electron-dense deposits found in the basement membrane of glomeruli in kidneys. Whether these are indeed immune complexes awaits further investigation. The epithelial cells that are in close proximity to the connective tissue in these lesions are sending down small pseudopods or large cytoplasmic extensions into the lamina propria, in consequence, *ultrastructurally* these carcinoma in situ lesions can already be considered microscopically invasive bladder cancers.

It has been reported that measurement of the desmosomal frequency in invasive cancers of the bladder by quantitative electron microscopy shows statistically fewer desmosomes per unit surface for invasive cancers than for tumors that are noninvasive (13). The figures were also compared with those for normal urothelial cells. There has been a recorded decrease in the number of desmosomes in invasive tumors which has contributed to reductions in cell-to-cell adhesiveness. The possibility is also raised that desmosomal frequency may be a predictor "growth potential" of transitional cell carcinoma arising in human urinary bladder. This report did not designate the location of the cells taken for measurement. Were they taken from the surface, the intermediary layer, or the advancing front of the tumor that was invading the connective tissue? If indeed they were taken from the luminal surface of the "invading tumors," it is conceivable that there was a decrease in desmosomal frequency. However, the most important part of any human bladder cancer, pathobiologically, is not its luminal surface, but that which is invading. All the micrographs that follow in this chapter refute the above findings. (Figs. 22–42). It appears that desmosomal frequency and other types of cellular attachments are not decreased, but may even be increased. The morphological characteristics of this increase and its absolute quantitation in nearly 100 surgical specimens of invading tumor are now being studied by means of stereological methods.

Within the cells that are contiguous with the lamina propria, there appear to be numerous surface infoldings and cytoplasmic filaments with focal condensations (Figs. 26–29). Microtubular elements are also seen, as well as numerous arrays of rough-surfaced endoplasmic reticulum and polyribosomes.

INVASIVE BLADDER CANCER AND HISTIOCYTES

Many of the squamous cell and urothelial tumors of the bladder exfoliate cells into the urine and consequently can be cytologically detected and the tumor treated early. In the tissues as well as the urine obtained from patients with these tumors, numerous histiocytes and/or inflammatory cells are detected. On a number of occasions, we have noted that if there is crushing or cautery artifact, a false diagnosis can be easily made for the presence or absence of tumor. Not only can the urothelial cells that are invading be destroyed beyond recognition but

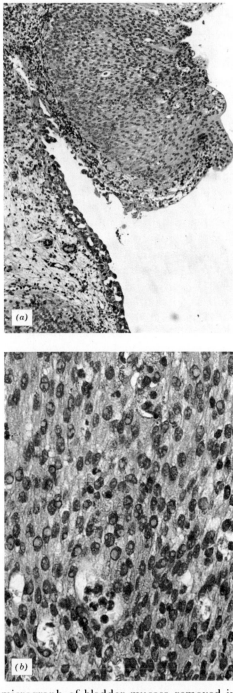

Figure 22. (*a*) Light micrograph of bladder mucosa removed in a cystoscopy fluid of glycine. Note the epithelial proliferation with small blood vessels in the upper portion of the picture. The lower portion contains a urothelium in which the pathology is consistent with carcinoma in situ. (×78) (*b*) View at higher power of upper portion of urothelium. Note şmall capillaries filled with blood cells scattered throughout urothelium. Nuclei are also vacuolated. (×333)

246

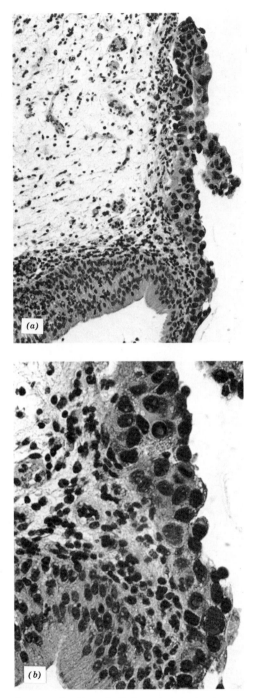

Figure 23. (a) Light micrograph of urothelial carcinoma in situ covering an inflamed and edematous lamina propria. The urothelium that is normal lines a cystitis cystica in the lower portion of the picture. (×136) (b) View at higher power demonstrating the increased nuclear size of the carcinoma in situ nuclei as compared with the normal urothelial nuclei in the lower portion of the micrograph. Both the neoplastic and normal nuclei are partially vacuolated. (×333)

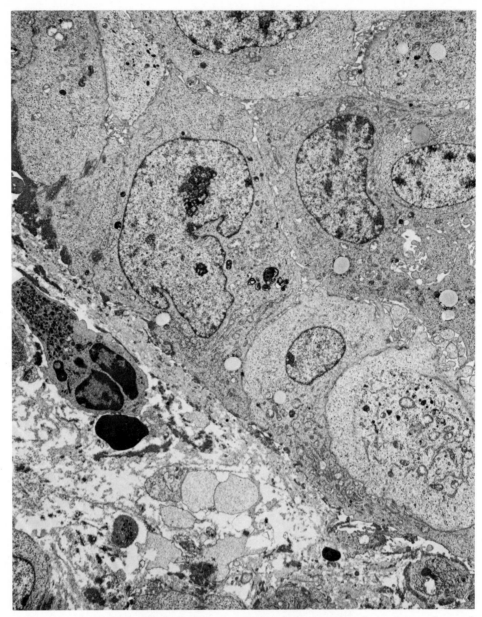

Figure 24. TEM of a carcinoma in situ. No distinct basement membrane is seen beneath the neoplastic cells which are sending microvilli into the surrounding lamina propria. Note an electron-dense material in the lamina propria next to these cells at left. There is a PMNL in the stroma. The tumor cells have numerous cell connections between them. (×3,450)

248

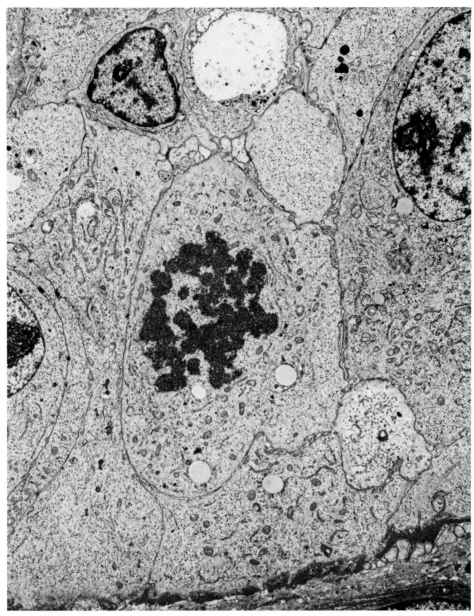

Figure 25. TEM of carcinoma in situ with a mitotic figure. Basal cells have numerous cell connections between them and are sending pseudopods through an electron-dense material into the surrounding lamina propria. (×3,875)

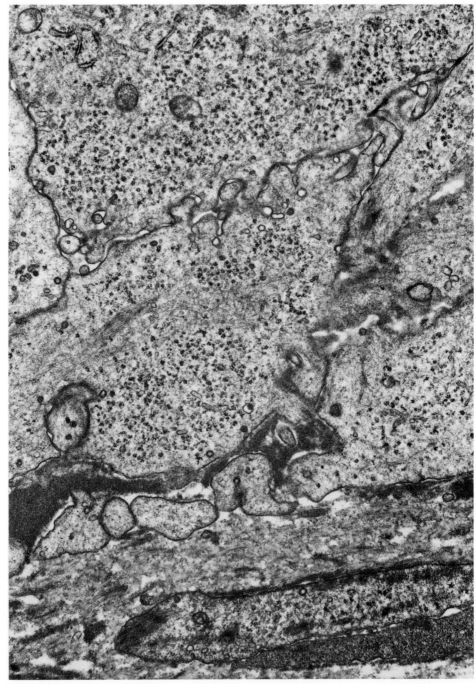

Figure 26. Figure 25 at higher magnification. Note filaments in cytoplasm with focal condensation, electron-dense material outside the cell, cell connections between the cells, and cytoplasmic pseudopodal extension into the surrounding lamina propria. (×9,125)

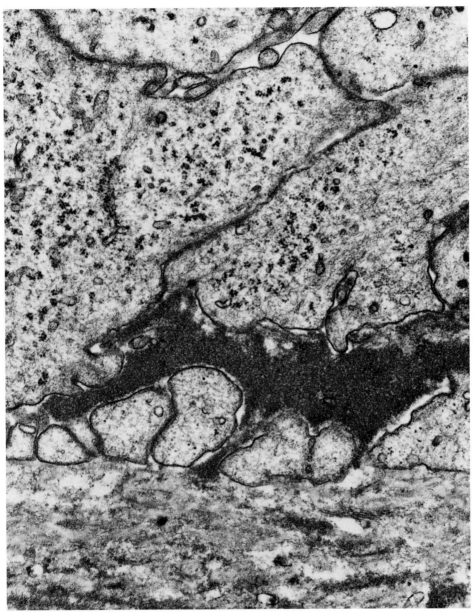

Figure 27. Figure 26 at higher magnification. Note the substructure to the electron-dense material outside the cell and the fibrillar cytoplasm, as well as the pseudopodal extension of the cell. (×20,750)

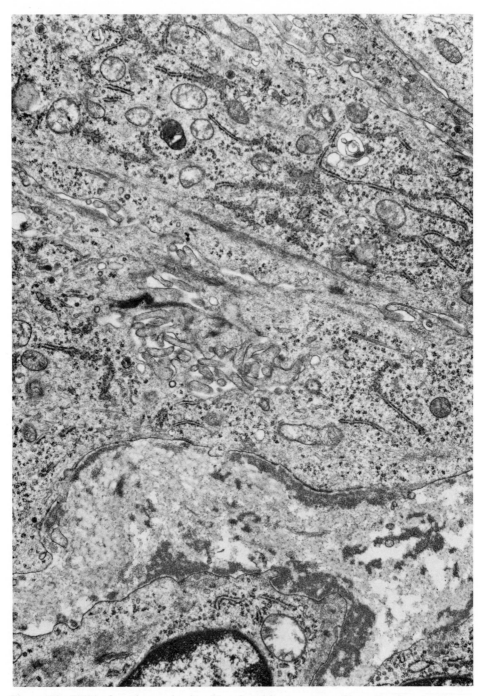

Figure 28. TEM of carcinoma in situ. Note hemidesmosomes of invading basal cells near electron-dense material. Note also fibrillar cytoplasm with focal condensations, microvilli, and intercellular connections. Some vesicles are seen in the cell at right. (×9,125)

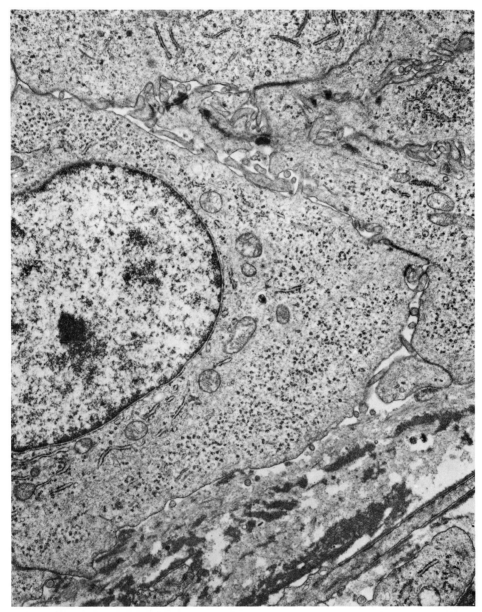

Figure 29. TEM of carcinoma in situ with numerous intercellular connections of various types in upper portion of picture. Note electron-dense material in the lamina propria in lower portion of picture. (×13,000)

also the multinucleated giant cells can be so distorted that they (histiocytes) are misdiagnosed as tumor cells. They are ubiquitous and seem to increase in numbers when there have been a) previous biopsies of the urinary bladder, b) prior radiotherapy to the bladder, c) chemotherapeutic instillation into the bladder, and d) invading tumor.

Figures 30 and 31 show extensive squamous cell change in the urothelium that abuts the lumen of the bladder. In the lamina propria beneath this squamous urothelium, there are marked numbers of inflammatory cells, some invading tumor, and multinucleated tumor cells. If these latter cells are distorted, they can easily be misinterpreted as tumor cells, especially in those clinical conditions when they appear in vast numbers. By electron microscopy, these cells are very distinctive and different from invading bladder tumor cells (Figs. 37-39). The syncytium of histiocytes or macrophages has several nuclei and a markedly vacuolated cytoplasm with numerous vesicles that are lined by either a single or double layer of membranes. The cytoplasm may contain mitochondria, dense bodies, and phagolysosomes, but the Golgi apparatus is hard to find (14).

These giant cells (Fig. 38) are almost always adjacent to collagen and/or tumor cells, but they do not appear to be fused to the latter. Their nuclei have prominent chromocenters, sometimes several, with margination of some of the chromatin at the nuclear membrane. There are also fine granular electron-dense bodies in their nuclei.

Another variant of these cells appears to be in close proximity to the epithelial tumor cells and, on light microscopy, is very easily ignored or misdiagnosed as part of the tumor cell cluster (Fig. 31). Electron microscopy of these clusters (Figs. 37 and 39) reveals that these cells have cytoplasm similar to that of the multinucleated cell (Fig. 38). These histiocytes or macrophages are not attached to the epithelial tumor cells (Figs. 37 and 39). The tumor cells have numerous cellular attachments of various sorts, including desmosomes. The macrophages (Fig. 37 and upper left corner of Fig. 39) are also found in the normal urothelium in very small numbers. These histiocytes or macrophages are morphologically identical to those found in lymph nodes when there is prominent sinus histiocytosis in the lymph nodes. What then is the function of these macrophages or nonattached cells? Are they processing tumor antigen, or are they a form of cell-mediated immunity? This is currently a subject of intense investigation.

PAPILLARY TUMORS

Scanning electron microscopy of papillary tumors reveals numerous microvilli on the surface cells, which are in contact with the urine, rather than the ridged pattern that is seen on the surface of normal, noninflamed urothelial cells (Fig. 4, insert). These surface tumor cells then cover, in a tight fashion, the intermediary and basal cells. The three different cells layers overlie a central fibrovascular core that is devoid of smooth muscle. Inflammatory cells, if they are present, are usually found surrounding the vasculature, and very rarely will they cross the basement membrane of the papillary tumor; however, in normal urothelium, it is not uncommon to find lymphocytes crawling in between the various cell layers

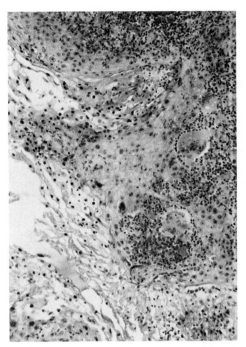

Figure 30. Squamous cells desquamating into lumen of bladder. There are numerous histiocytes, inflammatory cells, and invading tumor. (×125)

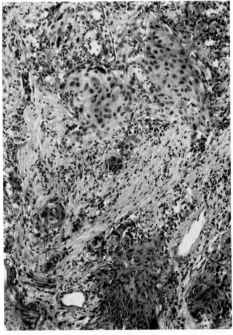

Figure 31. Undifferentiated and invading part of exophytic squamous cell tumor. Surrounding individual clusters of tumor are numerous dark cells. (×125)

255

Figure 32. SEM of a papillary portion of a bladder tumor in which there are cone-like projections of tumor cells surrounding a fibrovascular core. (×280)

(Fig. 16). The surface cells of well-differentiated papillary tumors are joined by junctional complexes that are identical to those noted in normal bladder epithelium. Scanning electron microscopy reveals the surface topography of well-differentiated papillary tumors (Figs. 32, 33a, 33b, 34a, and 34b). Some tumors have cone shapes, while others form continuous ridges of papillary fronds (Figs. 33a and 33b). The latter are most frequently discerned in ureteral and kidney pelvic transitional cell tumors; however, umbrella or cap cells are common to all exophytic well-differentiated papillary tumors that are more than seven cell layers thick. The cap cells override the surface cells in an almost noncontinuous fashion. They are very light in color as compared to the underlying and surrounding neoplastic surface cells. In Figures 34a and 34b, several of the cap cells are seen; a few of them attached to one or two sister cap cells.

When grades II-IV papillary tumors are examined by SEM, the cap cells are no longer seen, and an occasional surface cell is devoid of microvilli and is covered by an almost smooth-surfaced plasmalemma (Fig. 36). Another morphological phenomenon that becomes more apparent is extreme cytoplasmic variability in regard to the distribution numbers and morphology of various cytoplasmic components. Scanning electron microscopy of these undifferentiated grades II-IV papillary tumors and carcinoma in situ is illustrated in Figure 35. Here there are deep grooves between many of the adjacent surface cells. Transmission electron microscopy of such areas shows that the cellular attachment sites between the luminal surfaces of these surface cells are decreased

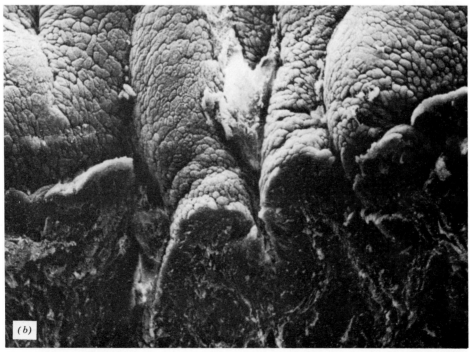

Figure 33. (*a*) SEM of ureteral tumor showing ridges of well-differentiated tumors of the urothelium instead of cone-like projections of tumor (×25.5) (*b*) Higher magnification of a similar ridge-like urothelial tumor of the renal pelvis. (×77)

Figure 34. (a) SEM of surface of well-differentiated papillary urothelial tumor demonstrating numerous cap cells that are white as compared to the darker surface cells. (×500) (b) SEM of portion at higher power. (×1,200)

258

Figure 35. SEM of surface cells in carcinoma in situ. Note deeper grooves or valleys between the neoplastic surface cells. No cap cells are seen, and the surface cells contain microvilli of various sizes. (×2,500)

Figure 36. SEM of surface of undifferentiated papillary tumor. Note disappearance of microvilli on the surface of some of the cells and a sparser distribution over some of the others. (×6,000)

259

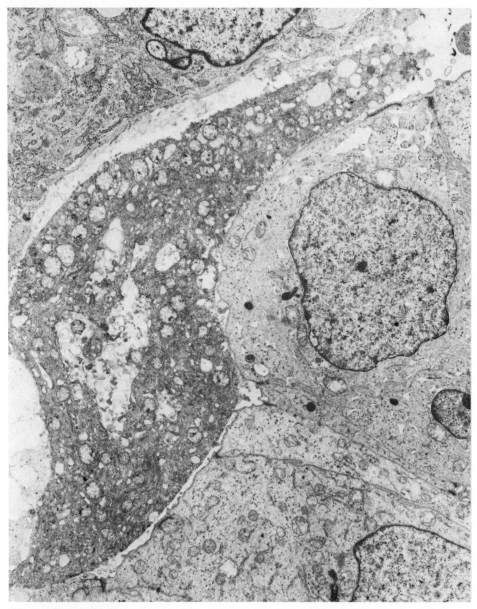

Figure 37. TEM of invading bladder tumor. Note fibroblast in upper portion of picture and a comma-shaped dark cell or histiocyte closely covering the advancing edge of several well-differentiated urothelial cancer cells. The cancer cells have numerous intercellular connections, some microvilli, and several mitochondria. (×5,750)

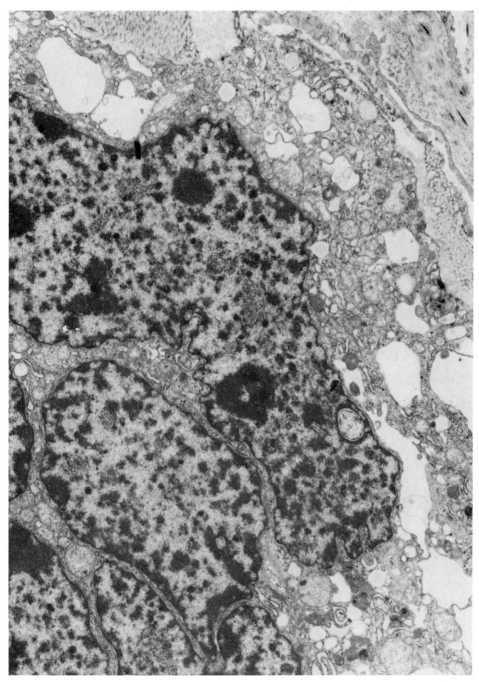

Figure 38. TEM of a multinucleated giant cell seen in Figure 30. This syncytium of histiocytes or macrophages contains at least four nuclei. It is adjacent to collagen at top of picture. (×5,250)

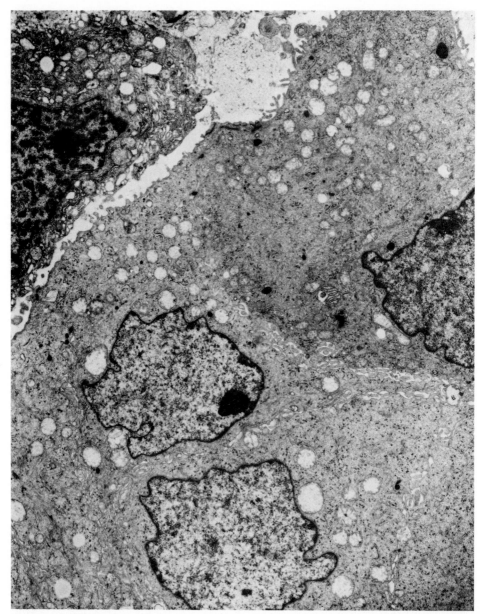

Figure 39. TEM of an invading bladder cancer adjacent to a macrophage seen at upper left. Note the numerous microvilli on surface of the cell and the numerous different types of intercellular connections between the various tumor cells. The cytoplasm contains swollen mitochondria, a fibrillar cytoplasm, and very little evidence of a Golgi apparatus. (×5,750)

in number. However, it seems that another pattern is seen by transmission electron microscopy when the invading section of the tumor is examined. Here various junctional complexes are seen in great numbers (Figs. 30–42).

The tripartite junctional complexes are seen in great abundance, and it is this author's interpretation from the material that has been examined in his laboratory that there is a reversal of polarity of differentiation of these cells, with the polarity of differentiation being greatest at the border of the advancing invading front of the tumor cells. This can be discerned very specifically in Figure 40, where the advancing front of a bladder tumor, taken from a stage C bladder cancer, is seen. In some of the cells, the advancing front is seen to have numerous microvilli and many tripartite junctional complexes next to the surface. Beneath the advancing tumor cells are a number of other tumor cells that bear a very close resemblance to the intermediary cells and the intercellular cisternae that are often seen between these cells. Note the blood vessels, collagen, elastic tissue, and fibroblasts in the connective tissue adjacent to the advancing front of the bladder tumor. The cytoplasm of this grade II-III bladder cancer shows swelling of the mitochondria with distorted and irregular cristae. This pattern is visible no matter how the bladder cancer sample is obtained, i.e., through a needle biopsy or from a surgical specimen. An abundance of filaments is present, and this is especially so if there is squamous cell change associated with the malignant urothelial cells. In the latter case, the filaments are often in the form of tonofilaments. Cytoplasmic surface villi, in some instances, have a glycocalyx or electron microscopic fuzz attached to them (Figs. 41 and 42).

Another interesting morphological finding is the appearance, in a very attenuated fashion, of vesicles that were found in normal cells and also asymmetrical membranes (Fig. 41). The tumor cells were examined from a stage C lesion. Perivesicle fat cells can be seen at lower left. Note the numerous swollen mitochondria and very fibrillar portions of the cytoplasm of this invading stage C urothelial cell cancer.

SUMMARY

The human urinary bladder offers the scientist interested in cancer a unique opportunity for monitoring the induction, conversion, and progression of the urothelium to a cancer that invades and kills. All this can be accurately observed by means of urinary cytology and electron microscopy as employed by the surgical pathologist. The urologist can be quickly alerted if the urothelium is at high risk for cancer. He can then visualize the exophytic urothelial growth through the cystoscope and treat it by some form of surgery, or other physical modality. No other human internal cancer can be as carefully watched both before and after its inflection point in its cancer growth curve as the human bladder carcinoma. Perhaps, with the aid of the TEM, SEM, and urinary cytology, the human bladder cancer problem can be better understood, and possibly cured before the end of this century.

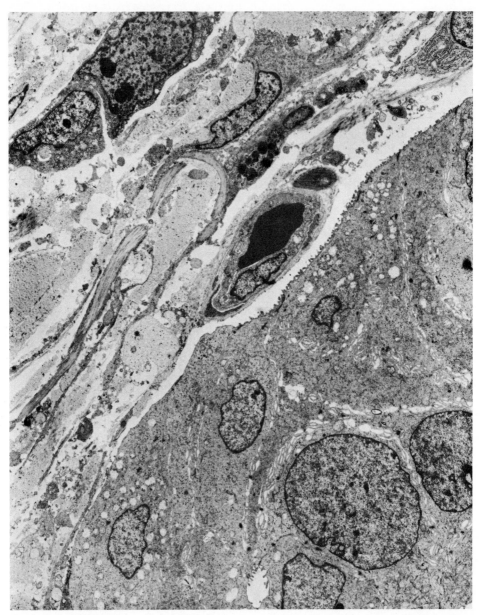

Figure 40. TEM of an invading bladder tumor. In the upper portion of the picture, note fibroblasts, elastic fibers, collagen, and a small capillary in close proximity to numerous cancerous urothelial cells. Numerous microvilli are seen on the surface of some of these cells, as in Figure 36. Other cells do not have these microvilli. However, many of the cells still retain numerous intercellular connections of various types. There is a reversal of the polarity of differential in these invading tumor cells. The greatest differentiation is now away from the luminal surface of the tumor. (×3,400)

264

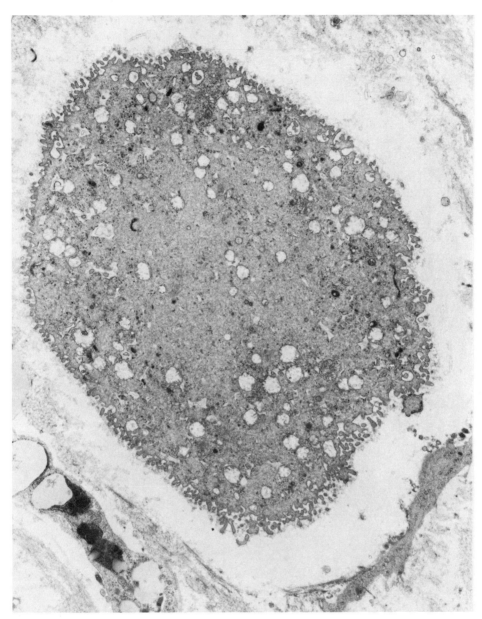

Figure 41. TEM of a stage C bladder cancer in contact with some fat cells in the perivesicle fat. Note that the surface of these cells contains many microvilli, with some indication of asymmetrical membranes and some vesicles being formed. The microvilli are surrounded by a glycocalyx electron-dense fuzz. The cytoplasm is very fibrillar and contains numerous distorted mitochondria. (×5,750)

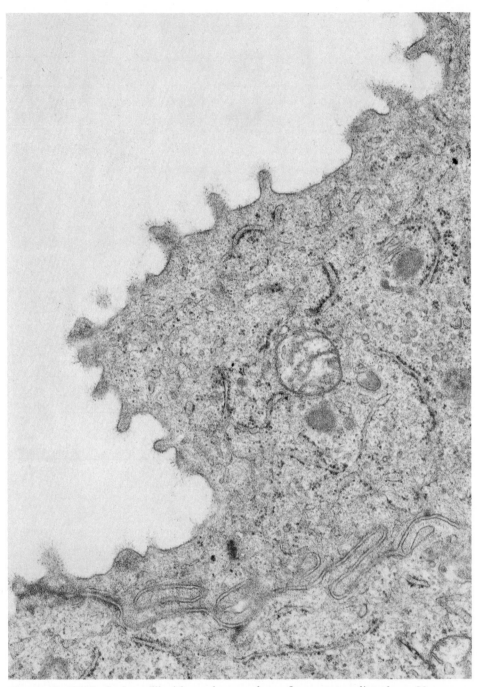

Figure 42. TEM of microvilli with an electron-dense fuzz surrounding them. Note some of the cell connections near the surface of these cells and the complex intercellular interdigitation. (×30,000)

REFERENCES

1. Monis B, Zambrano D: Ultrastructure of transitional epithelium of man. *Z Zellforsch* 87:101, 1968.

2. Fulkner MJ, Cooper EH, Tanaka T: Proliferation and ultrastructure of papillary transitional cell carcinoma of the human bladder. *Cancer* 27:71, 1971.

3. Battifora HA, Eisenstein R, Sky-Peck HH, et al: Electron microscopy and tritiated thymidine in gradation of malignancy of human bladder carcinomas. *J Urol* 93:217, 1965.

4. Battifora H, Eisenstein R, McDonald JH: The human urinary bladder mucosa. An electron microscopic study. *Invest Urol* 1:354, 1964.

5. Walker BE: Electron microscopic observations on transitional epithelium of the mouse urinary bladder. *J Ultrastruct* 3:345, 1960.

6. Staehelin LA, Chlapowski FJ, Bonneville MA: Lumenal plasma membrane of the urinary bladder. I. Three-dimensional reconstruction from freeze-etch images. *J Cell Biol* 53:73, 1972.

7. Firth JA, Hicks RM: Interspecies variation in the fine structure and cytochemistry of mammalian transitional epithelium. *J Anat* 116:31, 1973.

8. Koss LG: The asymmetric unit membranes of the epithelium of the urinary bladder of the rat. An electron microscopic study of a mechanism of epithelial maturation and function. *Lab Invest* 21:154, 1969.

9. Clark M, O'Connell KJ: Scanning and transmission electron microscopic studies of an exstrophic human bladder. *J Urol* 110:481, 1973.

10. Melicow MM, Hollowell JW: Intra-urothelial cancer: carcinoma in situ, Bowen's disease of the urinary system: Discussion of thirty cases. *J Urol:* 68:763, 1952.

11. Melicow MM: Histological study of vesical urothelium intervening between gross neoplasms in total cystectomy. *J Urol* 68:261, 1952.

12. Melamed MR, Vousta NG, Grabstald H: Natural history and clinical behavior of in situ carcinoma of the human urinary bladder. *Cancer* 17:1533, 1964.

13. Alroy J, Weinstein RS, Pauli BU: Relation of desmosomes to tumor invasiveness in human urinary bladder transitional cell carcinomas. Seventy-third Annual Meeting, American Association of Pathologists and Bacteriologists, Boston, 1976, p 31a.

14. Kern WH, Dermer GB, Tiemann, RM: Comparative morphology of histiocytes from various organ systems. *Acta Cytol* 14:205, 1970.

6
Gynecology

Alex Ferenczy, M.D.
Associate Professor of Pathology and Obstetrics and Gynecology
McGill University
Director, Division of Gynecologic Pathology
Jewish General Hospital
Montreal, Quebec, Canada

Ralph M. Richart, M.D.
Professor of Pathology
Columbia University College of Physicians and Surgeons
Director, Division of Obstetrics and Gynecologic Pathology
The Sloane Hospital for Women
New York, New York

The female reproductive system is the seat of an almost infinite variety of neoplastic and non-neoplastic conditions, most of which are relatively easy to diagnose on routine histological and histochemical examination. However, there are a certain number of neoplasms and hormone-dependent endometrial conditions which are difficult to classify and in which the study of ultrastructural characteristics is valuable for either the differential or the confirmatory diagnosis of the lesions. Also, electron microscopy provides valuable information on the histogenetic origin and development of some of the poorly understood genital neoplasms.

In this chapter, the specific fine structural features that are diagnostic or help in the understanding of the histogenesis of these conditions will be described and illustrated. To facilitate orientation to the chapter, the various neoplastic processes will be presented according to the various segments of the female genital tract, although some of the tumors, such as the clear cell adenocarcinoma and leiomyosarcoma, may be encountered in more than one portion of the reproductive system.

VULVA

Paget's Disease

Paget's disease is an uncommon vulvar malignancy in which the epidermis is invaded by large, pale cells referred to as Paget cells. The disease may be con-

Supported by Grant MA5137 from the Medical Research Council of Canada.

fused with intraepidermal malignant melanoma. In contrast to malignant melanoma, however, the Paget cells, whether they are of mammary or extramammary origin, are rich in PAS-alcian blue-positive, diastase-and hyaluronidase-resistant secretory products. Ultrastructurally, the secretory material is stored within membrane-bound, Golgi-derived granules (1–6). In addition, Paget cells are attached by desmosomes to the adjacent normal squamous cells and, occasionally, form glandular lumens, whereas melanoma cells lack these fine structural characteristics (Figs. 1a and b).

Although it is generally accepted that Paget's disease of the female mammary gland develops as a result of epidermotropic migration of underlying carcinoma, the origin of extramammary Paget cells, including those involving the vulva, remains controversial. When the disease is associated with an underlying sweat gland carcinoma, the Paget cells are thought to migrate from the subjacent carcinoma into the epidermis. When no underlying disease is present, the cells are thought to develop from either the malignant conversion of epidermal squamous cells (3) or the intraepidermal portion of the sweat gland apparatus (3), although some authors favor an extraepidermal derivation from either the

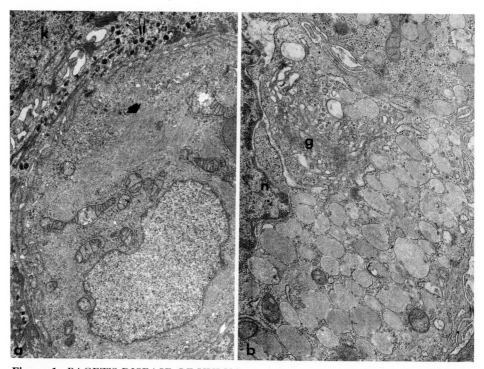

Figure 1. PAGET'S DISEASE OF VULVA

a. A nonsecretory Paget cell with interlacing bundles of microfilaments (arrow) and scattered mitochondria. Free and bound ribosomes are surrounded by epidermal squamous cells (k). The latter are rich in electron-dense tonofibrils and membrane-bound keratinosomes (double arrow). These cellular specializations are absent in Paget cells. (× 3,400) *b.* The perinuclear (n) region of a secretory Paget cell is shown, occupied by numerous Golgi-derived (g), membrane-bound secretory granules in all stages of development. The content of the granules is granulofilamentous acid mucin. (× 11,000)

eccrine (1,6,7) or apocrine (5) dermal sweat gland system. An eccrine, rather than apocrine, derivation is supported by the similar ultrastructural organization of secretory granules and the similar distribution of amylophosphorylase, leucine aminopeptidase, and oxidative enzymes in both Paget cells and the eccrine epithelium (7). In the normal apocrine glandular epithelium, the secretory granules are characteristically pleomorphic, and the cells have a dissimilar enzyme content.

VAGINA

Clear Cell Adenocarcinoma

In recent years a close association has been reported between the occurrence of clear cell adenocarcinoma of the vagina and cervix in young girls and women (8 to 30 years of age) and the intrauterine exposure to the nonsteroidal estrogen diethylstilbestrol (DES) as a result of its administration to their mothers for the treatment of high-risk pregnancies (8). In addition to the clear cell carcinomas, a variety of other stilbestrol-related but non-neoplastic abnormalities of the lower genital tract have been encountered. These include vaginal adenosis (the presence of glandular epithelium in the vagina), endocervical eversion on the cervical portio, and transverse fibrous ridges in the vagina and on the cervix (8). Histologically, most of the adenocarcinomas occurring in the DES progeny are of the clear cell type and contain tubules, solid nests, papillae, or cysts lined by cells with a water-clear cytoplasm and, occasionally, a hobnail cytonuclear architecture (8) (Fig. 2). The clear cytoplasm is due to abundant intracytoplasmic glycogen (10,11) (Figs. 2 and 3).

Morphologically similar, but hormonally unrelated, clear cell adenocarcinomas are observed in the endometrium and ovary and were regarded in the past as mesonephric in origin (12). However, the presently available morphological (10,11,13–18) and experimental (19) evidence favors a Müllerian origin for the clear cell cancers, whatever their site of origin in the female genital tract. A Müllerian origin is supported by the frequent association of clear cell carcinomas of the vagina and Müllerian-derived vaginal adenosis (8,9), the histological continuity between cervical clear cell carcinoma and normal endocervical epithelium, and the observation of morphologically identical neoplasms in the cervix (10), endometrium (7,9,10), and ovary (6,8,11). In addition, the DES-related vaginal clear cell carcinomas occur principally anteriorly and superficially, whereas the mesonephric carcinomas are found deep in the lateral wall of the vagina where histological continuity between the neoplastic growth and mesonephric remnants has occasionally been demonstrated (20). Of additional diagnostic value, the cells of mesonephric carcinomas are not clear and contain negligible amount of intracytoplasmic glycogen.

Sarcoma Botryoides

Vaginal sarcoma botryoides, a variant of embryonal rhabdomyosarcoma, is a highly aggressive malignant neoplasm that typically occurs in young children. The lesion is composed of a myxomatous stroma containing atypical spindle-

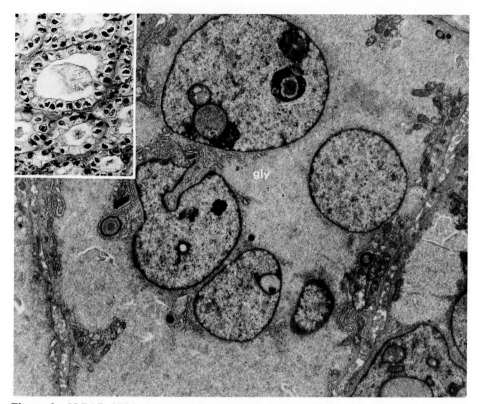

Figure 2. CLEAR CELL ADENOCARCINOMA OF VAGINA
The neoplastic cells contain pools of glycogen particles (gly) and islands of organelles. Multinucleation and intranuclear cytoplasmic inclusions are artifactual and are due to the tangential sectioning of a complexly convoluted nucleus and indented nuclear membranes, respectively. The lateral plasma membranes have numerous microvilli and a few desmosomes. (× 3,725) *Insert:* Clear cell adenocarcinoma of vagina. The tubules are lined by large neoplastic cells with clear cytoplasm, and the lumen contains secretory products. (Hematoxylin-eosin, × 150)

shaped and stellate cells, mitotic figures, a subepithelial cellular zone ("cambium layer"), and a few striated muscle cells. The rhabdomyoblasts are most easily identified by transmission electron microscopy, as many of the undifferentiated cells seen by light microscopy contain intracellular myofilaments in an abortive stage of differentiation forming Z bands and, to a lesser degree, A and I bands (21,22). Occasionally, cells with well-organized myofibrils with Z bands in register are found, indicating a degree of myogenic maturation. These latter cells are the ones that are recognized in hematoxylin-eosin (H&E)-stained sections as embryonal rhabdomyoblasts (Fig. 4).

Sarcoma botryoides must be differentiated from the benign polypoid rhabdomyoma of the vagina (23), which occurs in adults, grows slowly, and is composed of mature, connective tissue containing numerous striated rhabdomyoblasts. The rhabdomyoblasts in the benign rhabdomyoma, however, have a high degree of myofibrillar differentiation with easily identifiable Z, A, and I bands.

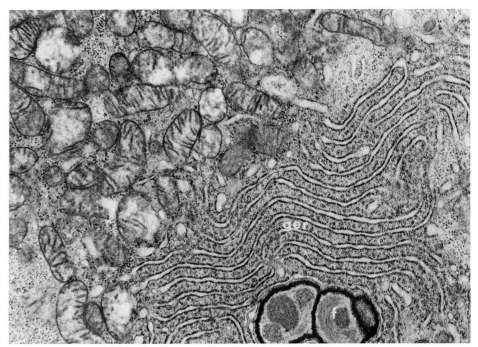

Figure 3. CLEAR CELL ADENOCARCINOMA OF VAGINA
Detailed view of neoplastic clear cells with stalks of parallel membranes of granular endoplasmic reticulum (ger) in close association with masses of glycogen particles and mitochondria. Bundles of microfilaments (f) are surrounded by crown-like aggregates, presumably nuclear ribosomes. (× 11,625)

These mature ultrastructural features, coupled with the clinical and histological characteristics, serve to distinguish the benign rhabdomyoma from the malignant vaginal sarcoma botryoides (23).

CERVIX

Undifferentiated Carcinomas

By light microscopy alone, neoplastic growths that contain no features of differentiation are often difficult to classify, and it is frequently difficult to determine even whether the neoplasm is of mesenchymal or epithelial origin. Similarly, the question of the parent cell type generally cannot be resolved. Ultrastructurally, the presence of junctional complexes, such as desmosomes, zonula occludens, and zonula adherens, between adjacent cells is characteristic specialization of epithelia and serves to distinguish carcinomas from sarcomas. Detection of cytoplasmic tonofilaments and tonofilament-desmosomal complexes between neoplastic cells and interdigitating microvilli identifies a histologically undifferentiated neoplasm as squamous cell carcinoma (Fig. 5a). Poorly formed abortive glandular lumina with microvilli or mucin-containing secretory granules are

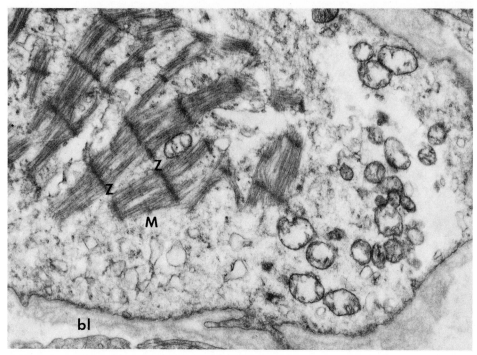

Figure 4. SARCOMA BOTRYOIDES OF VAGINA

Embryonal rhabdomyosarcoma cell with well-differentiated myofilaments forming sarcomeres delimited by Z bands (Z). The M bands are visible in a few sarcomeres. A basal lamina-like (bl) material seems to be deposited along the cell membrane. (× 19,500) (*Courtesy of L. Overbeck, ref. 22*)

features characteristic of adenocarcinoma of endocervical origin (Fig. 6). Of 18 malignant cervical neoplasms classified histologically as undifferentiated and examined by electron microscopy, 10 were of squamous cell origin and 2 of glandular origin. The remaining 6 lesions were devoid of specific fine structural features that could be used to identify their cell of origin (24).

Spindle Cell Squamous Cell Carcinoma (Pseudosarcoma)

In most instances, no difficulty is encountered in the diagnosis of invasive squamous cell carcinoma of the cervix uteri. In some cases, however, the squamous cells assume a distinctive spindle-shaped configuration, grow in interlacing bundles or with a whorled arrangement, and simulate a sarcoma. The squamous, epithelial derivation of the neoplastic cells is relatively easy to identify at the ultrastructural level. The distinctive features include desmosomal-tonofilamentous attachments between the neoplastic cells and intracytoplasmic bundles of tonofilaments (Fig. 5b). A basal lamina is sometimes produced by these fusiform squamous neoplastic cells, and specialized cellular characteristics are comparatively less abundant and less well developed (25) than in the conventional invasive squamous cell carcinoma. However, provided adequate sampling

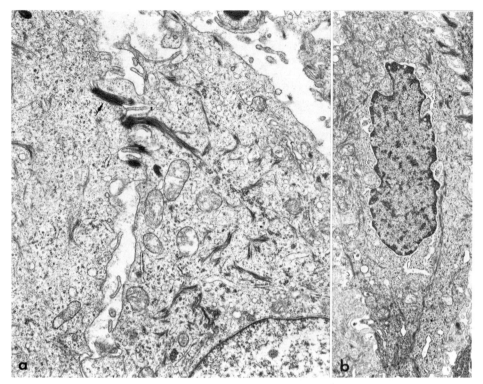

Figure 5. POORLY DIFFERENTIATED SQUAMOUS CELL CARCINOMA OF CERVIX.
a. Electron microscopic appearance of cells from histologically undifferentiated carcinoma of the cervix containing bundles of aggregated tonofilaments and desmosome-tonofilament attachment complexes (arrow). These features, together with dilated intercellular spaces and interdigitating microvilli, are diagnostic of carcinoma of squamous cell origin. (× 12,000) *b. Spindle-shaped, Pseudosarcomatous, Squamous Cell Carcinoma of Cervix.* With the exception of a fusiform cytonuclear configuration, the neoplastic cells are ultrastructurally similar to those found in the non-spindle-shaped cell variety of squamous cell carcinoma. (× 3,250)

is available, their presence can be demonstrated. Although the mechanisms of development of the fibroblastoid squamous cells and their peculiar growth patterns are not understood, their general ultrastructural organization closely resembles that of squamous neoplastic cells cultured in vitro (26).

ENDOMETRIUM

The Endometrium in Infertility

The primary purpose of histological dating of endometrial tissues is to provide the clinician with as accurate a diagnosis as possible of the functional status of patients complaining of infertility. The correct morphological interpretation of the changes the endometrium undergoes through the menstrual cycle presup-

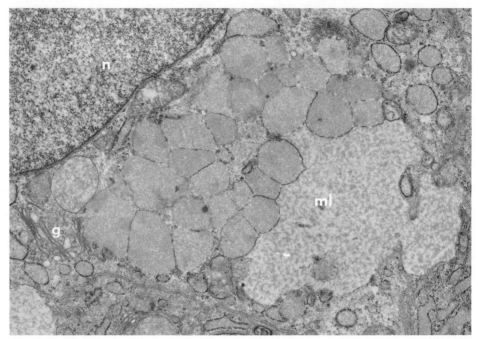

Figure 6. POORLY DIFFERENTIATED ADENOCARCINOMA OF ENDOCERVIX
Electron microscopy of a cervical neoplasm that histologically and histochemically was classified as undifferentiated carcinoma. The mucinous nature of the lesion is evidenced by finding occasional neoplastic cells with large collections of perinuclear membrane-bound secretory granules with finely granular filamentous content. Many of the granules are fused together by intercoalescence, and rupture of their limiting membranes results in the escape of mucin within the cytoplasmic matrix leading to mucinous lakes (ml). The formation of mucinous granules is initiated in the perinuclear (n) ribosome-ergastoplasm-Golgi (g) complex, and the mitochondria provide the necessary energy for mucinogenesis. ($\times$ 12,000)

poses a thorough knowledge of the histology of the normal and abnormal endometrium. In most instances these changes can be determined in routine hematoxylin-eosin-stained sections when adequate clinical information is available and the endometrial sample is properly prepared. There are a number of cases, however, in which, despite a normal length cycle, the question of whether ovulation has occurred cannot be determined by histological study.

In these conditions, a fine structural evaluation of the epithelial cells in the presumed secretory phase may resolve the issue. At the electron microscopic level, ovulation is followed by the development of a nucleolar channel system (NCS) (27,28) (Fig. 7). The differentiation of this system occurs in response to rising levels of postovulatory progesterone (29,30) and is regarded as diagnostic that ovulation has occurred (28). The nucleolar channel system apparently develops from the nuclear membrane (27) and is seen as early as the first post-ovulatory day, reaches a maximum on the 21st day of the cycle, and disappears by the 25th cycle day. The significance of the NCS is presently unknown, but its

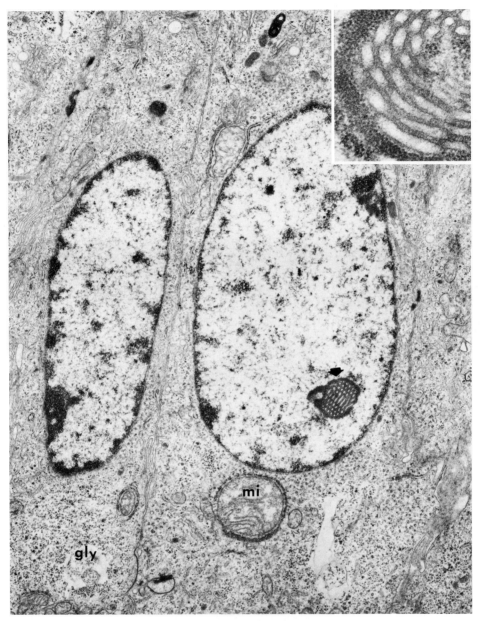

Figure 7. NORMAL POSTOVULATORY ENDOMETRIUM

Nucleolar channel system (arrow) associated with subnuclear glycogen particles (gly) and giant mitochondria (mi) in endometrial gland cells represent fine structural features indicating that ovulation has occurred. (× 5,875) *Insert:* High-power view of a nucleolar channel system. It is tangentially sectioned, containing channels in circular and tubular profile. The channels are embedded in a dense amorphous matrix, and the peripheral portion of the channel system is delimited by a row of electron-dense, 150-Å granules resembling ribosomes. (× 46,000)

277

exclusive presence in the postovulatory period suggests that it is related in some way to the implantation of the blastocyst (28). The failure to identify the NCS in early or mid luteal phase endometria implies anovulation. The presence of intracytoplasmic glycogen deposits and giant mitochondria without NCS should not be considered evidence of ovulation, for the development of these organelles can be seen in vivo or induced in vitro by estrogens alone (29,30).

Adenomatous Hyperplasia and Endometrial Carcinoma

By using the electron microscope, the difficult problem of distinguishing between adenomatous hyperplasia (Fig. 8) and early endometrial carcinoma (Fig. 9), referred to by many as carcinoma in situ, can be resolved. When a systematic evaluation of the ultrastructure of these endometrial tissues is undertaken, it can be seen that there are several alterations that distinguish hyperplastic from malignant neoplastic cells. Cells in *adenomatous hyperplasia* contain a striking increase in ciliary shafts (Fig. 10), and the nonciliated cells have abundant and

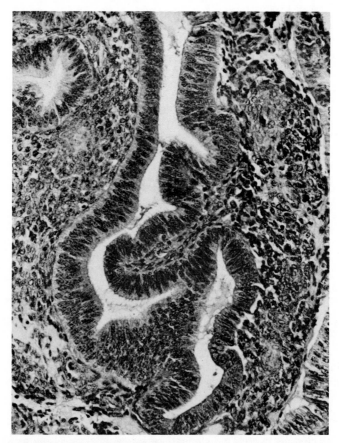

Figure 8. ADENOMATOUS HYPERPLASIA
A voluminous endometrial gland with projections into the dense, hyperplastic stroma is shown. The lining cells are tall columnar with numerous cilia, and the pseudostratified, elongated nuclei have dense, granular chromatin. (Hematoxylin-eosin, × 220)

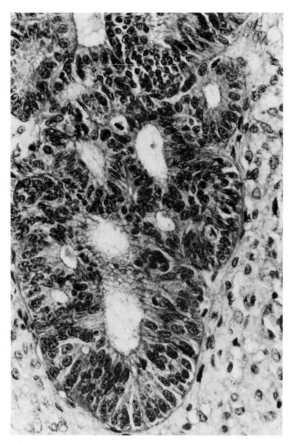

Figure 9. EARLY ENDOMETRIAL ADENOCARCINOMA (CARCINOMA IN SITU)
In this condition, the glands characteristically have a cribriform pattern, which is produced by the interanastomosis of intraluminal papillary projections. The cells lining the glands contain pleomorphic nuclei with irregular, granular chromatin and prominent nucleoli. The cytoplasm has characteristically a pink, eosinophilic and finely granular appearance. Mitoses are frequent and cilia inconspicuous. (Hematoxylin-eosin × 400)

well-developed surface microvilli associated with bundles of peri- and subnuclear, 80-Å thick microfilaments (24,31,32) (Fig. 11a), and free and bound ribosomes, supranuclear mitochondria, and lysosomes (Fig. 11b). In addition, aggregates of glycogen granules are found in most hyperplastic epithelial cells (11). The changes observed in hyperplastic epithelium are presumably a reflection of long-standing estrogenic stimulation and are similar to those found in normal cyclic proliferative endometrium. However, the estrogen-dependent ultrastructural characteristics are considerably more numerous and better developed in hyperplastic epithelial cells than in their normal, cyclic counterpart (24,31,32).

In the early, well-differentiated adenocarcinoma cells (24,32), variously called severe adenomatous hyperplasia, atypical adenomatous hyperplasia, dysplasia, or carcinoma in situ (Fig. 9) there is a striking decrease in the estrogen-dependent ciliary apparatuses, as well as in the length of surface microvilli and primary lysosomal activity, whereas the Golgi-mitochondrial-granular endoplasmic re-

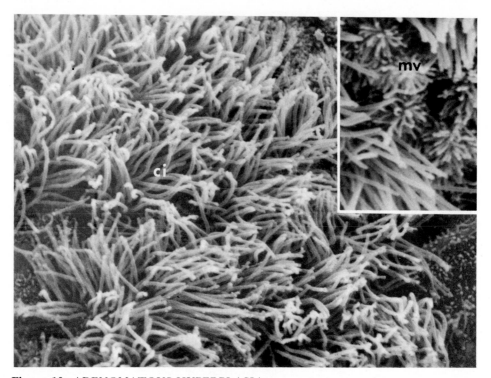

Figure 10. ADENOMATOUS HYPERPLASIA
The endometrial surface in adenomatous hyperplasia contains numerous ciliated cells (ci) alternating with nonciliated microvillous cells. The number of ciliary shafts and microvilli exceed that found in the normal proliferative endometrium or cystic glandular hyperplasia. (×6,000) *Insert:* Surface microvilli (mv) are arranged in promontories. (×8,000)

ticulum (GER) complex is increased, as is the synthesis of glycoproteins. Mitochondria and membranes of the GER have numerous bizarre, pleomorphic forms, and bundles of microfilaments have a disorganized, often whorled arrangement in a peri- and subnuclear location (Fig. 12). The plasma membranes near the basal lamina are markedly convoluted, and intracytoplasmic glycogen is sparse or absent. These cytoplasmic changes are accompanied by nuclear aneuploidy. *In well-differentiated, invasive adenocarcinoma,* the neoplastic glandular epithelium closely resembles that of the in situ variety with a comparatively increased mitotic activity, organellar pleomorphism, autophagocytic lysosomal activity, and juxtanuclear bundles of microfilaments (11,32–35). *Most poorly and undifferentiated carcinomas* contain abortive glandular lumen formation, and perinuclear microfilaments are arranged in concentric bundles (11,24,32–34) (Figs. 13a and b).

Although microfilaments vary quantitatively from one tumor to another, as well as from one neoplastic cell to another, they represent the most constant ultrastructural feature of endometrial carcinomas and are present even in the least differentiated, anaplastic growth of endometrial origin (11,25,32). The microfilaments present in the neoplastic cells differ in distribution, configuration, and number from those found in normal proliferative and hyperplastic

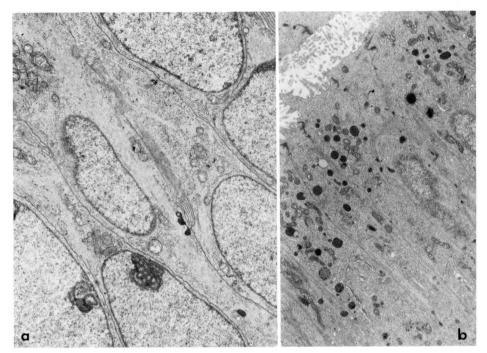

Figure 11. ADENOMATOUS HYPERPLASIA

a. The glandular cells are rich in perinuclear bundles of microfilaments, a characteristic feature of proliferative, hyperplastic, and neoplastic endometrial epithelium. Organelles and nuclear chromatin are regular in distribution and configuration. The lateral plasma membranes are generally straight. (×5,125) *b.* Membrane-bound primary lysosomes, presumably rich in acid hydrolases, are numerous in hyperplastic gland cells. (×3,075)

endometrium. In the former, they are abundant, have a peri- and subnuclear location, and are arranged in a whorle-like pattern. In the latter, the microfilaments are less numerous and form narrow bundles extending from the bases to the supranuclear region (11). Additional, though less specific, fine structural alterations include pleomorphic nuclei and nucleoli, complexly convoluted plasma membranes, interdigitating microvilli, an increase in autophagocytic activity, and a marked decrease in desmosomal attachments between adjacent neoplastic cells. The cells, regardless of the degree of dedifferentiation, produce a basal lamina, although in the poorly differentiatted forms, it is often fragmented.

Endometrial Stromal Sarcomas

Endometrial stromal sarcomas are composed of cells resembling the endometrial stroma. They invade the myometrium and its lymphatic and venous vasculature and characteristically have irregular margins. They are divided into low-grade sarcoma, or endolymphatic stromal myosis (ELSM) (Fig. 14a), and a highly malignant endometrial stromal sarcoma. The former has few mitoses, whereas the latter shows a high mitotic rate. Electron microscopy confirms the endometrial stromal cell origin of stromal sarcomas (11,36,37).

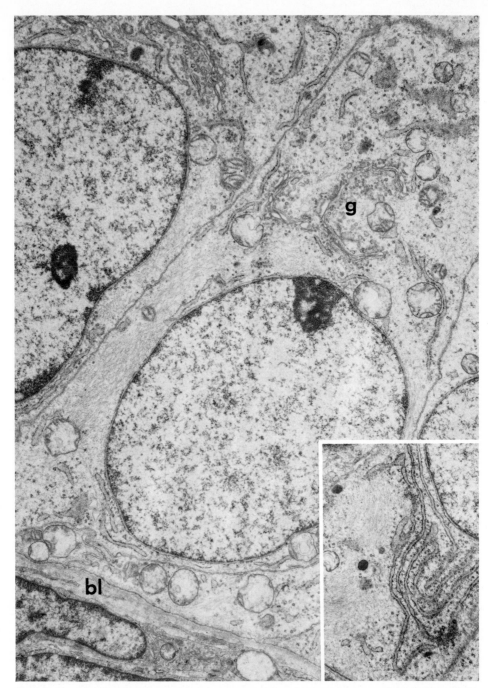

Figure 12. EARLY ENDOMETRIAL ADENOCARCINOMA (CARCINOMA IN SITU)
In contrast to hyperplastic endometrial cells, in early neoplasia large quantities of microfilaments are deposited concentrically in a subnuclear and perinuclear location. The Golgi (g) is well developed with peripheral secondary vesicles, and mitochondria are distended. The nuclei have a round contour and slightly irregular chromatin pattern. A basal lamina (bl) separates the neoplastic cells from the subjacent fibroblasts. (×9,050) *Insert:* Note bizarre configuration of membranes of GER. Cytoplasmic eosinophilia is probably due to an increase in free ribosomes intermingling with bundles of microfilaments. (×8,000)

282

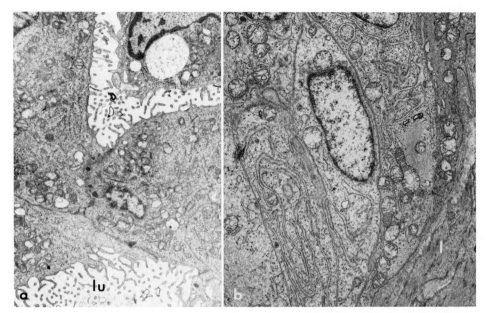

Figure 13. POORLY DIFFERENTIATED ADENOCARCINOMA OF ENDOMETRIUM

a. The neoplastic cells have abundant organelles, and the surface microvilli project into a small glandular lumen (lu). (×2,700) *b.* Invasive neoplastic cells with bundles of microfilaments, pleomorphic mitochondria, and cisternae of GER are separated from the adjacent smooth muscle cells (1) by a well-developed basal lamina. (×3,300)

Endolymphatic Stromal Myosis (ELSM)

The cells in ELSM resemble the stromal cells of the normal proliferative endometrium and vary in their differentiation, as do the stromal cells in the early, mid, and late proliferative phases of the cycle (11,36,37). The majority of the neoplastic stromal cells contain all the fine structural characteristics of fibroblasts engaged in active collagen synthesis (Fig. 14b), including a well-developed GER, free ribosomes, mitochondria, and Golgi complexes associated with abundant intracytoplasmic microfilaments (tropocollagen). The cytoplasmic membrane demonstrates peripheral condensation of microfibrils and complex microvillous projections. Akin to normal endometrial stromal cells (11), the neoplastic cells occasionally contain single ciliary apparatuses (37). Typical of cells of fibroblastic origin, they are uniformly devoid of a covering basal lamina. In many low-grade stromal sarcomas, there are scattered, well-circumscribed, hyaline nodules encased by neoplastic stromal cells. Ultrastructurally, these structures are composed predominantly of cross-striated precollagen fibers and are apparently produced by neoplastic stromal cells rich in tropocollagen fibrils (11).

It has been suggested that the endometrial stromal sarcomas have a histological appearance similar to that of the hemangiopericytomas seen in other parts of the body (38). Indeed, both neoplasms are highly vascular, and the cells are frequently associated with small and larger vessels. However, at the ultrastructural level, endometrial stromal sarcoma cells contain neither intracytoplasmic myofilaments nor abundant plasma membrane micropinocytic vesicles and pro-

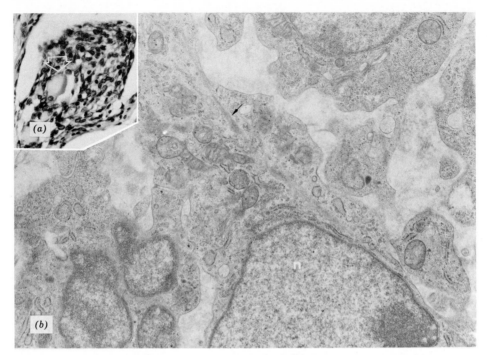

Figure 14. ENDOLYMPHATIC STROMAL MYOSIS

a. Typical of this type of neoplasm, masses of spindle-shaped stromal cells associated with hyalin nodules (arrow) are found within the myometrial vessels. (Hematoxylin-eosin, ×180) *b.* The narrow cytoplasm contains numerous perinuclear (n) free and bound ribosomes, Golgi, and mitochondria resembling normal endometrial stromal cells of the proliferative phase of the menstrual cycle. Many of the cells also have bundles of microfilaments, presumably tropocollagen fibers (arrow). The general ultrastructural organization of ELSM cells is consistent with active intracellular fibrillogenesis. The plasma membranes exhibit microvillous projections and are uniformly devoid of desmosomes and basal lamina. (×6,300)

duce no basal lamina, features that are characteristic of both normal and neoplastic pericytes (39,40).

Endometrial Stromal Sarcoma

The lesion in high grade endometrial stromal sarcoma is composed of a comparatively more pleomorphic and disorganized cell population than that of ELSM, rendering the histogenetic classification difficult and often uncertain. Ultrastructurally, the fibroblastic nature of the malignant cells can be recognized by the presence of abundant intracytoplasmic microfilaments associated with numerous GER filled with flocculent material, free ribosomes, mitochondria, and Golgi apparatus (Fig. 15). The cytoplasmic membrane contains complex microvillous processes that are occasionally attached by focal microfilamentous thickenings resembling macula adherens. In general, the neoplastic cells do not form basal lamina, although in a few cells an often interrupted, poorly formed basal lamina-like structure is seen (11). Focal, peripheral condensation of microfilaments and rare pinocytic vesicles occur along the cytoplasmic membrane.

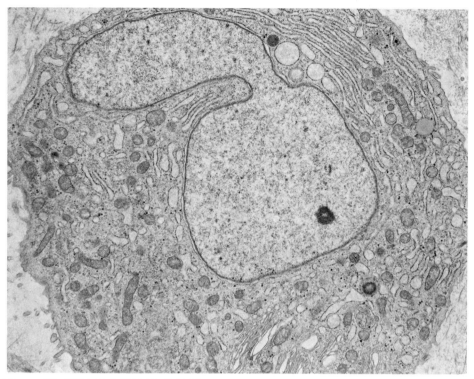

Figure 15. ENDOMETRIAL STROMAL SARCOMA
In high-grade lesions, the cytoplasm of neoplastic cells is packed with dilated GER, Golgi, free ribosomes, and mitochondria. Scattered microfilaments and microtubules are encountered. The plasma membrane has a few pinocytic vesicles and numerous short microvillous processes. (×6,125)

In contrast to the well-differentiated, low-grade stromal sarcoma cells of ELSM, in the poorly differentiated variety, formation of extracellular collagen is depressed. As a result, the extracellular space is composed predominantly of amorphous ground substance containing occasional bundles of pre- and mature collagen fibers. Similar ultrastructural characteristics were reported in moderately to poorly differentiated fibrosarcomas of extragenital sites (41).

MYOMETRIUM

Cellular Leiomyoma

Cellular leiomyomas are highly cellular smooth muscle tumors with or without nuclear atypia and containing less than five mitoses per 10 high-power fields. These lesions have a benign clinical course. At the ultrastructural level (42), the neoplastic cells have all the features that characterize well-differentiated smooth muscle cells in general (Fig. 16). These include abundant, and diffusely distributed, 80Å-thick myofilaments scattered with spindle-shaped densities, the plasma membrane dense bodies (which are thought to represent attachment sites

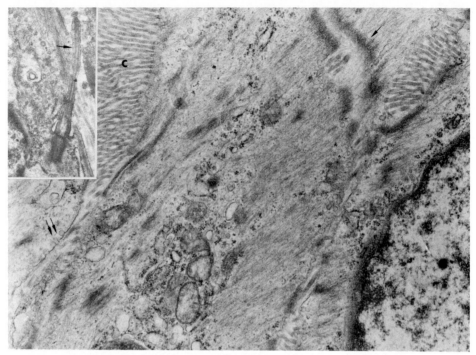

Figure 16. CELLULAR LEIOMYOMA OF UTERUS
In addition to nuclear enlargement, smooth muscle cells of cellular leiomyoma possess
more abundant, conically arranged myoplasmic organelles than common leiomyoma
cells. There is abundant production of myofilaments with the characteristic spindle-
shaped densities. Bundles of myofilaments converge into marginal dense bodies (arrow).
The myoplasmic membrane has rows of micropinocytic vesicles (double arrow), and the
extracellular space is rich in mature collagen fibers (c). A basal lamina, which is generally
produced by smooth muscle cells, is not apparent in this micrograph. (×27,200) *Insert:*
Single cilia are more common in cellular than ordinary leiomyoma. A mature, perinuclear
cilium (arrow) with ciliary vesicles and a basal body is projecting into the extracellular
space. (×25,000)

of myofilaments), micropinocytic vesicles, and a well-developed basal lamina.
The cells in cellular leiomyomas differ from the conventional leiomyoma cells in
that the former contain comparatively larger nuclei, prominent nucleoli, and
increased myoplasmic organelles such as free ribosomes, mitochondria, and
Golgi apparatus. Single cilia occur more frequently, and intracytoplasmic glycogen is less conspicuous than in the ordinary leiomyoma (42) (Fig. 16—insert).
The presently available ultrastructural observations suggest that the cellular
leiomyoma represents a variety of common uterine leiomyoma, rather than an
intermediate form between the leiomyoma and the leiomyosarcoma (42).

Leiomyoblastoma

This peculiar neoplasm, also named "bizarre leiomyoblastoma," "myoid tumors,"
or "epithelioid (bizarre) leiomyoma," is composed of round, epithelioid, rather
than spindle-shaped, smooth muscle cells. Characteristically, the epithelioid cells

are sensitive to delayed fixation, and intracytoplasmic vacuolization resembling signet ring cells is frequently present (Fig. 17a). The nuclei are round, the cytoplasm is devoid of myofibrils by light microscopy, and histochemical tests to demonstrate myofilaments are not useful. Leiomyoblastomas are encountered more frequently in the gastric wall than in other sites, but several cases of uterine leiomyoblastoma have been reported, as well as lesions at other body locations. Leiomyoblastomas generally have a benign course, but large lesions or those with high mitotic rates may exhibit malignant behavior with metastasis.

Previous electron microscopic studies of leiomyoblastomas (43–45) have unequivocally demonstrated their smooth muscle origin, and myofilaments in periodic arrangements, micropinocytic vescles, and often a basal lamina are their diagnostic features (Figs. 17b and 18). These ultrastructural markers, however, are less well developed in leiomyoblastoma cells than in mature smooth muscle cells, presumably reflecting their cellular immaturity. Another distinctive ultrastructural feature of leiomyoblastoma cells, which is not observed in typical smooth muscle cells or the conventional benign neoplastic smooth muscle cells, is the presence of occasional microtubular, microfilamentous, intracytoplasmic crystalloid structures (44) (Fig. 18-insert). The failure to find ultrastructural alterations in freshly fixed specimens that would correspond to the cytoplasmic vacuoles in light microscopy suggests that this histological feature is an artifact due to delayed fixation, rather than to the molecular constitution of various tissue fixatives.

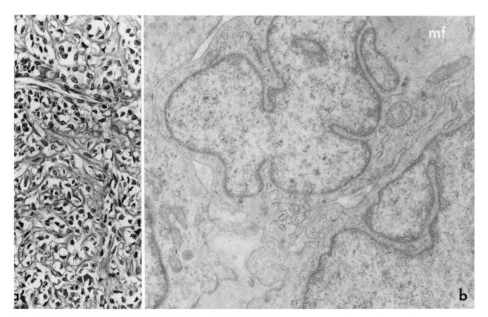

Figure 17. LEIOMYOBLASTOMA OF UTERUS

a. The cells have an epithelioid appearance with vacuolated cytoplasm resembling signet ring cells. (Hematoxylin-eosin, ×140) *b.* Ultrastructurally, the myogenic origin of tumor cells is evidenced by the presence of short bundles of myofilaments (mf) with spindle-shaped densities. The myoplasm is polygonal rather than fusiform and contains scattered organelles. In this case, the cells do not produce a basal lamina. (×11,400)

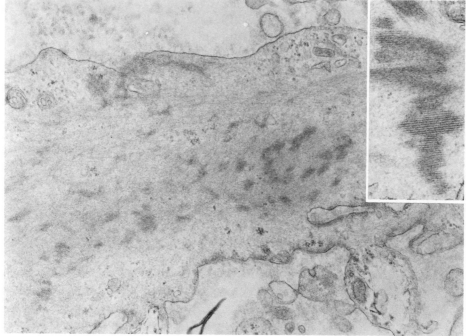

Figure 18. LEIOMYOBLASTOMA OF UTERUS
Cellular detail of bundles of myofilaments with periodic spindle-shaped densities. The latter are composed of 50 to 60-Å-thick actin filaments that are interconnected by 20-Å-thick collateral fibers. (×22,500). *Insert:* Leiomyoblastomas often contain clusters of crystal-like microtubulomicrofilamentous strucures. (×19,000)

Leiomyosarcoma

The leiomyosarcoma is among the most common malignant sarcoma of the uterus. The lesion generally grows diffusely and has infiltrating margins. Histologically, the neoplastic cells have various degrees of differentiation, ranging from well-differentiated spindle-shaped smooth muscle cells with blunt-ended nuclei to highly pleomorphic undifferentiated mesenchymal cells. In the absence of parametrial invasion or distant metastases, the histological diagnosis of leiomyosarcoma is based on evaluating its mitotic activity. The presence of five or more mitoses per 10 high-power fields is generally regarded as indicative of malignant behavior.

Ultrastructurally (41–43,46), malignant smooth muscle cells may be identified by the presence of myofilaments with focal periodicity. In contrast to normal or neoplastic benign smooth muscle cells, the smooth muscle sarcomas contain myofilaments appearing as short bundles, often deposited in a haphazard fashion (Figs. 19a and b). The most undifferentiated mesenchymal cells are totally devoid of myofilaments. Micropinocytic activity, plasma membrane densities, and basal lamina formation are reduced, as compared to both cellular and ordinary leiomyoma. The cytoplasmic membranes are generally straight and occasionally juxtaposited by small desmosome-like attachments. In keeping with the high metabolic turnover of malignant cells in general, free ribosomes,

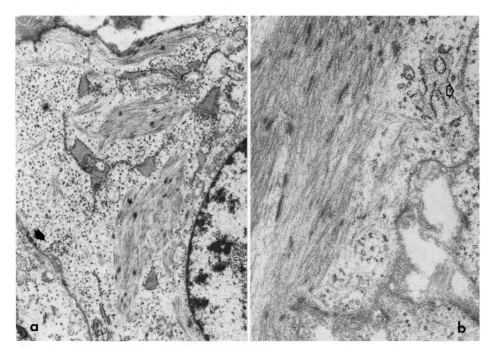

Figure 19. LEIOMYOSARCOMA OF UTERUS
a. Poorly differentiated malignant smooth muscle cells are characterized by haphazardly oriented short bundles of microfilaments with spindle-shaped densities and plasma membrane dense bodies. The GER is cystically dilated, and the plasma membranes are straight, reinforced by a few macula adherences (arrow). (×12,800) *b.* Myofilaments with focal spindle-shaped densities and micropinocytic vesicles (arrow) are shown at higher magnification. (×18,500)

mitochondria, and pleomorphic GER are conspicuous in many cells. The single cilia that are seen in the benign myomas are rarely seen in the leiomyosarcomas. The nuclei are prominent, with indented nuclear membranes, and contain pleomorphic nucleoli.

OVARY

Serous Cystadenocarcinoma

Serous cystadenocarcinomas are the most common epithelial malignancies of the ovary. The well-differentiated and moderately differentiated papillary forms are easy to identify in histological sections because the neoplastic lining epithelium retains its similarity to tubal epithelium, including the presence of ciliated and nonciliated secretory cells. In addition to these types of cells, others containing glycogen, microfilaments, or apical mucous secretory granules, resembling endometrial- and endocervical-type cells, respectively, may be recognized at the electron microscopic level (11,47,48). In the adenocarcinomas, classified histologically as poorly differentiated serous carcinomas, the specialized cellular features found in the better differentiated lesions are lost, including ciliary shafts. Glandular lumen formation is abortive; the lateral cytoplasmic membranes are convoluted with numerous interdigitating microvilli; and autophagocytic cytolysosomal bodies may be numerous (Fig. 20). A characteristic ultrastructural feature of these lesions is the presence of collections of small electron-dense, membrane-bound bodies, chiefly in a suprabasal location (11) (Fig. 20—insert). They presumably represent lysosomal or some types of secretory granules.

Mucinous Cystadenocarcinoma

Histologically, mucin-producing ovarian carcinomas range from papillary cystadenomas of borderline malignancy to poorly differentiated, partly solid, partly cystic neoplasms. The diagnosis of the borderline variety relies on the histological evaluation of epithelial stratification, mitotic activity, and nuclear pleomorphism and an absence of stromal invasion. Transmission and scanning electron microscopic studies of the mucinous cystadenomas (49) reveal that the lining

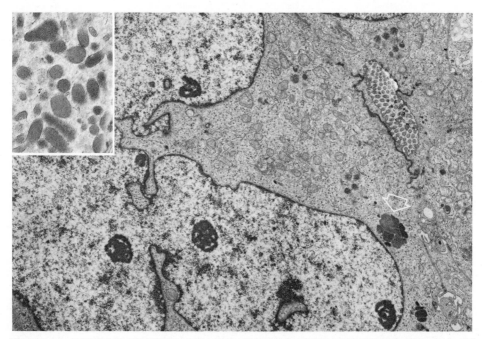

Figure 20. POORLY DIFFERENTIATED SEROUS CYSTADENOCARCINOMA OF OVARY
The cells have prominent nuclei with indented nuclear membranes and several enlarged nucleoli. The cytoplasm contains numerous organelles and a collection of membrane-bound electron-dense bodies (arrow). The latter are considered characteristic of "serous-type" cells, for they are often present in normal and neoplastic tubal and ovarian serous epithelium. The cells demonstrate attempt at glandular lumen formation. The above cellular organization and absence of special cellular functions such as mucinogenesis favor a serous origin for these neoplastic cells. (×3,600) *Insert:* Higher magnification of membrane-bound electron-dense granules. (×17,500)

epithelium is composed of endocervical cells alone or a mixture of endocervical and colonic adenoma-type cells. In the borderline lesions, the neoplastic epithelium is made of admixtures of endocervical and colonic-type cells, the latter being similar, fine structurally, to those found in colonic carcinomas in situ (50)

The epithelium in the well-differentiated papillary mucinous cystadenocarcinomas, as well as in mucinous lesions of lesser differentiation (24,50), is exclusively of the intestinal type, and typical colonic-type goblet cells in various stages of development, nonsecretory abortive cells, and argentaffin cells are seen (Fig. 21). The fine structural organization of the intestinal-type cells found in these ovarian neoplasms has a close resemblance to adenomatous (51) and malignant (52) colonic epithelium, matched for degree of epithelial differentiation (51). In contrast to the ovarian lesions, however, mature goblets cells are rarely found in primary malignant neoplasms of gastrointestinal origin. The conspicuous presence, or absence, of goblets cells may therefore be regarded as a diagnostic criterion in distinguishing between primary and secondary ovarian mucin-producing adenocarcinomas, especially when both ovarian and intestinal lesions occur simultaneously. This diagnostic criterion, however, seems to be of no help

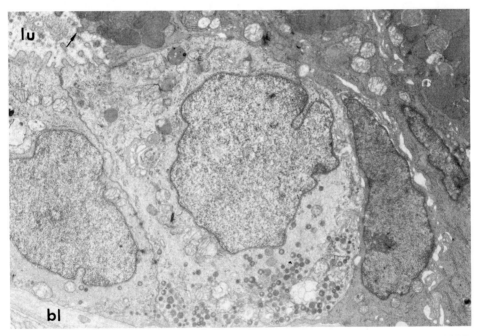

Figure 21. WELL-DIFFERENTIATED, PAPILLARY MUCINOUS CYSTADENO-CARCINOMA OF OVARY
The glandular lumen (lu) lining cells are rich in polymorphic, subnuclear neurosecretory granules filled with electron-dense substance, presumably serotonine. The cytoplasm is packed with fine filaments and produces a basal lamina (bl). These cells are recognized as argentaffin cells of midgut derivation. The cells adjacent to the argentaffin cells are filled with acid-mucin-containing granules with coarse, granulofilamentous content and stubby surface microvilli (arrow). ($\times$3,800)

in the poorly differentiated mucinous carcinomas, since both the ovarian and colonic varieties contain very few goblet cells.

Mucin production in poorly differentiated ovarian carcinomas is profoundly reduced (24,50), precluding their identification by light microscopy. At the ultrastructural level, however, a certain proportion of cells retain some of the submicroscopic features that are characteristic of intestinal epithelium, and their presence will serve to identify the neoplasm as mucinous in origin (Figs. 22a and b).

Endometrioid Carcinoma

Endometrioid carcinoma is the second most common primary neoplasm of the ovary. The histological diagnosis is dependent upon a resemblance to uterine endometrial adenocarcinoma, frequently including the presence of foci of squamous metaplasia (Fig. 23a). Fine structural examination is of considerable diagnostic value (11,53) since the ultrastructural appearance of the endometrioid carcinomas differs significantly enough from those of the mucinous or serous neoplasms to allow for a definitive diagnosis, even in poorly differentiated

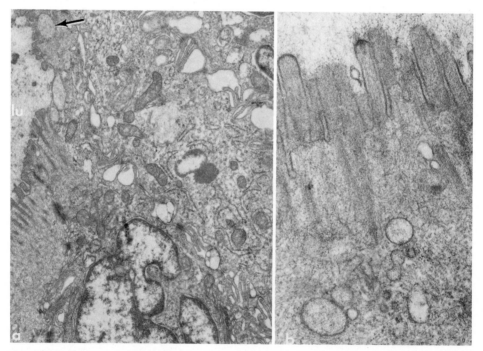

Figure 22. POORLY DIFFERENTIATED MUCINOUS CYSTADENOCARCINOMA OF OVARY

a. The mucinous nature of this lesion is recognized by the presence of intestinal-type luminal (lu), microvillous "brush border" supported by microfilaments and a few mucin-containing secretory granules (arrow). (×4,400) *b.* View of short, irregular blunt-ended intestinal-type surface microvilli at higher power. As in the colonic epithelium, the microvillous processes are covered by fine branching filamentous glycocalyx and supported by microfilaments extending from the subjacent cytoplasmic substance. (×35,600)

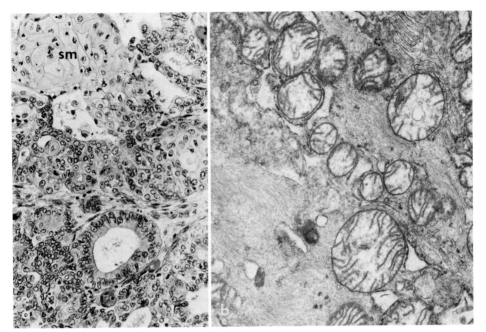

Figure 23. MODERATELY DIFFERENTIATED ENDOMETRIOID CARCINOMA OF OVARY

a. The neoplastic glands and associated squamous metaplasia (sm) are identical with adenocanthomatous lesions of endometrial origin. (Hematoxylin-eosin, ×180) *b.* Akin to adenocarcinoma of endometrium, the neoplastic cells of endometrioid carcinoma are rich in microfilaments and the mitochondria are pleomorphic. (×10,875)

forms. The distinctive features of endometrioid carcinoma, regardless of the degree of differentiation, include conspicuous bundles of cytoplasmic microfilaments, closely apposed mitochondria with parallel membranes of granular endoplasmic reticulum, and an absence of mucin secretory granules or subnuclear, electron-dense membrane-bound lysosomes (Fig. 23b). The value of differentiating endometrioid carcinoma from other malignant neoplasms of the ovary lies in the fact that the endometrioid tumors have a comparatively better prognosis.

Clear Cell Adenocarcinoma

The type of lesion in clear cell adenocarcinoma is a histogenetic variant of endometrioid tumors, sometimes referred to as mesonephroma. As in clear cell carcinoma of the vagina, cervix, and endometrium, the lesion in the ovary is composed of tubules with luminal secretions or solid sheets of large epithelial cells with a clear cytoplasm. Both histochemically and ultrastructurally (11,13,15,18), the cells are packed with glycogen granules (Figs. 2 and 3) which are associated with well-developed GER, Golgi apparatus, and numerous mitochondria. Some of the neoplastic cells contain clusters of membrane-bound electron-dense bodies, with a polygonal contour and interconnected by hexagonal microtubules. These structures contain lipid bodies and needle-shaped crystalline material (Fig. 24). Similar structures have been observed in glycogen-rich, neoplastic clear cell carcinoma of the endometrium (17) and are regarded as lysosomal bodies undergoing partial crystallization.

Sex Cord-Stromal Tumors

Neoplasms derived from the ovarian cortical stroma are often hormonally active, leading either to feminization or virilization. There are wide variety of histological growth patterns reproducing with variable success the specialized cellular elements of the developing female- and/or male-oriented gonads. Consequently, a confusing nomenclature has evolved, and the histogenetic origin of this category of ovarian neoplasms remains uncertain. On the basis of their histological, and often hormonal, manifestations, they have been variously classified as granulosa-theca cell tumors, Sertoli-Leydig cell tumors, lipid cell tumors, sex cord-mesenchymal tumors, mesenchymomas, and stromal cell tumors. Electron microscopic examination (11) of these lesions has provided circumstantial evidence for a common histogenetic origin—the specialized ovarian stromal fibroblast. This common origin is supported by ultrastructural, cytochemical (54), and radioautographic (55) studies which have demonstrated the postnatal differ-

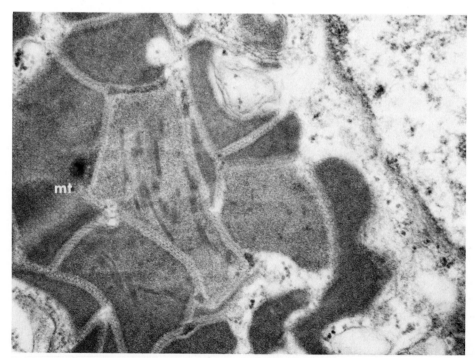

Figure 24. CLEAR CELL ADENOCARCINOMA OF OVARY
Peculiar electron-dense bodies with polyhedral contour, interconnected by hexagonal microtubules (mt), are observed in some of the neoplastic clear cells. Within the amorphous matrix are needle-shaped crystallin-like structures. (×65,000)

entiation of both human and rodent ovarian stromal cells into granulosa and theca cells. The production of biologically active steroids occurs in those cells that possess the ultrastructural and enzymatic distribution characteristic of normal steroidogenic cells of the ovary or testis (11).

The most common stromal tumors are the granulosa-theca cell tumor, Sertoli-Leydig cell tumor, hilus cell tumor, lipid cell tumor, and the luteoma of pregnancy. Most can be readily diagnosed by light microscopy alone when they are well differentiated, but the accurate diagnosis of some of the poorly differentiated sarcomatoid-type lesions can be enhanced by electron microscopy.

Granulosa-Theca Cell Tumor

Histologically, a number of pattern variants occur in the granulosa-theca cell tumors. They have been designated well-differentiated micro- and macrofollicular, moderately differentiated trabecular, and poorly differentiated sarcomatoid types. In the majority of cases, fibromathecomatous areas are present and vary from fusiform fibroblast-like cells to round, lutenized, theca-like cells. The latter are rich in vesicular, agranular endoplasmic reticulum (AER), Golgi apparatus, free ribosomes, pleomorphic mitochondria with tubular cristae, and lipid bodies. This type of cell is considered the site of steroid hormone production (11).

By means of ultracytoenzymological studies, the AER has been shown to contain 3-β-hydroxysteroid dehydrogenase, glucose-6-phosphate dehydrogenase, and various aromatizing enzymes in a perilipidic and perimitochondrial location, suggesting that the AER is the site of steroidogenesis. The well-developed mitochondrial system provides a source of energy for hormonal biosynthesis, and the lipid droplets represent sites of storage of steroid hormone precursors such as cholesterol and its esters (11). Although nonluteinized, neoplastic granulosa cells are devoid of such subcellular organization, in vitro tumor biosynthetic studies (56) have suggested that in neoplastic granulosa cells, steroid synthesis does occur with the production of androstenedione and estrogens.

Ultrastructurally (11,56–60), neoplastic granulosa cells closely resemble the preovulatory granulosa cells of the follicular apparatus. The two most distinctive features are deep nuclear indentations (Fig. 25a) and the formation of Call-Exner bodies (Fig. 25b). The latter are produced in the normal follicle by central cellular degeneration which leads to cystic intercellular spaces that characteristically are delimited by a basal lamina (11). These structures may be found ultrastructurally even in the least differentiated (sarcomatoid) granulosa cell tumor. Since Call-Exner bodies are seen only in granulosa cells, they constitute the major diagnostic criterion for this group of ovarian neoplasms. The specific cytoplasmic elements that are present include abundant microfilaments, often arranged in a whorled pattern, small, elongated, or tubular mitochondria, free and bound ribosomes, and lipid bodies. The cytoplasmic membrane has well-developed interdigitating microvilli, and the cells are attached to each other by small desmosomes. In contrast to normal granulosa cells, the neoplastic counterparts do not contain annular tight junctional nexuses (11). A fine basal lamina surrounds clusters of cells but not individual cellular units. The formation of glands is not a feature of granulosa cells. Apparent transitional forms between granulosa cells and fibrothecomatous cells have been described.

The poorly differentiated, sarcomatoid granulosa cell tumors have fine struc-

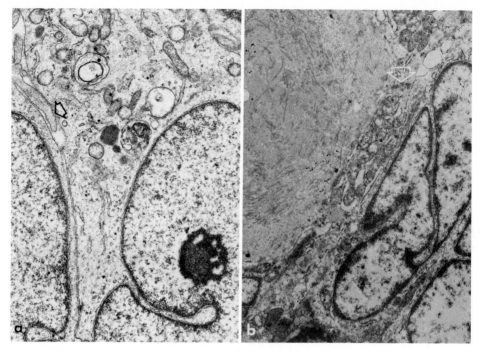

Figure 25. GRANULOSA-THECA CELL TUMOR OF OVARY
a. The nuclei of neoplastic granulosa cells typically contain deep cytoplasmic indentations that correspond to the characteristic "coffee bean" nuclear pattern on light microscopy. Within the cytoplasm are scattered small tubular mitochondria and bundles of microfilaments. The plasma membranes have several microvilli (arrow) projecting into the extracellular space and have infrequent desmosomal attachments. (×12,300) *b.* A Call-Exner body, consisting of a round extracellular space filled with basal lamina and procollagen connective tissue fibers. Characteristically, the lining cells have tight junctional complexes (arrow). (×12,000)

tural features that are similar to those found in the better differentiated, epithelial cells, except that their cytoplasm is spindle-shaped, the nucleocytoplasmic ratio is increased, and cytoplasmic microfilaments are less numerous (Fig. 26). With the exception of enlarged nuclei and nucleoli, malignant, metastatic granulosa cells contain no distinctive characteristics that serve to differentiate them from nonmetastatic granulosa cells (60).

Sertoli-Leydig Cell Tumor
Sertoli-Leydig cell tumors are the most common virilizing neoplasms of the ovary, but may also be nonfunctional or even feminizing. The epithelial components recapitulate to a varying extent the cells of the testis at different stages of development. Histologically, the tumors are classified into three groups, according to their degree of differentiation. The well-differentiated form consists of tubules lined by cells resembling Sertoli cells. The moderately differentiated or intermediate type consists of immature Sertoli cells growing in cords resembling embryonic or fetal testicular sex cords. The poorly differentiated or sarcomatoid type is composed predominantly of interlacing bundles of spindle-shaped cells

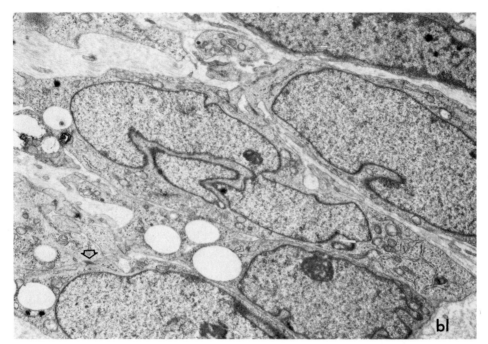

Figure 26. POORLY DIFFERENTIATED, SARCOMATOID GRANULOSA CELL TUMOR OF OVARY

Deeply indented nuclear membranes, increased nucleocytoplasmic ratio, relatively poor organellar content, lipid bodies, and microfilaments, together with occasional Call-Exner bodies, are diagnostic features of sarcomatoid granulosa cell tumors. The cells, despite their spindle-shaped appearance, retain their epithelial differentiation, as evidenced by the formation of desmosomal attachments (arrow) and a basal lamina (bl). (×4,400)

resembling the urogenital mesenchymal cells of the primitive gonad. The latter form is identified as a Sertoli-Leydig cell tumor only when cords of cells associated with Leydig cells are demonstrated. The stromal component of these tumors contains polyhedral, eosinophilic cells, interpreted as Leydig cells, which occasionally contain crystals of Reinke. Virilization is due to androgen production by the neoplastic Leydig or luteinized stromal cells (11). In vitro biosynthetic studies and analysis of venous blood draining the tumors have shown that both testosterone and androstenedione are produced through a steroid pathway similar to that which operates in the normal adult testis. Sertoli-Leydig cell tumors of the intermediate and sarcomatoid types may contain mucinous epithelium, cartilage, bone, smooth and skeletal muscle, and adipose tissue.

The ultrastructural features of virilizing Sertoli-Leydig cell tumors (11,61–66) are principally those associated with male-oriented cells, including Sertoli cells (11,66), Leydig cells (66), and cells lining the ductuli efferentes (64), although female-oriented granulosa-theca cells (63) and undifferentiated ovarian stromal cells (61,63,64) have also been described. The variations in interpretation are largely due to the degree of differentiation of the neoplasms studied; well- and moderately well-differentiated lesions resemble the Sertoli cells of immature, fetal testis (11,66), whereas poorly differentiated tumor cells lack most of the cellular features that are characteristic of either male- or female-oriented gonadal cells. In addition, some of the poorly differentiated lesions have been classified as Sertoli-Leydig cell tumors, mainly because the patients became virilized rather than because of the findings on morphological examination.

Sertoli-Leydig cell tumors may be distinguished from granulosa-theca cell tumors by identifying true tubular lumens formed by the neoplastic cells. The lumens are lined by epithelial cells, attached to each other by desmosomes and tight junctional complexes, and the luminal plasma membrane contains infrequent microvilli and occasional ciliary apparatuses (67). Like normal human Sertoli cells (68), some of the neoplastic tubular cells also have intraluminal apical cytoplasmic protrusions, consistent with an apocrine type of secretory activity. The peripheral portion of the cells is separated from the stroma by a basal lamina. The tubules differ from the Call-Exner bodies of the granulosa-theca cell tumor in that in the Call-Exner bodies the "lumen" is lined by a basal lamina and contains fibrillar material and extruded cellular debris. Although true lumens are more frequently found in well-differentiated and intermediate types of Sertoli-Leydig cell tumors, they can be found in an abortive stage in the lesser differentiated neoplasms as well (64).

The epithelial cells of the tubular and trabecular types of lesions exhibit features of the immature Sertoli cells observed in the testicular feminization syndrome (69), rather than of truely neoplastic testicular Sertoli cells (70). These features include a polyhedral cellular contour, abundant cytoplasmic substance with scattered GER and AER, microfilaments, microtubules, mitochondria, glycogen particles, lipid and lipofuscin granules, and tightly apposed plasma membranes (Fig. 27). In general, the nuclei are devoid of deep cytoplasmic indentation. A distinct nucleolus with a prominent, central pars amorpha and a peripheral fibrillar nucleolonema is sometimes seen (66). Well-developed intranuclear sphaeridia (nuclear bodies) with a dense, central, granular mass and a clear, external fibrillar halo—typically present in human Sertoli cells (71)—have not been reported in the epithelial cells of Sertoli-Leydig cell tumors. The sphaeridia correspond both in size and morphology to cytoplasmic ribosomes and are believed to be associated with high intranuclear protein synthesis.

The morphology of Sertoli cells differs from that of normal or neoplastic granulosa cells. The latter have a greater nuclear to cytoplasmic ratio, indented nuclei, prominent aggregates of microfilaments, and larger intercellular spaces. The stromal component of the tumors may help in differentiating Sertoli-Leydig cell from granulosa-theca cell tumors only when crystals of Reinke (11,62) or abundant glycogen particles (66) are found in the polyhedral stromal cells (Fig. 28). These features are present in normal and neoplastic Leydig cells, but not in stromal lutein cells of the ovary.

Hilus Cell Tumor

Hilus cell tumors are among the least frequent neoplasms of the ovary and are usually associated with virilization, although occasionally feminization is observed. The virilization is due to high levels of androgens, principally testosterone and its precursors, dehydroepiandrosterone and androstenedione, all of which are secreted by the neoplastic cells.

Both histologically and ultrastructurally (11,72,73), neoplastic hilus cells are

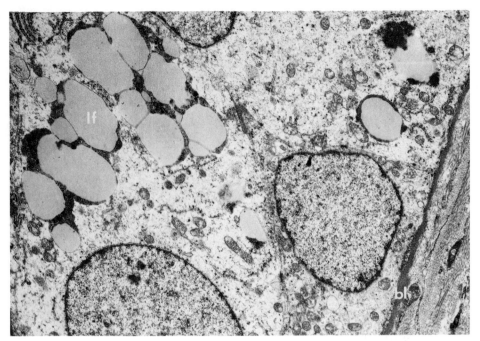

Figure 27. WELL-DIFFERENTIATED, TUBULAR SERTOLI-LEYDIG CELL TUMOR OF OVARY

The tubular epithelial cells resemble immature, testicular Sertoli cells, including round, nonindented nuclei, abundant clear cytoplasmic matrix with small mitochondria, stalks of granular GER, and lipofuscin granules (lf). In the tubular and trabecular (intermediate) forms, the epithelial cells produce more desmosomes and often a thickened basal lamina (bl). (×7,000)

similar to normal (61) and hyperplastic (74) hilus cells, as well as normal (11) and neoplastic (61,75) stromal lutein cells (lipid cell tumor) and normal and neoplastic testicular Leydig cells (76). The common fine structural features are abundant tubular and vesicular AER, pleomorphic mitochondria with tubular cristae, prominent Golgi, lipid, lipochrome pigment bodies, and lysosomes.

However, Reinke crystals or their precursor forms are considered the pathognomonic feature for the identification of neoplastic ovarian hilus cells (Figs. 29a and b). Although Reinke crystals have been identified in the Leydig cell element of a few Sertoli-Leydig cell tumors (11,62), they have not been encountered in granulosa-theca cell and lipid cell tumors, or in luteomas (77). Since hilus cell neoplasms with Reinke crystals have never been observed to behave in a malignant fashion, their identification is considered indicative of benignancy.

The rod-shaped Reinke crystals of light microscopy appear ultrastructurally as protein-containing, intracytoplasmic or intranuclear inclusions, composed of aggregates of six 300-Å-wide hexagonal microcylinders surrounding one central microcylinder (11,73). The fine structural morphology of the microcylinders of mature Reinke crystals closely resembles that of the microtubular rosettes, or elementary bodies (11,73), which are believed to represent precursor elements of

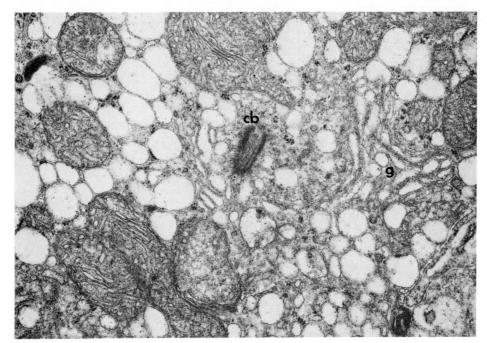

Figure 28. WELL-DIFFERENTIATED, TUBULAR SERTOLI-LEYDIG CELL TUMOR OF OVARY
The stromal component of the lesion has clusters of steroidogenic cells with abundant vesicular, agranular endoplasmic reticulum, hypertrophic Golgi (g), and pleomorphic mitochondria with tubular cristae. A tangentially sectioned centriolar body (cb) is seen. (×27,750)

crystal development. The microtubular rosettes presumably correspond to the eosiniphilic, hyalin spheres found in hilus cell tumors at the light microscopic level. Since crystals of Reinke are identified as true crystals by electron microscopy, the term crystal rather than crystalloid should be used.

At the present time, the precise significance of Reinke crystals as related to cellular dynamics is unknown. However, they are not unique to either ovarian hilus or testicular Leydig cells, for similar structures have been described in perivascular fibroblasts of normal human adrenal glands (78), virus-infected or chemically treated cells, and in the sexual antebrachial organ of the prosimian primate (Lemur Catta).

Another distinctive feature of hilus cells is the presence of abundant intracytoplasmic glycogen particles. These are comparatively reduced in other steroidogenic cells of ovarian origin.

Luteoma of Pregnancy
Pregnancy luteoma is considered to be a hyperplastic, rather than a neoplastic, proliferation of ovarian lutein cells. The condition is induced by high levels of chorionic gonadotropin during gestation and after termination of pregnancy. This lesion spontaneously involutes. The luteoma cells are capable of producing

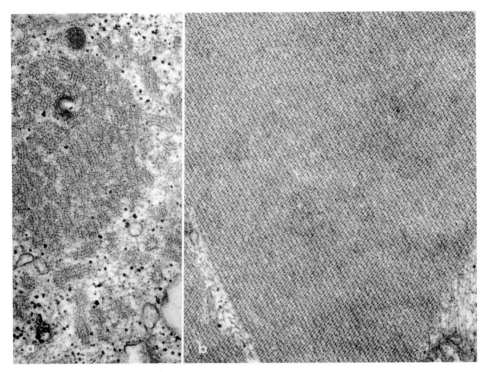

Figure 29 HILUS CELL TUMOR OF OVARY

a. Some of the neoplastic hilus cells are packed with striated, elementary, tubular inclusions, considered to be precursors to crystals of Reinke. These inclusions are composed of six hexagonal microtubules arranged in a rosette-like pattern around one central microtubule. (×23,400) *b.* Electron microscopic appearance of crystals of Reinke with a geometric contour. The crystals are made of juxtaposed hexagonal microtubules. (×45,250)

high levels of androstenedione, which may lead to masculinization of both the mother and her infant. A morphologically and steroidochemically similar lesion is encountered in postmenopausal women, as well as in mice (79).

Histologically, the lesion is made up of large, polygonal cells with a granular eosinophilic cytoplasm. Characteristically, large eosinophilic intracytoplasmic "colloid" bodies are observed in some of the luteoma cells. Similar structures are found in the granulosa lutein cells of the gestational corpus luteum (11). The nuclei are vesicular, with a single, prominent nucleolus. A fine reticulin network surrounds individual cells or groups of cells.

Ultrastructurally, the luteoma cells (11,77) are rich in vesicular AER, Golgi, pleomorphic mitochondria with tubular cristae and electron-dense granules, and, typically, annular nexuses (Figs. 30a and b). The "colloid" bodies seen histologically presumably correspond to giant, perinuclear, membrane-bound lysosomes with a homogenous amorphous content (11) (Fig. 30a). These structures are in all respects similar to those found in both the granulosa and theca lutein cells of the corpus luteum. The cells are devoid, however, of the prominent microvillous processes seen in granulosa lutein cells. The subcellular and

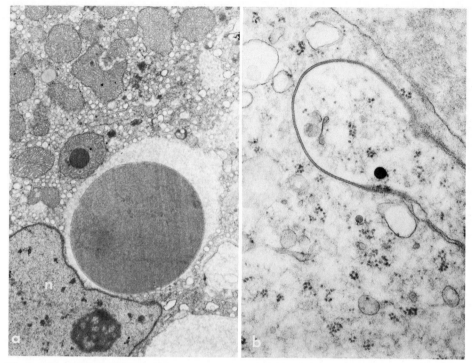

Figure 30. GESTATIONAL LUTEOMA OF OVARY
a. The cytoplasmic substance is filled with a myriad of Golgi-derived vesicular AER associated with pleomorphic mitochondria. Characteristically, in this type of lesion lipid bodies are scant or absent. A giant, perinuclear (n), membrane-bound "colloid" body is shown. The neighboring mitochondria contain a spherical electron-dense granule. (×9,250) *b.* High magnification of a pentalaminar, tight junctional annular nexus with a central core. The distance between the two thick membranes of nexus is about 20 to 40 Å. (×46,250)

biochemical features of the luteoma of pregnancy suggest that it derives either from the luteinized theca interna or granulosa of the gestational corpus luteum, or from ovarian stromal cells.

Dysgerminoma

The dysgerminoma is an uncommon malignant ovarian tumor occurring in children and young adults, and in dysgenetic gonads. Like germ cell tumors arising in other sites, the dysgerminoma is highly sensitive to ionizing irradiation. In general, dysgerminomas are similar histologically and histochemically to the testicular seminoma. However, several distinctive features differentiate the ovarian from the testicular germ cell tumors. The dysgerminoma is usually composed of a homogeneous population of pale, large cells, with vesicular nuclei and prominent nucleoli, resembling primitive embryonal germ cells. The testicular seminoma tends to be more heterogeneous in the cell pattern, varying from the poorly differentiated anaplastic seminoma to the well-differentiated spermatocytic form (80). Characteristically, the stroma of the dysgerminoma is often obscured by lymphoplasmocytic infiltrates and contains foreign body, or sarcoid-like, giant cell granulomatous foci. Occasionally, steroidochemically active, luteinized, stromal cells are observed (81).

Ultrastructurally, three different types of neoplastic cells can be found in pure germinomas (81–84). These cells resemble primitive germ cells (82,83), oogonia (81), moderately differentiated seminoma tumor cells (84–86), and those described in a pinealoma (87) and a thymic germ cell tumor (88). The principal, or type I, tumor cells have large nuclei in which the nucleolus has a typical widespread ribbon or lace-like pattern (Fig. 31). The cytoplasm is poor in organelles and lacks glycogen and cytoplasmic microvillous processes. This cell type corresponds to the clear germinal tumor cells seen on light microscopy and resembles the primitive ovarian oogonium. The type III cell has masses of glycogen granules, an intracytoplasmic and peripheral, microfilamentous web, stacks of GER or annulate lamellae, and prominent cytoplasmic evaginations (Fig. 32). This cell type resembles the normal embryonic germinal cells. The microfilaments and cytoplasmic microvilli are structures associated with cellular ameboid motion, a phenomenon that occurs during germ cell migration from the endoderm of the yolk sac to the urogenital ridges. The type II cell contains less glycogen and microvillous processes, but more free and bound ribosomes than

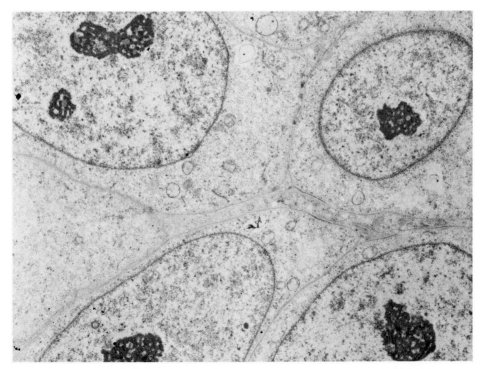

Figure 31. DYSGERMINOMA OF OVARY
The neoplastic germ cells are recognized by their prominent nucleoli which are often arranged in a ribbon-like pattern. Cells are poor in organelles and have clear cytoplasm. These are the more mature type 1 cells. The plasma membranes are tightly apposed, but desmosomal attachments are generally poorly developed or absent. (×6,625)

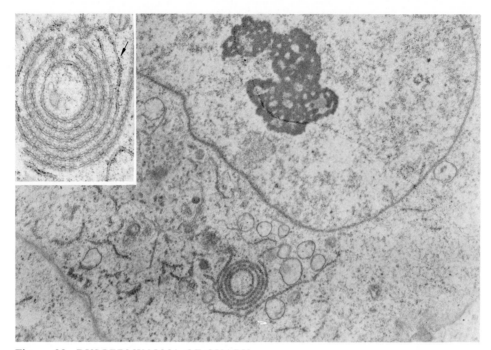

Figure 32. DYSGERMINOMA OF OVARY
Type 3 dysgerminoma cell in which the organelles, including a concentrically arranged annulate lamella, demonstrate a perinuclear distribution, reminiscent of Balbiani's vitelline body of primary ovocytes. (×8,750) *Insert:* Higher magnification of annulate lamella with periodically spaced diaphragms in close association with membranes of GER (arrow). (×19,375).

the type III cell and is regarded as an intermediate form between type I and III cells. Desmosomes, usually without tonofibrils and zonula adherens, are not a constant feature of either dysgerminomas or seminomas. The basal lamina that surrounds clusters of cells often appears thickened. Although dense granules, interpreted as proacrosomal granules (85), have been reported in a few seminomas, other characteristic features of germ cell, such as "intermitochondrial cement" or crystalloids of Lubarsch, have not been found.

CONCLUSION

The results obtained from studies of the ultrastructure of neoplasms of the female reproductive system demonstrate the satisfactory role of electron microscopy in diagnostic gynecologic pathology. By electron microscopy, histologically undifferentiated cancers may be classified into epithelial tumors (carcinomas) and mesenchymal tumors (sarcomas), and subclassified according to their histogenetic differentiation, i.e., adenocarcinoma, squamous cell carcinoma, leiomyosarcoma, endometrial stromal cell sarcoma, etc. In addition, electron microscopy helps in differentiating early endometrial adenocarcinoma from hyperplastic lesions that may be mistaken for neoplasia, and noninvasive borderline mucinous lesions of the ovary from their frankly malignant counterparts. The comparative fine structural evaluation of normal and neoplastic cells of the genital tract provides new and complementary data on the cellular origin and pathogenetic development of genital malignancies with controversial histogenesis, such as the clear cell adenocarcinoma of the vagina, mucinous neoplasms of the ovary, and ovarian sex cord-stromal tumors.

In view of the valuable information obtained by electron microscopy with respect to improved pathological diagnosis and understanding of the pathogenesis of genital neoplasms, electron microscopy should be integrated into all well-established gynecologic pathology services. We believe that the promotion of such diagnostic units is increasingly needed to improve contemporary gynecologic practice and patient care.

REFERENCES

1. Ferenczy A, Richart RM: Ultrastructure of perineal Paget's disease. *Cancer* 29:1141, 1972.
2. Yshibashi Y, Jager G: Elektronenmikroskopische Untersuchung einnes extramammaren Morbus Paget. *Arch Klin Exp Dermatol* 234:293, 1969.
3. Fetherston WC, Freidrich EG Jr: The origin and significance of vulvar Paget's disease. *Obstet Gynecol* 39:735, 1972.
4. Demopoulos RI: Fine structure of the extramammary Paget's cell. *Cancer* 27:1202, 1971.
5. Neilson D, Woodruff JD: Electron microscopy in in situ and invasive vulvar Paget's disease. *Am J Obstet Gynecol* 113:719, 1972.
6. Koss LG, Brockunier A Jr: Ultrastructural aspects of Paget's disease of the vulva. *Arch Pathol* 87:592, 1969.
7. Belcher RW: Extramammary Paget's disease. Enzyme histochemical and electron microscopic study. *Arch Pathol* 94:59, 1972.
8. Herbst AL, Robboy SJ, Scully RE, et al: Clear-cell adenocarcinoma of the vagina and cervix in girls: An analysis of 170 registry cases. *Am J Obstet Gynecol* 118:713, 1974.
9. Herbst AL, Kurman RJ, Scully RE, et al: Clear-cell adenocarcinoma of the genital tract in young females. Registry report. *N Engl J Med* 287:1259, 1972.
10. Silverberg SG, DeGiorgi LS: Clear cell carcinoma of the vagina. A clincal, pathologic and electron microscopic study. *Cancer* 29:1680, 1972.
11. Ferenczy A, Richart RM: Female reproductive system, in *Dynamics of Scan and Transmission Electron Microscopy.* New York, John Wiley, 1974.
12. Hameed K: Clear cell "mesonephric" carcinoma of uterine cervix. *Obstet Gynecol* 32:564, 1968.
13. Okagaki T, Richart RM: "Mesonephroma ovarii (hypernephroid carcinoma)." Light microscopic and ultrastructural study of a case. *Cancer* 26:453, 1970.
14. Silverberg SG, DeGiorgi LS: Clear cell carcinoma of the endometrium. *Cancer* 31:1127, 1973.
15. Silverberg SG: Ultrastructure and histogenesis of clear cell carcinoma of the ovary. *Am J Obstet Gynecol* 115:394, 1973.
16. Roth LM: Clear cell adenocarcinoma of the female genital tract. *Cancer* 33:990, 1974.
17. Rorat E, Ferenczy A, Richart RM: The ultrastructure of clear cell adenocarcinoma of endometrium. *Cancer* 33:880, 1974.
18. Salazar H, Merkow LP, Walter WS, et al: Human ovarian neoplasms: Light and electron microscopic correlations. II. The clear cell tumor. *Obstet Gynecol* 44:551, 1974.
19. Forsberg JG: Late effects in the vaginal and cervical epithelia after injections of diethylstilbestrol into neonatal mice. *Am J Obstet Gynecol* 121:101, 1975.
20. Hart WR, Norris HJ: Cervix adenocarcinoma of mesonephric type. *Cancer* 29:106, 1972.

21. Overbeck L: Die Ultrastruktur des Sarcoma Botryoides oder Traubeusarkoms der Vagina beim King. *Z Geburtsh Gynaekol* 167:13, 1967.

22. Overbeck L: Elektronenmikroskopische Untersuchungen des embryonalen Rhabdomyosarkoms. *Frank Z Pathol* 77:49, 1967.

23. Leone PG, Taylor HB: Ultrastructure of a benign polypoid rhabdomyoma of the vagina. *Cancer* 31:1414, 1973.

24. Ferenczy A: The ultrastructural morphology of gynecologic neoplasms. *Cancer* 38:463, 1976.

25. Lightiger B, Mackay B, Tessmer CF: Spindle-cell variant of squamous carcinoma. A light and electron microscopic study of 13 cases. *Cancer* 26:1311, 1970.

26. Wilbanks GD: In vitro studies on human cervical epithelium, benign and neoplastic. *Am J Obstet Gynecol* 121:771, 1975.

27. More IAR, Armstrong EM, McSeveney D, et al: The morphogenesis and fate of the nucleolar channel system in the human endometrial glandular cell. *J Ultrastruct Res* 47:74, 1974.

28. Gore BZ, Gordon M: Fine structure of epithelial cells of secretory endometrium in unexplained primary infertility. *Fertil Steril* 25:103, 1974.

29. Clyman M: Electron microscopic changes produced in the human endometrium by norethindrone acetate with ethinyl estradiol. *Fertil Steril* 14:352, 1963.

30. Kohorn EI, Rice SI, Gordon M: In vitro production of nucleolar channel system by progesterone in human endometrium. *Nature* (London) 228:671, 1970.

31. Richart RM, Ferenczy A: Endometrial morphologic response to hormonal environment. *Gynecol Oncol* 2:180, 1974.

32. Ferenczy A: Ultrastructural pathology of the uterus, in Wynn RM (ed): *Biology of the Uterus.* New York, Plenum Press, 1977, p 545.

33. Gompel C: Ultrastructure of endometrial carcinoma: Review of fourteen cases. *Cancer* 28:745, 1971.

34. Thrasher TV, Richart RM: An ultrasturctural comparison of endometrial adenocarcinoma and normal endometrium. *Cancer* 29:1713, 1972.

35. Cavasos F, Lucas FV: Ultrastructure of the endometrium, in Norris HJ, Hertig AT, Abell MR (eds): *The Uterus.* International Academy Pathology Monograph. Baltimore, Williams and Wilkins, 1973, p 136.

36. Komorowski RA, Garancis JC, Clowry LJ: Fine structure of endometrial stromal sarcoma. *Cancer* 26:1042, 1970.

37. Akhtar M, Kim PY, Young I: Ultrastructure of endometrial stromal sarcoma. *Cancer* 35:406, 1975.

38. Greene RR, Gerbie AB: Hemangiopericytoma of the uterus. *Obstet Gynecol* 3:150, 1954.

39. Silverberg SG, Willson MA, Board JA: Hemangiopericytoma of the uterus: An ultrastructural study. *Am J Obstet Gynecol* 110:397, 1971.

40. Battifora H: Hemangiopericytoma. Ultrastructural study of five cases. *Cancer* 31:1418, 1973.

41. Tannenbaum M: Ultrastructural pathology of human renal cell tumors, in Sommers SC (ed): *Pathology Annual.* New York, Appleton-Century-Crofts, 1970, p 249.

42. Ferenczy A, Richart RM, Okagaki T: A comparative ultrastructural study of leiomyosarcoma, cellular leiomyoma and leiomyoma of the uterus. *Cancer* 28:1004, 1971.

43. Kay S, Still WJS: A comparative electron microscopic study of a leiomyosarcoma and bizarre leiomyoma (leiomyoblastoma) of the stomach. *Am J Clin Pathol* 52:403, 1969.

44. Cornog JL: The ultrastructure of leiomyoblastoma. *Arch Pathol* 87:404, 1969.

45. Hajdu SI, Erlandson RA, Paglia MA: Light and electron microscopy studies of a gastric leiomyoblastoma. *Arch Pathol* 93:36, 1972.

46. Tobon H, Murphy AE, Salazar H: Primary leiomyosarcoma of the vagina. Light and electron microscopic observations. *Cancer* 32:450 1973.

47. Roberts DK, Marshall RB, Wharton JT: Ultrastructure of ovarian tumors. I. Papillary serous cystadenocarcinoma. *Cancer* 25:947, 1970.

48. Gondos B: Electron microscopic study of papillary serous tumors of the ovary. *Cancer* 27:1455, 1971.

49. Fenoglio CM, Ferenczy A, Richart RM: Mucinous tumors of the ovary. Ultrastructural studies of mucinous cystadenomas with histogenetic considerations. *Cancer* 36:1709, 1975.

50. Fenoglio CM, Ferenzy A, Richart RM: Mucinous tumors of the ovary. Ultrastructural features of mucinous cystadenocarcinoma. *Am J Obstet Gynecol* 125:990, 1976.

51. Kaye GI, Fenoglio CM, Pascal RR, et al: Comparative electron microscopic features of normal, hyperplastic and adenomatous human colonic epithelium. Variations in cellular structures relative to the process of epithelial differentiation. *Gastroenterology* 64:926, 1973.

52. Fischer ER, Sahrkey D: The ultrastructure of colonic polyps and cancer with special reference to the epithelial inclusion bodies of Leuchtenberger. *Cancer* 15:160, 1962.

53. Cummins PA, Fox H, Langley FA: An electron microscopic study of the endometrioid adenocarcinoma of the ovary and a comparison of its fine structure with that of normal endometrium and of adenocarcinoma of the endometrium. *J Pathol* 113:165, 1974.

54. Hertig AT, Adams EC: Studies on the human oocyte and its follicle. I. Ultrastructural and cytochemical observations on the primordial follicle stage. *J Cell Biol* 34:647, 1967.

55. Peters H, Pederson T: Origin of follicle cells in the infant mouse. *Fertil Steril* 18:309, 1967.

56. MacAulay MA, Welicky I, Schulz RA: Ultrastructure of a biosynthetically active granulosa cell tumor. *Lab Invest* 17:562, 1967.

57. Pedersen PH, Larsen JF: Ultrastructure of a granulosa cell tumour. *Acta Obstet Gynecol Scand* 49:105, 1970.

58. Gondos B, Monroe SA: Cystic granulosa cell tumor with massive hemoperitoneum. Light and electron microscopic study. *Obstet Gynecol* 38:683, 1971.

59. Bjersing L, Frankendal B, Angstrom T: Studies on a feminizing ovarian mesenchymoma (granulosa cell tumor). I. Aspiration biopsy cytology, histology and ultrastructure. *Cancer* 32:1360, 1973.

60. Gondos B: Ultrastructure of a metastatic granulosa-theca cell tumor. *Cancer* 24:954, 1969.

61. Kempson RL: Ultrastructure of ovarian stromal cell tumors. *Arch Pathol* 86:492, 1968.

62. Berendsen PB, Smith EB, Abell MR, et al: Fine structure of Leydig cells from an arrhenoblastoma of the ovary. *Am J Obstet Gynecol* 103: 192, 1969.

63. Jensen AB, Fechner RE: Ultrastructure of an intermediate Sertoli-Leydig cell tumor. A histogenetic misnomer. *Lab Invest* 21:527, 1969.

64. Murad TM, Mancini R, George J: Ultrastructure of a virilizing ovarian Sertoli-Leydig cell tumor with familial incidence. *Cancer* 31:1440, 1973.

65. Kalderon AE, Tucci JR: Ultrastructure of a human chorionic gonadotropin- and adrenocorticotropin-response functioning Sertoli-Leydig cell tumor (type I). *Lab Invest* 29:81, 1973.

66. Roth LM, Cleary RE, Rosenfield RL: Sertoli-Leydig cell tumor of the ovary with an associated mucinous cystadenoma. An ultrastructural and endocrine study. *Lab Invest* 31:648, 1975.

67. Ramzy I, Boss C: Sertoli cell tumors of ovary. *Cancer* 38:2447, 1976.

68. Schulze C: On the morphology of the human Sertoli cell. *Cell Tiss Res* 153:339, 1974.

69. Ferenczy A, Richart RM: The fine structure of the gonads in the complete form of testicular feminization syndrome. *Am J Obstet Gynecol* 113:339, 1972.

70. Able ME, Lee JC: Ultrastructure of a Sertoli cell adenoma of the testis. *Cancer* 23:481, 1969.

71. Obregon EB, Esponda P: Ultrastructure of the nucleus of human Sertoli cells in normal and pathological testes. *Cell Tiss Res* 152:467, 1974.

72. Green JA, Maqueo M: Histopathology and ultrastructure of an ovarian hilar cell tumor. *Am J Obstet Gynecol* 96:478, 1966.

73. Merkow LP, Slifkin M, Acevedo HF, et al: Ultrastructure of an interstitial (hilar) cell tumor of the ovary. *Obstet Gynecol* 37:845, 1971.

74. Hameed K: Brenner tumor of the ovary with Leydig cell hyperplasia. A histologic and ultrastructural study. *Cancer* 30:945, 1972.

75. Koss LG, Rothschild EO, Fleisher M, et al: Masculinizing tumor of the ovary, apparently with adrenocortical activity. A histologic, ultrastructural and biochemical study. *Cancer* 23:1245, 1969.

76. Dadoune JP, Galian P, Steg A, et al: Leydig-cell adenoma of the testis. A histological, ultrastruc-

tural, histoenzymatic and biochemical study of a case. *Arch Anat Pathol* (Paris) 15:322, 1967.

77. Garcia-Bunuel R, Brandes D: Luteoma of pregnancy: Ultrastructural features. *Hum Pathol* 7:205, 1976.

78. Magalhaes MC: A new crystal-containing cell in human adrenal cortex. *J Cell Biol* 55:126, 1972.

79. Roth LM, Sternberg WH, Huseby RA, et al: Transplantable luteoma of the mouse. An ultrastructural and biochemical study. *Lab Invest* 27:115, 1972.

80. Rosai J, Khodadoust K, Silber I: Spermatocytic seminoma. II. Ultrastructural study. *Cancer* 24:103, 1969.

81. Lynn RA, Varon HH, Kinglsey WB, et al: Ultrastructural and biochemical studies of estrogen secretory capacity of a "nonfunctional" ovarian neoplasm (dysgerminoma). *Am J Pathol* 51:639, 1967.

82. Overbeck L, Philipp E: Die Ultrastruktur des Disgerminoms im Ovar. Zugleich eim Beitrag zur Histogenese des Tumors. *Z Geburtsh Gynaekol* 170:125, 1969.

83. Kay S, Silverberg SG, Schatzki PF: Ultrastructure of an ovarian dysgerminoma: Report of a case featuring neurosecretory-type granules in stromal cells. *Am J Clin Pathol* 58:458, 1972.

84. Jensen KH, Kempson RL: The ultrastructure of gonadoblastoma and dysgerminoma. *Hum Pathol* 5:79, 1974.

85. Pierce GB Jr: Ultrastructure of human testicular tumors. *Cancer* 19:1963, 1966.

86. Holstein AF, Korner F: Light and electron microscopical analysis of cell types in human seminoma. *Virchows Arch Pathol Anat Histol* 363:97, 1974.

87. Ramsey HJ: Ultrastructure of a pineal tumor. *Cancer* 18:1014, 1965.

88. Levine GD: Primary thymic seminoma—a neoplasm ultrastructurally similar to testicular seminoma and distinct from epithelial thymoma. *Cancer* 31:729, 1973.

7

Ultrastructural Changes of Peripheral Nerve

Peter W. Lampert, M.D.
Department of Pathology
University of California, San Diego
La Jolla, California

Sydney S. Schochet, Jr., M.D.
Department of Pathology
University of Texas Medical Branch
Galveston, Texas

INTRODUCTION

Morphological studies of nerve biopsies include quantitative histometric evaluations, studies of teased fiber preparations, and electron microscopic examination. In this chapter the diagnostic significance of electron microscopy is emphasized. We will describe and illustrate ultrastructural changes in human and experimental neuropathies and discuss the findings according to primary or predominant involvement of either axons, myelin sheaths, Schwann cells, or interstitial tissue.

The sural nerve at the level of the ankle is most commonly chosen for biopsy examination (1,2). It is a readily accessible sensory nerve that can be partially resected (fascicular biopsy) with minimal discomfort to the patient. The surgical removal of the nerve or its fascicles should be accomplished with the least mechanical distortion. The resected nerve is immersed in phosphate-or cacodylate-buffered 2.5% glutaraldehyde for at least 4 hours. Mincing of unfixed nerve should be avoided because it produces severe artifacts of myelin sheaths and axons. After fixation, the nerve is cut into small blocks for further processing, i.e., several rinses in buffer and postfixation in osmium. Light microscopic evaluation of one-micron-thick, cross and longitudinal sections from all blocks precedes thin sectioning for electron microscopy. Thick sections stained with paraphenylenediamine (3) can be used for histometric studies in regard to number and size of myelinated and unmyelinated nerve fibers. Teased fiber preparations are obtained by dissecting the osmium-fixed nerve in glycerin or medium-hard epoxy (1,4).

309

NORMAL PERIPHERAL NERVE

Peripheral nerve trunks consist of nerve fascicles surrounded by the *epineurium,* a condensation of connective tissue composed of fibroblasts and collagen fibers (Fig. 1). The epineurium is traversed by large arteries, veins, and lymphatics that run longitudinally between fascicles. Each nerve fascicle is enclosed by the *perineurium,* a sheath composed of several layers of flattened cells (Fig. 2). A very thick basal lamina covers the surface of perineural cells, which are further characterized by abundant micropinocytotic vesicles, cytoplasmic filaments, and tight junctions between cells. Functionally, the perineurium acts as a perifascicular diffusion barrier to macromolecular proteins (5,6). The *endoneurium* consists of the intrafascicular tissue between myelinated and unmyelinated nerve fibers. Collagen fibers and fibroblasts may subdivide the fascicles into compartments. Occasional mast cells are encountered (7). Endoneurial vessels consist of capillaries and venules lined by tightly joined endothelial cdlls. Lymphatics are absent (8). Functionally, the endothelial lining of endoneurial vessels constitues a diffusion barrier to proteins, as well as to some nonproteinaceous substances of low molecular weight (5,9). Corpuscles of obscure significance (Renaut bodies) composed of concentric arrays of attenuated cytoplasmic processes of fibroblasts, collagen fibers, and amorphous and finely fibrillar material are occasionally observed in the endoneurium adjacent to the perineural sheath (10).

In the sural nerve, small unmyelinated axons measuring 0.2 to 3 μm in width outnumber myelinated axons by a 4 to 1 ratio (2). The size of the myelinated axons range from 2 to 14 μm, with a bimodal distribution showing peaks at 4 and 11 μm (2). A single Schwann cell and its surrounding basal lamina enclose several unmyelinated axons (Fig. 3). Longitudinal sections show a continuous row of Schwann cells that cover unmyelinated axons without interruption, due to overlapping processes of adjacent Schwann cells. In contrast, a single myelinated axon is enclosed by one Schwann cell per myelin segment that may vary in length from 0.3 to 1.5 μm. Gaps or nodes of Ranvier separate myelin segments (Fig. 4). The length of the nodal gap is inversely proportional to the diameter of the axon and varies from about 0.3 to 2 μm. At nodes of Ranvier, axons are in contact with extracellular space. The nodes are bridged by the basal lamina that forms a continuous sheath around single myelinated axons. This tubular investment of nerve fibers by the continuous basal lamina is called the neurilemmal sheath (11).

The *Schwann cell* has an elongated nucleus oriented along the length of the axon. The cytoplasm contains mitochondria, ribosomes, rough and smooth endoplasmic reticulum, filaments, microtubules, occasional glycogen granules, myelin-like inclusions (Elzholz bodies), and π granules of Reich (Figs. 5–7). The latter consist of membrane-bound lamellar bodies associated with electron-dense amorphous material. These rod-shaped, metachromatically staining bodies are associated with acid phosphatase activity suggestive of their lysosomal nature (12). A narrow rim of Schwann cell cytoplasm always surrounds the myelin sheath. The narrow space between the plasma membranes of overlapping tongues of Schwann cell cytoplasm is called the mesaxon (Fig. 6).. The *myelin sheaths* are formed and maintained by Schwann cells. They consist of myelin lamellae that are compactly wrapped around axons. The number of lamellae or the thickness of the sheath is proportional to the diameter of the axon. A myelin lamella is formed by the fusion of the inner surface of the plasma membrane of

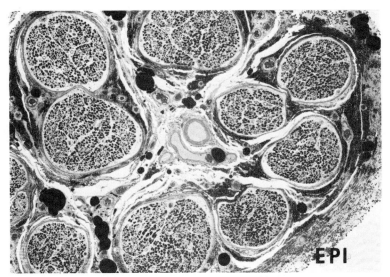

Figure 1. Sural nerve containing nerve fascicles surrounded by the epineurium (EPI), which is composed of condensed connective tissue. Large vessels are located in the center of the nerve between fascicles. (Paraphenylene diamine, ×40)

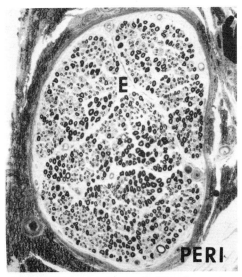

Figure 2. Nerve fascicle of sural nerve surrounded by the perineurium(PERI), which is composed of layers of flattened cells that act as a diffusion barrier. Myelinated and unmyelinated axons are located within the endoneurium (E), which also contains vessels, fibroblasts, mast cells, and collagen fibers. (Paraphenylenediamine, ×120)

311

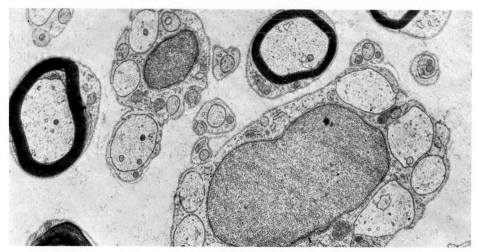

Figure 3. Endoneurium containing myelinated and unmyelinated axons. Several unmyelinated axons are embedded within the same Schwann cell. In contrast, only one myelinated axon is enclosed by a single Schwann cell. (×12,000)

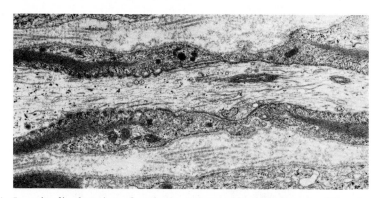

Figure 4. Longitudinal section of a myelinated axon showing a node of Ranvier. (×8,000)

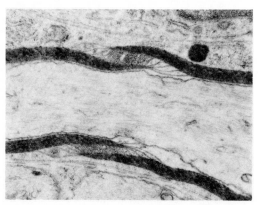

Figure 5. Longitudinal section of a myelinated axon showing a Schmidt-Lanterman incisure in the myelin sheath where major dense lines open and enclose Schwann cell cytoplasm. (×12,000)

312

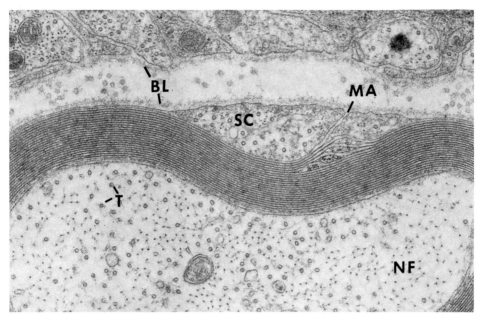

Figure 6. Cross section of myelinated axon. The axoplasm contains neurofilaments (NF) and tubules (T), as well as endoplasmic reticulum and mitochondria. The myelin sheath is cut through the outer portion of a Schmidt-Lanterman incisure where major dense lines split to enclose Schwann cell cytoplasm. A basement lamina (BL) covers the surface of the Schwann cell bridging the outer mesaxon (MA). (×38,500)

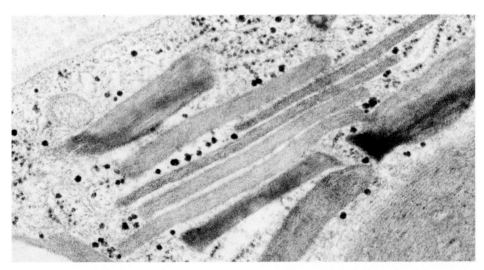

Figure 7. Lamellated electron dense bodies (π granules of Reich) and glycogen granules in Schwann cell in the sural nerve from a patient with chronic neuropathy. (×54,000)

the Schwann cell, resulting in a very dense osmophilic layer or major dense line flanked by less or minor dense lines. The spiral wrapping and compaction of myelin lamellae around the axon produce a sheath that reveals regularly spaced major and minor dense lines with a periodicity of about 13 to 17 nm, depending on methods of fixation and embedding. A split of the minor or intraperiod line may be observed at high resolution (13). In cross section, the outer- and inner-most myelin lamellae can be shown to connect with the plasma membrane of the Schwann cell (Fig. 6). At nodes of Ranvier, myelin lamellae terminate in tongues of Schwann cell cytoplasm, most of which adhere to a series of ridges on the surface of the axon (14). Stacks of desmosome-like densities linking adjacent terminal tongues of Schwann cell cytoplasm may be observed at some nodes (15). Longitudinal sections further demonstrate the presence of Schmidt-Lanterman incisures (Fig. 5), the number being proportional to the thickness of the sheath. In these regions, each major dense line opens and encloses a portion of Schwann cell cytoplasm (16). The *axon* is confined by a surface membrane, the axolemma. The axoplasm contains neurotubules and filaments (Fig. 6) measuring 25 and 10 nm in width, respectively. Suggestive evidence for their involvement in axoplas-mic transport mechanisms has been published (17). Other axoplasmic organelles comprise mitochondria, vesicles of smooth endoplasmic reticulum, and occa-sionally a lamellar electron-dense body. Vesicles with dense granular cores occur in autonomic nerve fibers (18).

AXONAL CHANGES

Although Schwann cells and myelin sheaths undergo changes when axons de-generate, electron microscopy can reveal early alterations that are limited to the axoplasm and thus indicative of a primary neuronal or axonal disturbance. It is further possible to recognize patterns of axoplasmic alterations similar to those consistently developing in transected nerves at proximal and distal axonal stumps (reactive axonal enlargement), in distal degenerating axons (Wallerian degeneration), and in regenerating axonal sprouts. Less clearly defined are changes seen in neuropathies associated with metabolic or nutritional distur-bances of axons or neurons. In these conditions, morphological changes similar to those seen in reactive, degenerating, and regenerating axoplasm may occur, and often predominate in the most distal parts of nerves (dying back neuropathies). In order to emphasize the difference between this type of axonal degeneration and that of Wallerian degeneration, we discuss these neuropathies under the heading of axonal dystrophy. Finally, there are neuropathies in which axons are lost or decreased in size without visible axoplasmic alterations. The term axonal atrophy is reserved for these conditions.

Reactive Axonal Enlargement

Mechanical damage to the axon, such as ligation, crush, or transection of nerves, causes axonal enlargements at or near the site of injury (19,20). Resected seg-ments of nerve fibers develop bulbous swellings at both ends (20). Axoplasmic organelles, i.e., mitochondria and vesicular and tubular elements of smooth endoplasmic reticulum, as well as lamellar electron-dense bodies derived from degenerating mitochondria, accumulate in the enlarged stumps of transected axons (21,22,23). Similar changes develop in nerve fibers adjacent to areas of necrosis produced by vascular or inflammatory lesions. Axoplasmic organelles also accumulate in axons next to a region in which axoplasmic flow has been arrested by agents that disrupt microtubules, such as vincristine (24). Intact myelin sheaths and Schwann cells may surround reactive enlargements at a distance away from the area of trauma, ischemia, or inflammation (Fig. 8). Unmyelinated axons react to injury in the same manner (25).

Wallerian Degeneration

In 1850, Waller observed that the distal part of a transected nerve degenerates (26). By light microscopy, the disintegrating nerve fibers are recognized by finding linear rows of myelin and axonal fragments contained within the

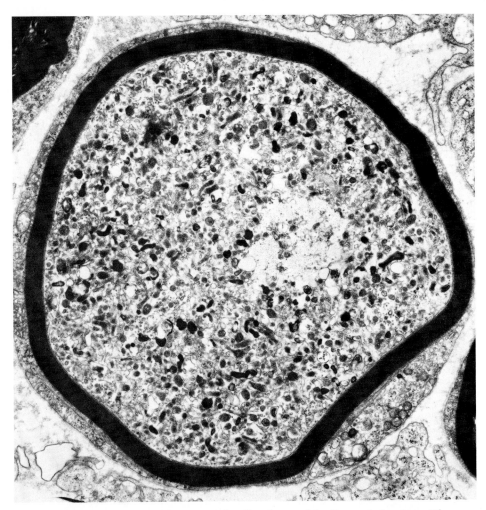

Figure 8. Reactive axonal enlargement in allergic neuritis of rat sciatic nerve. The axoplasm is filled with mitochondria, lamellar electron-dense bodies, and vesicular elements. (×12,000)

neurilemmal tube. Ultrastructural studies demonstrate the accumulation of axoplasmic organelles in axonal stumps prior to visible changes in the more distal parts where neurotubules and filaments may still be visible in some axons 48 hours after transection (21). Thereafter, most axons show either hydropic swelling or condensation of disintegrating axoplasm (Fig. 9). The axolemma ruptures, and neurofilaments, tubules, and endoplasmic reticulum undergo granular dissolution. Mitochondria remain recognizable for longer periods but eventually also dissolve. The myelin sheaths around degenerating axons collapse and break up into lamellar ovoid configurations that are enclosed within Schwann cell cytoplasm (Fig. 10). These changes are first visible in paranodal regions and are associated with an increased formation of Schmidt-Lanterman incisures (27). Schwann cells rapidly proliferate and engulf myelin and axonal debris within membrane-bound vacuoles or autophagosomes (23,28). The proliferating cells are rich in ribosomes and granular endoplasmic reticulum. At later stages, macrophages are noted in the endoneurium. Their derivation from hematogenous mononuclear cells rather than from Schwann cells has been demonstrated (29).

In advanced stages, columns of Schwann cells enclosed by a persistent basal lamina (Büngner's bands) represent the only yet revealing vestige of a degenerated nerve fiber. Several months after Wallerian degeneration, the Schwann cell processes of Büngner's bands reveal rather clear cytoplasm containing filaments, occasional microtubules, and glycogen granules (30,31). Schwann processes are also seen wrapped around bundles of collagen. The finding of Büngner's bands

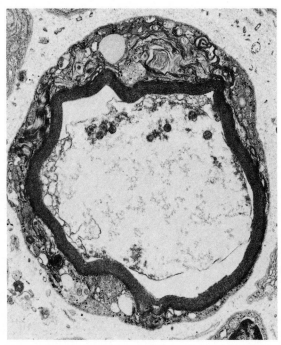

Figure 9. Degenerating axon showing hydropic swelling of the disintegrating axoplasm in sural nerve from a patient with a chronic neuropathy. (×6,000)

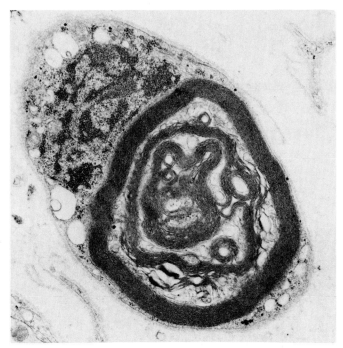

Figure 10. Degenerated axon showing collapse and disruption of the surrounding myelin sheath. Sural nerve from a patient with a chronic neuropathy. (×16,000)

(Fig. 11), i.e., Schwann cell processes within a common basal lamina, is diagnostic of a completely degenerated nerve fiber. Any condition that causes an interruption of axons or a loss of the parent nerve cell will leave such revealing traces. Remnants of this type occur in sural nerves of older patients (32). The finding of collapsed or disintegrating myelin sheaths around clear, watery, or dark, clumped axons containing disintegrating axoplasmic organelles (Figs. 9 and 10) suggests a more recent, destructive injury to either axons or neurons.

Axonal Regeneration

After nerve transection, regenerating axons sprout from proximal nerve stumps and proceed to grow along supporting stroma (33,34,35). Effective regeneration is accomplished if sprouts grow within the perineural sheaths, along or within Büngner's bands, which guide the fibers to appropriate end organs. The identification of regenerating axons may be difficult. The sprouts often contain an increased number of neurofilaments (Fig. 12), and close to their tips or growth cones, they may show more numerous mitochondria, vesicular profiles, and electron-dense bodies. Regenerating axons in experimental neuropathies have been described to contain stacks of endoplasmic reticulum (Fig. 13) and, occasionally, giant mitochondria with abundant cristae (35). The myelin sheaths around regenerated fibers rarely regain the same thickness and internodal length as seen around normal fibers (36). In neuropathies that are not associated with disruptive lesions of the peri- and endoneurium, regeneration proceeds

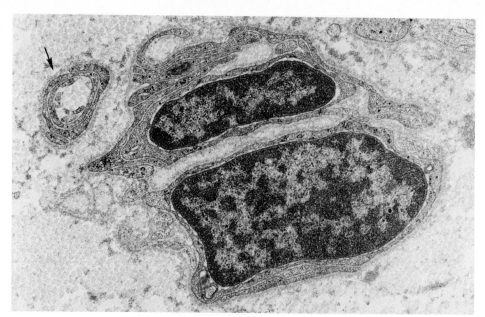

Figure 11. Denervated Schwann cell cords (Büngner's bands) in sural nerve of a patient with chronic neuropathy. The "resting" Schwann cells are rich in filaments. Also note Schwann cell processes encircling collagen (arrow). (×20,000)

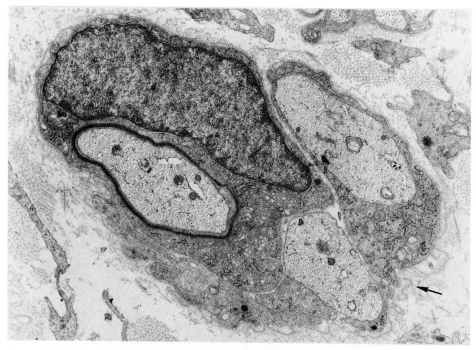

Figure 12. Regenerating axonal sprouts surrounded by a common, ruffled basement lamina (arrow) in sciatic nerve of rat with alloxan diabetic neuropathy. The proliferating Schwann cells are rich in ribosomes and granular endoplasmic reticulum. One of the axonal sprouts is only partially covered by Schwann cell cytoplasm, and another has already acquired a thin myelin sheath. Note increased numbers of neurofilaments in the axoplasm. (×8,000)

318

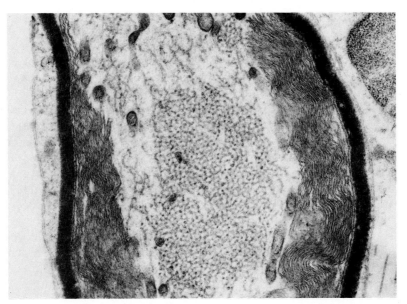

Figure 13. Proliferation of endoplasmic reticulum in a regenerating axon derived from rat dorsal root after transection. (×8,000)

more efficiently since the regenerating sprouts are confined within Büngner's bands. The Schwann cells within Büngner's bands proliferate and begin to wrap around each advancing axon. Some of the regenerating axons may only be partially covered by Schwann cell cytoplasm, whereas others may already have acquired a thin myelin sheath. The activated Schwann cells reveal abundant cytoplasm, rich in ribosomes and granular endoplasmic reticulum. The finding of a bundle of small axons, each surrounded by a proliferated Schwann cell and all confined within the persistent neurilemmal sheath of a Büngner's band, is diagnostic of regenerating axons (Fig. 12).

Axonal Dystrophy

Under this heading, we lump neuropathies in which axonal degeneration proceeds more slowly as compared to the rapidly developing axoplasmic disintegration of Wallerian degeneration that occurs after interruption of axons or acute necrosis of neurons. Metabolic, nutritional, and toxic disturbances of axons or nerve cells have been shown to cause such neuropathies, which frequently involve predominantly the distal parts and terminal ends of axons, with little or no visible changes of proximal axons or nerve cells (37). The concomitant involvement of Schwann cells in some of these neuropathies suggests that the metabolic disturbance is not limited to axons. Diabetic, uremic, and alcoholic neuropathies may show these features (38,39). Neuropathies associated with hypothyroidism (40), intoxications by isoniazid (41), acrylamide (42), vinca alkaloids (43), tri-ortho-cresyl-phosphate (44,45), or vitamin deficiencies (46,47,48) are other examples. In diseases of as yet unknown inborn errors of metabolism, such as giant axonal neuropathy (49) and infantile neuroaxonal dystrophy (50), the dystrophic axoplasmic alterations are morphologically most striking.

Because of the variety of ultrastructural changes that predominate in the degenerating axons of these neuropathies, a multitude of different factors must play a role in their pathogenesis. An impairment of axoplasmic flow has been postulated as the most likely cause for the selective involvement of distal axonal segments. Morphologically, the neuropathies show a spectrum of changes. Some axons demonstrate features indistinguishable from Wallerian degeneration or reactive axonal changes (Fig. 14). Others reveal alterations similar to those seen in regenerating axons. Often a particular axoplasmic organelle is predominantly affected. A striking proliferation of endoplasmic reticulum has been described in tri-orth-cresyl-phosphate neuropathy (44,45) and in thiamine deficiency (46). An accumulation of neurofilaments accounts for the enormous enlargement of axons in giant axonal neuropathy (Fig. 15). In this regard, it is of interest that intoxication with vinca alkaloids and hexacarbon compounds produces similar aggregates of neurofilaments, causing an arrest of axoplasmic flow (24,37,51). Various mitochondrial alterations have been described. The intramitochondrial deposition of glycogen is most commonly observed (39,40,50,52,53). Glycogen collects between the outer and inner mitochondrial membrane and eventually forms large membrane-bound aggregates that almost completely fill the width of

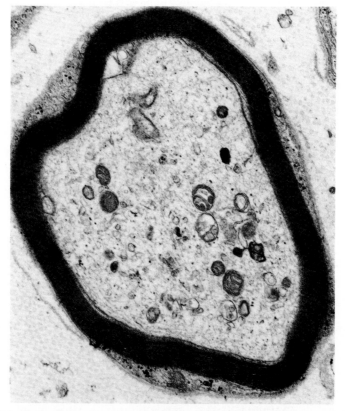

Figure 14. Myelinated axon containing an increased number of filaments, mitochondria, vesicular structures, and dense bodies in the sural nerve from a patient with chronic neuropathy. Axoplasmic changes of this type occur in degenerating and reactive dystrophic axons. (× 16,000)

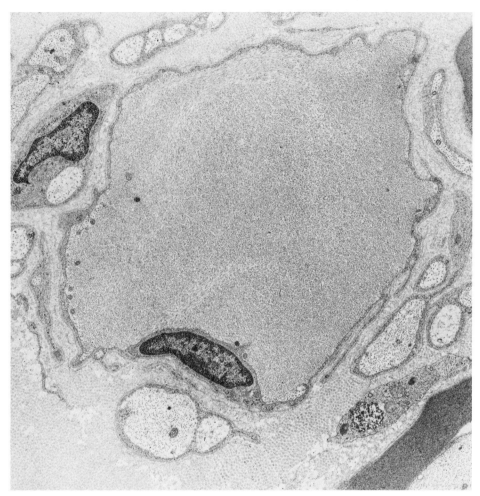

Figure 15. Demyelinated axonal enlargement densely packed with neurofilaments in sural nerve from a child with giant axonal neuropathy. (×6,000)

an axon as here illustrated in alloxan diabetic neuropathy (Fig. 16). In this condition we also observed the intra-axonal deposition of polyglucosans (Fig. 17) that are known to be stored in the liver and muscle in glycogenesis type IV (54) and in neurons in myoclonus epilepsy (55). Intramitochondrial lipid deposits are common in dystrophic axons in vitamin E deficiency (56). The changes in infantile neuroaxonal dystrophy are most bizarre, consisting of axonal enlargements filled with layered loops of membranes, branched tubular profiles, abnormal mitochondria, and lipid inclusions (50).

Axonal Atrophy

Small atrophic axons occur in chronic neuropathies characterized by a slowly progressive loss of nerve cells or axons. These conditions are distinct from Wallerian and dystrophic axonal degeneration by the lack or rare visualization of

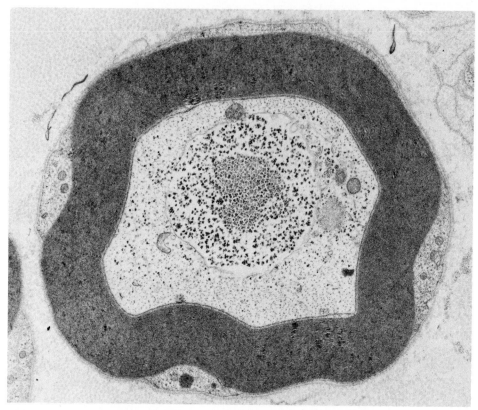

Figure 16. Intra-axonal accumulation of glycogen in rat sciatic nerve in alloxan diabetic neuropathy. (×12,000)

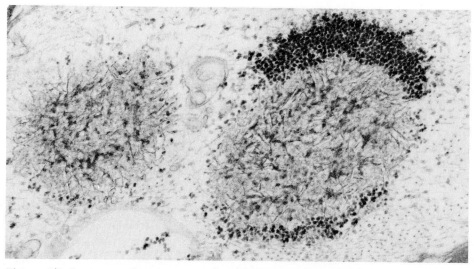

Figure 17. Intra-axonal aggregates of polyglucosan deposits closely associated with glycogen granules in rat sciatic nerve in alloxan diabetic neuropathy. (×28,000)

disintegrating axons. They represent inherited disorders that are classified according to neuronal type or nerve involved. Electron microscopy is useful by demonstrating the sequelae of axonal degeneration, namely Büngner's bands, remnants of basal laminae, increased amounts of endoneurial collagen, and concentric layers of Schwann cell processes around longitudinally oriented collagen fibers (Fig. 11) or around single atrophic, myelinated, or unmyelinated axons. Axons of small diameter surrounded by a sheath that is thicker than expected for the size of the axon suggest axonal atrophy, as described in Friedreich's ataxia (57). Demyelinated and remyelinated axons with short internodes can be demonstrated in these chronic neuropathies (57). There is evidence suggesting that segmental demyelination develops subsequent to axonal atrophy (39). Concentric layers of Schwann cell processes frequently develop in conditions in which repeated demyelination and remyelination occur. Cycles of insidous axonal degeneration, demyelination, abortive regeneration, and remyelination may well occur in slowly progressive neuropathies and might account for concentric Schwann cell layers. However, such hypertrophic changes of Schwann cells (onion bulbs) more commonly indicate a disorder of Schwann cells, as described below.

CHANGES OF MYELIN SHEATHS

Different modes of myelin sheath disintegration are distinguished by light microscopy. Diffuse, focal, or perivascular demyelination with or without associated cellular infiltrates may be observed. Longitudinal sections and teased fiber preparations can reveal continuous, paranodal, or segmental demyelination of individual fibers. Electron microscopic examination adds further information by demonstrating initial changes in myelin sheaths or myelin supporting Schwann cells (primary demyelination), or in axons (secondary demyelination).

Primary Segmental Demyelination

Injuries affecting directly either the myelin sheath or its supporting Schwann cell may cause segmental demyelination. Primary alterations of myelin sheaths with little or no initial changes of Schwann cells have been described in experimental neuropathies after intoxication with hexachlorophene (58), after treatment with AY 9944, an inhibitor of cholesterol biosynthesis (59), and in experimental allergic neuritis (60).

Hexachlorophene intoxication of rats leads to paralysis due to the formation of large vacuoles within myelin sheaths of both central and peripheral nerve fibers (Figs. 18 and 19). The vacuoles develop between myelin lamellae after a split of the intraperiod or minor dense line of myelin sheaths. Paranodal regions are spared initially, but in chronic hexachlorophene neuropathy, the entire myelin segment is involved and eventually disintegrates. Thick myelin sheaths around large axons in proximal parts of nerves and in spinal roots are predominantly affected. The pathogenesis is related to the preferential binding of hexachlorophene to lipids (61). The changes are reversible, i.e., recovering animals show few or no intramyelinic vacuoles, and there are segments of thin myelin sheaths suggestive of remyelination (62).

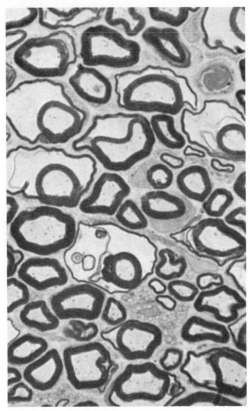

Figure 18. Hexachlorophene neuropathy. Note the presence of large intramyelinic vacuoles in the myelin sheaths of rat sciatic nerve. (×600)

Segmental demyelination occurs in developing peripheral nerves when suckling rats are treated with AY 9944, which inhibits cholesterol biosynthesis, resulting in an accumulation of 7-dehydroxycholesterol, a cholesterol precursor (59). Myelin sheaths disintegrate in the presence of intact axons and Schwann cells. Similar segmental and paranodal demyelination may also occur spontaneously but to a lesser degree during normal maturation of peripheral nerves (63). It has been suggested that such myelin disintegration in early postnatal weeks, particularly after treatment with AY 9944, is related to unstable myelin in which 7-dehydrocholesterol substitutes for cholesterol in the lipid cholesterol complex of myelin (59).

In experimental allergic neuritis, induced by sensitization to peripheral nerve myelin, segmental demyelination develops after initial, focal lysis of myelin lamellae (60,64). Early alterations consist of increased permeability of endoneurial vessels and infiltrates of mononuclear cells that pass through the walls of venules. The invading cells then reach out for myelinated axons, traverse the neurilemma, penetrate the outer mesaxon, separate the Schwann cell from the outer myelin lamella (Figs. 20 and 21), and after having surrounded the sheath, cause lysis of myelin lamellae (Fig. 22). At sites of myelinolysis, the lamellae either dissolve abruptly or are transformed into vesicular arrays of membranes.

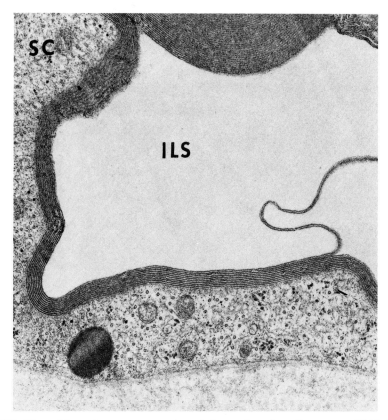

Figure 19. Hexachlorophene neuropathy. A large interlamellar space (ILS) has developed within the myelin sheath by the separation of myelin lamellae. The surrounding intact Schwann cell (SC) contains a lysosome. (×20,000)

After the focal dissolution of myelin lamellae, macrophages pass through the defect of the sheath and remove the remaining compact remnants of the internodal myelin segment. Single and multiple lamellae or the entire sheath are stripped by the invading phagocytes (Fig. 23–25). Finally, macrophages filled with myelin debris move out of the neurilemmal sheaths (Fig. 26), leaving behind a demyelinated but intact axon surrounded by Schwann cell processes that remyelinate the denuded axonal segment. The findings suggest that myelin sheaths, rather than the myelin supporting Schwann cells, are the target of the allergic reaction and that mononuclear cells are required to cause demyelination in allergic neuritis. In man, similar observations indicative of a diagnosis of allergic neuritis have been reported in idiopathic inflammatory polyradiculoneuropathy or the Landry-Guillain-Barré syndrome (65–68). The mechanisms responsible for triggering autosensitization against peripheral myelin in man are still obscure. The frequent association of antecedent infections, particularly with viruses that incorporate host membrane antigens in their envelopes, has been stressed (68).

Various injuries capable of destroying Schwann cells result in segmental demyelination. Some of the initial Schwann cell changes are diagnostic of specific

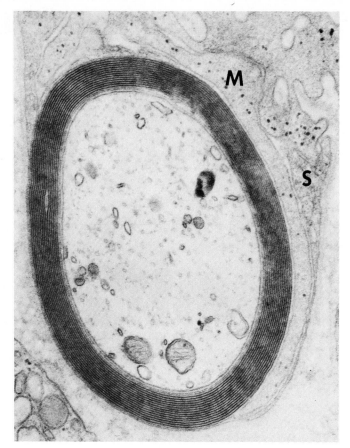

Figure 20. Penetration of mononuclear cell (M) between Schwann cell (S) and myelin sheath in experimental allergic neuritis of rat sciatic nerve, (×25,000)

diseases, as described below. The myelin sheaths related to degenerating Schwann cells disintegrate, beginning at paranodal regions. Macrophages traverse the neurilemmal sheath and strip both myelin and Schwann cell debris from the axon (69). The peeling of individual myelin lamellae or of thick layers of compact sheath by phagocytes represents a nonspecific change occurring whenever Schwann cells or their related myelin sheath are irreversibly damaged.

Paranodal Demyelination

Partial paranodal myelin disintegration followed by remyelination has been described, indicating that damage to a Schwann cell or its sheath does not always result in loss of the entire internodal segment. Paranodal changes of myelin sheaths have been explored in nerve injuries following mild nerve crush (70) and pneumatic tourniquet compression (71). If Schwann cell damage is limited to distal cytoplasmic portions, without involvement of nucleus and perikaryal cytoplasm, the cell survives in a truncated state and myelin disintegration occurs only at paranodal regions. Similar findings have been reported in diphtheric neuropathy (72). Early changes after administration of diphtheria toxin consist of swelling of Schwann cell processes that fill the nodal gap. Later the outer

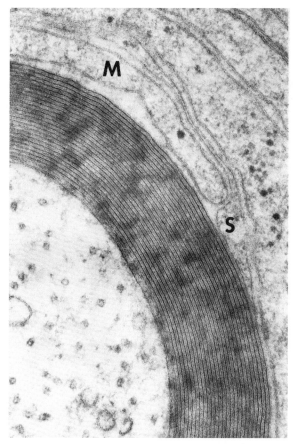

Figure 21. Advancing tongue of mononuclear cell (M) separating the Schwann cell (S) from its myelin sheath in experimental allergic neuritis of rat sciatic nerve. (×60,000)

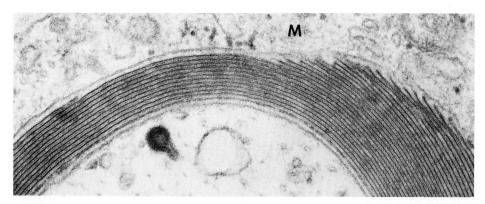

Figure 22. Lysis of myelin lamellae in contact with mononuclear cell (M) in experimental allergic neuritis of rat sciatic nerve. (×85,000)

327

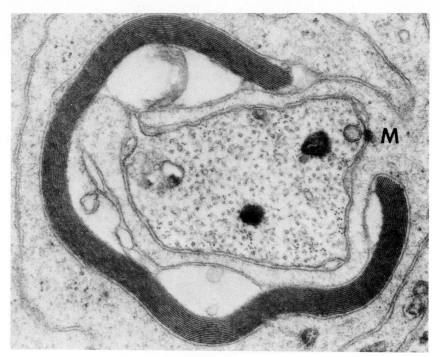

Figure 23. Macrophage (M) penetrating through a gap of the myelin sheath caused by focal myelinolysis. The remnant of the compact sheath is peeled off the axon by the macrophage. Allergic neuritis of rat sciatic nerve. (×25,000)

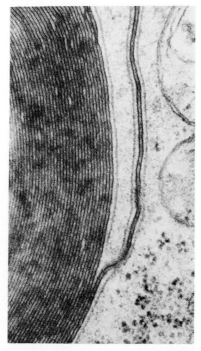

Figure 24. The cytoplasmic tongue of a macrophage is peeling two myelin lamellae off a damaged sheath in allergic neuritis of rat sciatic nerve. (×50,000)

328

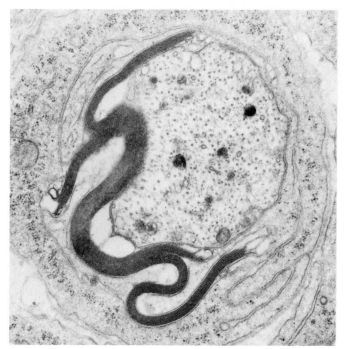

Figure 25. Partial removal of myelin sheath remnant by macrophage processes in allergic neuritis of rat sciatic nerve. (×20,000)

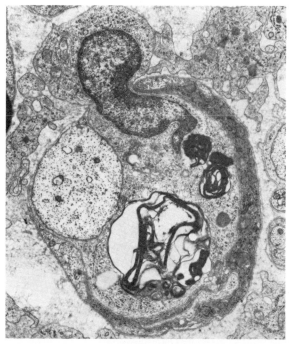

Figure 26. Completely demyelinated axon in allergic neuritis of rat sciatic nerve. Myelin debris is removed by a macrophage. Darker cytoplasmic processes beneath the neurilemma most likely represent Schwann cell processes. (×6,000)

myelin lamellae separate and detach from the axolemma, and the Schwann cell processes retract, causing a widening of the node. The inner myelin lamellae fragment, leaving remnants of membranes attached to the axolemma. During recovery, Schwann cells grow over the damaged paranodal region and remyelinate the widened node (73). Changes limited to paranodal regions mainly occur around large fibers with long internodal myelin segments.

Secondary Demyelination

Axonal degeneration is accompanied by either continuous or segmental loss of myelin sheaths. In Wallerian degeneration, myelin sheaths collapse and fragment into lamellar ovoids that are engulfed by the surrounding, proliferating Schwann cells. By light microscopy, such degenerating nerves show neurilemmal tubes, each containing a continuous row of myelin droplets. The presence of disintegrating axonal remnants, or the lack of visible axoplasm among the myelin debris, is easily recognized by electron microscopy of transverse sections, indicating myelin breakdown secondary to Wallerian degeneration.

In contrast, secondary segmental loss of myelin may occur around dystrophic or atrophic axons, suggesting that abnormal but still intact axons can affect myelin sheaths (39,46,57). In neuropathies with predominant involvement of distal nerve fibers, continuous rows of myelin ovoids consistent with Wallerian degeneration have been observed in the distal parts of nerves, whereas segmental demyelination occurred in the proximal portions (74,75). In uremic neuropathy, distal parts of the sural nerve at the level of the ankle have been compared with proximal portions from the midcalf region (39). Degenerating fibers with continuous linear myelin disintegration were abundant in the distal parts and rare in proximal portions. Teased fiber preparations revealed the presence of segmental demyelination around intact axons with decreased diameter. The paranodal and segmental demyelination was not random, but affected certain fibers along their course. The findings suggested that a decrease in axonal volume might induce this type of secondary segmental demyelination.

Remyelination

Regenerating axons and denuded axons after segmental and paranodal demyelination are remyelinated by Schwann cells in a fashion similar to that observed during normal myelination (76,77). Proliferating Schwann cells derived from adjacent cells by mitosis wrap around axons and form myelin lamellae by compaction of their cytoplasmic processes. The Schwann cell cytoplasm around remyelinating axons contains an increased amount of ribosomes and granular endoplasmic reticulum. Remyelinated axons rarely regain the same thickness and length of their former myelin segment. Remyelinated axons recognizable by thin myelin sheaths have been described many months after experimental demyelination (78). The length of remyelinated internodes remains shorter than the original myelin segments. A relationship between the length of a demyelinated segment and the number and length of remyelinated internodes exists (79). Short demyelinated segments or widened nodal gaps such as seen following paranodal demyelination are remyelinated by short intercalated myelin segments. Long stretches of demyelinated axons, as well as regenerated axons, are covered by longer remyelinated segments. Furthermore, remyelinated internodes of about the same length form around regenerated axons, whereas internodes of varying length develop around demyelinated fibers (80). The number and length of intercalated, remyelinated segments may thus provide clues regarding preceding axonal regeneration, or paranodal or segmental demyelination. There is also evidence suggesting that small paranodal defects are repaired without formation of intercalated myelin segments (73). Schwann cell cytoplasm and outer myelin lamellae of the affected myelin segment have been observed to project over the widened node, occasionally even overriding the Schwann cell of the adjacent internode (81).

SCHWANN CELL CHANGES

Alterations of Schwann cells frequently accompany or precede disturbances of myelin sheaths or axons. Nonspecific degenerative, reactive, and dystrophic changes, as well as alterations diagnostic of specific diseases, are recognized. Another reaction peculiar to Schwann cells and nerves subjected to repeated injuries leads to the formation of onion bulbs composed of concentric layers of Schwann cell processes.

Acute, *degenerative* changes consisting of hydropic cytoplasmic swelling have been described following a variety of injuries, e.g., in metabolic disturbances as seen in diabetic neuropathy (53) or after intoxication with lead (69) and diphtheria toxin (72). The swelling is associated with a loosening of myelin lamellae and the development of large intramyelinic vacuoles (Fig. 27). Another

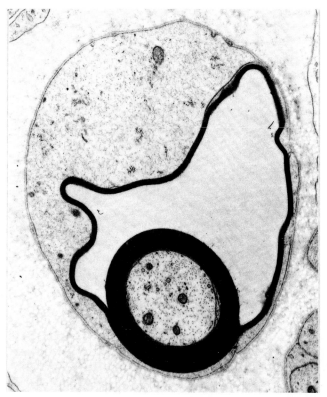

Figure 27. Hydropic degeneration of Schwann cell associated with intramyelinic accumulation of fluid in chronic lead neuropathy of rat sciatic nerve. (×8,000)

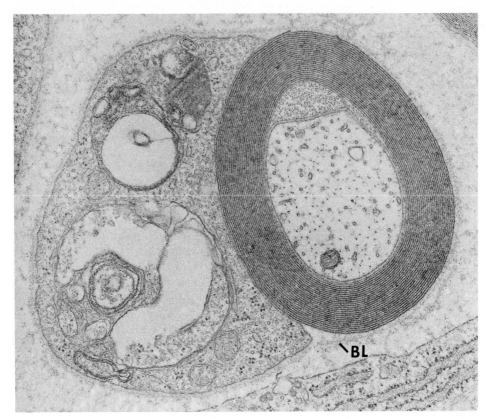

Figure 28. Vacuolar degeneration of Schwann cell in tellurium neuropathy of rat spinal roots. Note the retraction of the degenerated Schwann cell from the myelin sheath, leaving the outer myelin lamella in direct contact with the basal lamina (BL). (×40,000)

alteration consists of a focal degradation of cytoplasm followed by the formation of membranous whorls around the altered cytoplasm, as seen in tellurium neuropathy (82). The degenerating cytoplasm of Schwann cells retracts, leaving the outer myelin lamella in direct contact with the basal lamina (Fig. 28). Myelin disintegration secondary to the degeneration of Schwann cells begins in paranodal regions and at Schmidt-Lanterman incisures and proceeds to segmental demyelination (82).

Reactive Schwann cell changes have been studied during Wallerian degeneration (28–31). Early changes seen 12 to 24 hours after nerve transection consist of an increase in the number of Schmidt-Lanterman incisures, particularly in paranodal regions. The myelin sheath fragments and the lamellar debris are engulfed by proliferating Schwann cells. The myelin fragments are degraded within autophagosomes, i.e., membrane-bound cytoplasmic compartments with lysosomal properties (23). A striking increase in Schwann cells can be demonstrated in the degenerating nerves during the first week after transection (83). The proliferating cells are rich in ribosomes and granular endoplasmic reticulum. They may also show an increased amount of glycogen granules. At later stages, cytoplasmic filaments become more prominent at the expense of other organelles. In advanced stages, clear cytoplasmic processes rich in filaments are confined within the persistent neurilemmal tube. These columns of "resting"

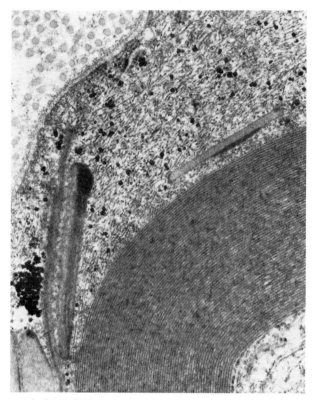

Figure 29. Accumulation of filaments and glycogen granules in a Schwann cell of the sural nerve of a patient with chronic neuropathy. (×30,000)

Schwann cells (Büngner's bands) may again revert to proliferating cells rich in ribosomes and granular endoplasmic reticulum when in contact with regenerating axonal sprouts. Similar changes occur in Schwann cells during demyelination and remyelination of axons. The finding of an increased amount of granular endoplasmic reticulum, filaments, and glycogen granules is therefore representative of a nonspecific reactive change (Figs. 29 and 30).

In *dystrophic* Schwann cells, an increased amount of normal or abnormal organelles is encountered. The πgranules of Reich, which consist of perinuclear, rod-shaped, lamellar lipid inclusions with a periodicity of about 5 to 6 nm (Fig. 7), are more frequent in Schwann cells of older patients and in persons with chronic neuropathies (84). Lamellar lipid deposits with a smaller periodicity of about 4 nm occur in Schwann cells of unmyelinated fibers and in paranodal regions of myelinated axons in diabetic neuropathy (85). Lipid droplets of varying electron density have been described in the cytoplasm of Schwann cells in neuropathies associated with hypothyroidism (40). In the latter, a striking accumulation of glycogen granules has also been stressed. Crystalline intramitochondrial inclusions consisting of 5-nm-wide, orderly arranged filaments have been observed in demyelinating neuropathies associated with altered lipid metabolism, such as Refsum's disease (86) and metachromatic leukodystrophy (87).

Some ultrastructural findings are diagnostic of specific diseases. In cutaneous

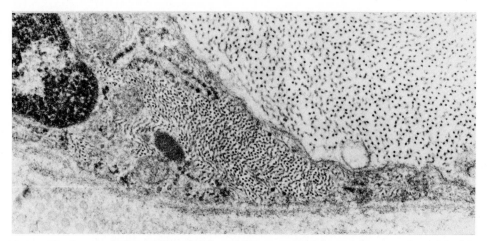

Figure 30. Accumulation of filaments in a Schwann cell of the sural nerve of a child with giant axonal neuropathy. (×60,000)

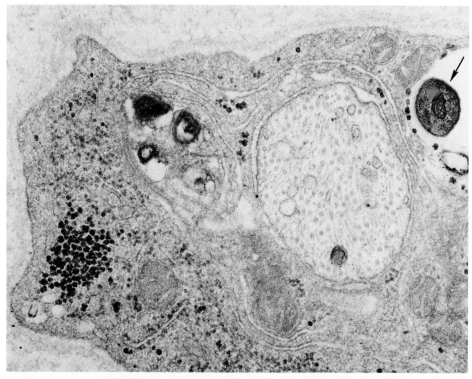

Figure 31. Demyelinated axon in cutaneous nerve from a patient with leprosy. Note glycogen granules and *Mycobacterium leprae* (arrow) in Schwann cell cytoplasm. (×60,000)

nerve biopsy specimens from suspected cases of leprosy, bacilli may be found within Schwann cells (Fig. 31). *Mycobacterium leprae* is known to show an affinity for Schwann cells and to cause segmental demyelination (88). Specific cytoplasmic inclusions diagnostic of metachromatic leukodystrophy (89,90), Krabbe's globoid cell leukodystrophy (91), or ceroid lipofuscinosis (92) have been described. In metachromatic leukodystrophy, complex, sulfatide inclusions consisting of zebra bodies and granular and lamellar structures, as well as round bodies of "tuff-stone" appearance, are encountered (Figs. 32 and 33). The myelin sheaths of the affected Schwann cells in metachromatic leukodystrophy often show loose outer myelin lamellae during early stages of demyelination (Fig. 34). In Krabbe's disease, the Schwann cells contain characteristic randomly distributed, curved, crystalline, and tubular profiles of galactocerebrosides within a clear matrix (Figs. 35,36). Ceroid lipofuscinosis can be diagnosed by finding typical inclusions composed of curvilinear bodies (Figs. 37 and 38). Characteristic lamellar spicules similar but not identical to the inclusions in Krabbe's disease

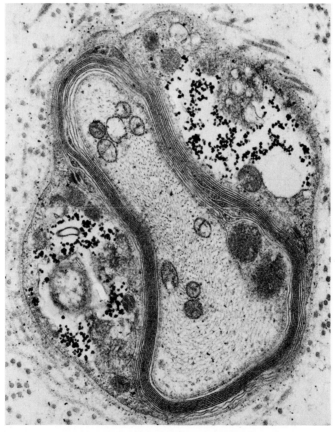

Figure 32. Myelinated axon of an intramuscular nerve from a patient with metachromatic leukodystrophy. Note collection of glycogen and lipid bodies in Schwann cell cytoplasm. A loosening of myelin lamellae is apparent. (×20,000)

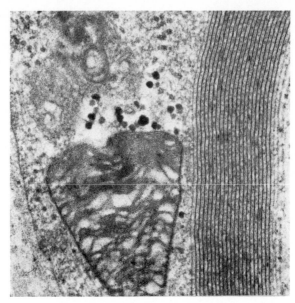

Figure 33. Glycogen granules and lamellated inclusion bodies in the Schwann cell cytoplasm of an intramuscular nerve from a patient with metachromatic leukodystrophy. (×50,000)

have been described in Schwann cells in adrenoleukodystrophy (93). In type II glycogenosis (Pompe's disease), Schwann cells may store glycogen in membrane-bound vacuoles (94).

Onion bulbs, characterized by multiple, concentric layers of Schwann cell processes, represent a histological reaction indicative of repetitive nerve injuries. They are the hallmark of hypertrophic neuropathies in Dejerine-Sottas and Refsum's disease (86, 95–97) but also occur in other conditions associated with repeated demyelination and remyelination such as diabetic neuropathy (98). The pathogenesis of onion bulb formation has been explored following nerve compression by repeated tourniquet applications (99) and in experimental lead neuropathy (69) in which Schwann cell degeneration and segmental demyelination occur. The basal lamina or neurilemma around demyelinated axons persists (Fig. 39). Proliferating Schwann cells derived from dividing cells of adjacent internodal segments wrap around the denuded axons and form new myelin sheaths (Fig. 40). Other cells grow along the persistent basal lamina creating an outer ring of Schwann cell processes. Repeated injury may cause necrosis of either the Schwann cell that remyelinated the central axon or the Schwann cell that grew along the original basal lamina (Fig. 41). Another persistent basal lamina remains after the disintegration of the second order of Schwann cells and serves as an additional scaffolding for the next wave of proliferating cells. Repeated demyelination and remyelination or degeneration and regeneration of Schwann cells eventually leads to numerous concentric layers of Schwann cell processes, as observed in Dejerine-Sottas hypertrophic neuropathy (Fig. 42). A remyelinated axon or longitudinally arranged bundles of collagen fibers are

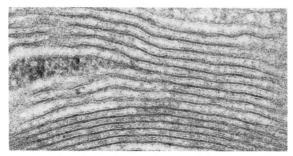

Figure 34. Even separation of lamellae of myelin sheath in metachromatic leukodystrophy. (×120,000)

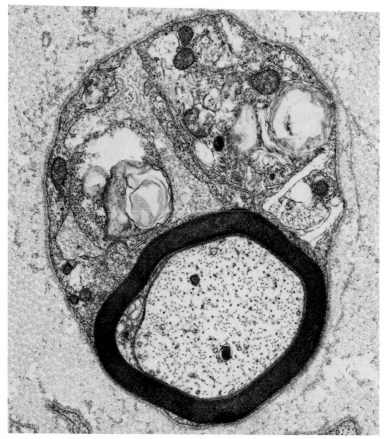

Figure 35. Myelinated axon in sural nerve from a patient with Krabbe's leukodystrophy. The Schwann cell is filled with cerebroside deposits. (×14,000)

337

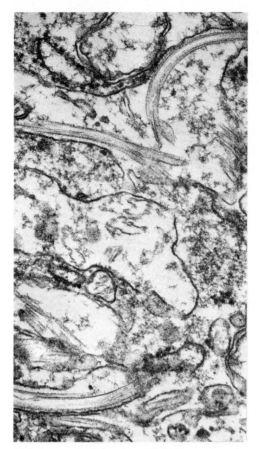

Figure 36. Curved crystalline inclusions in Schwann cell cytoplasm in Krabbe's leukodystrophy. (×60,000)

found in the center of a large onion bulb. Regenerating axons are occasionally encountered in the center or in the concentric peripheral layers of onion bulbs.

CHANGES OF INTERSTITIAL TISSUE

Nerve fibers may be damaged by infectious, vascular, or neoplastic disorders of interstitial tissue, but rarely is ultrastructural confirmation required. Electron microscopy contributes by demonstrating changes in the amount of collagen, number of fibroblasts, mast cells, and macrophages, and particularly also in regard to alterations of endoneurial vessels.

Nerves of older patients contain more fibroblasts and collagen fibers (2). Occasional remnants of basal lamina, most likely derived from degenerating Schwann cells, are encountered. There may also be an occasional macrophage containing lipid inclusions. Mast cells are more numerous (7). These cells have been shown to increase in number in distal degenerating nerve segments during Wallerian degeneration (7). They are also more prominent in chronic neuropathies, par-

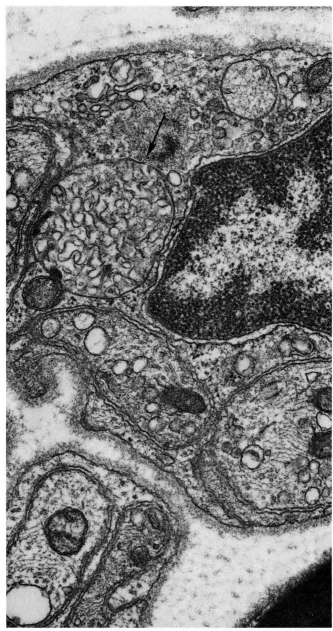

Figure 37. Unmyelinated axons in sural nerve of a patient with ceroid lipofuscinosis. The Schwann cell contains an inclusion filled with curvilinear bodies (arrow). (×37,000)

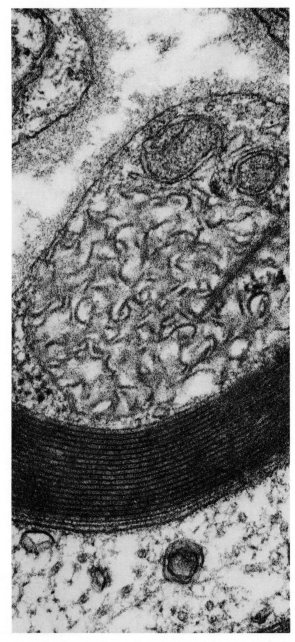

Figure 38. Schwann cell of a myelinated axon containing an inclusion filled with curvilinear bodies. Sural nerve of a patient with ceroid lipofuscinosis. (×85,000)

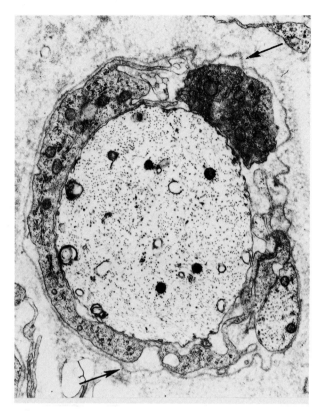

Figure 39. Demyelinated axon partially covered by proliferating Schwann cell processes that are located between the axon and the surrounding ruffled basal lamina that persisted after the degeneration of the Schwann cell that originally enclosed the axon. Chronic lead neuropathy of rat sciatic nerve. (×12,000)

ticularly those of the hypertrophic variety (100). In chronic neuropathy due to systemic amyloidosis, deposits of amyloid fibrils can be demonstrated around epineurial vessels, between perineurial cells, and as patchy deposits within the endoneurium (Figs. 43 and 44). Amyloid fibrils mingle with collagen fibers and attach to the basal lamina of vessels and Schwann cells (101, 102).

Endoneurial vessels may show ultrastructural changes of endothelial cells, basement membrane, and pericytes. In inflammatory conditions, such as allergic neuritis (60), vessels become permeable to proteins that leak between separated endothelial junctions. Hematogenous mononuclear cells penetrate between endothelial cells and through the basement membrane. The invading cells may also lodge beneath or within endothelial cells that show proliferative changes consisting of abundant cytoplasm filled with ribosomes and granular endoplasmic reticulum. Similarly activated endothelial cells have been observed in chronic lead neuropathy (69). In Fabry's disease, typical glycolipid inclusions can be demonstrated in endothelial, perithelial, and perineurial cells but, interestingly, not in Schwann cells (103). Thickening and reduplication of the vascular basal lamina of endoneurial vessels also suggest preceding endothelial damage (Fig. 45). Such

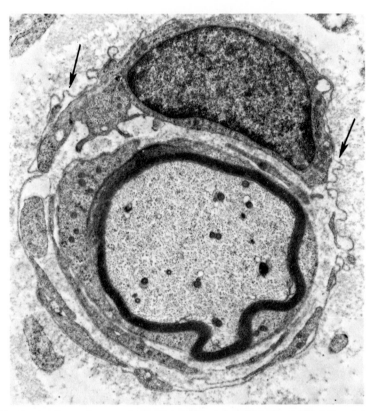

Figure 40. Remyelinated axon located within a persistent, ruffled basal lamina ensheathment (arrow). Schwann cell processes line the inner aspect of the persistent basal lamina and form a concentric layer of cellular processes around the remyelinated axon. Chronic lead neuropathy of rat sciatic nerve. (×18,000)

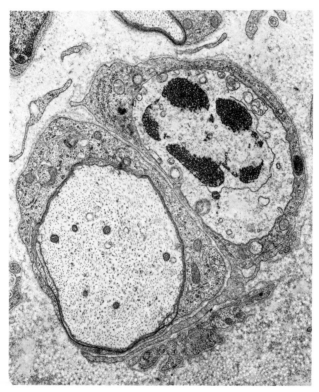

Figure 41. Beginning remyelination of an axon by a proliferated Schwann cell. Remnants of a necrotic Schwann cell are found enclosed within an adjacent cell that is also confined within the same original neurilemmal tube. Chronic lead neuropathy of rat sciatic nerve. (×8,000)

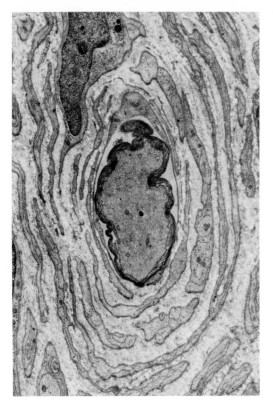

Figure 42. Onion bulb formation composed of concentric layers of Schwann cell processes around a remyelinated axon in sural nerve of a patient with Dejerine-Sottas neurpathy. (×6,000)

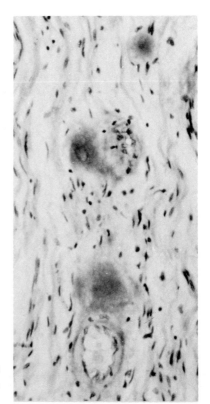

Figure 43. Light micrograph of anyloid deposits in sural nerve of patient with amyloid neuropathy. (×150)

344

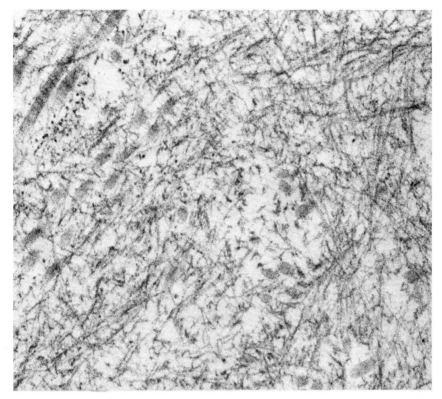

Figure 44. Amyloid filaments in endoneurium of sural nerve from patient with amyloid neuropathy. (×100,000)

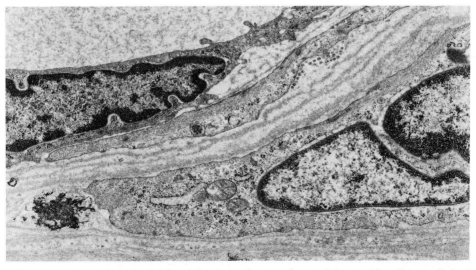

Figure 45. Reduplication of basal lamina of an endoneurial vessel in alloxan diabetic neuropathy of rat sciatic nerve. (×12,000)

supernumerary basement membranes have been conspicuous in nerves and other organs in human diabetes (85,104) and in experimental alloxan diabetic neuropathy (53). The presence of endoneurial edema characterized by wide separation of myelinated and unmyelinated nerve fibers is further suggestive of a vascular disorder. Since the perineurium acts as an effective diffusion barrier to macromolecules, the exudation of a protein-rich fluid into the endoneurium, as observed in inflammatory neuropathies (60), could result in edema harmful to nerve fibers either directly due to pressure or indirectly by causing ischemia. Prolonged or repetitive edema secondary to an alteration of the blood nerve barrier is frequently discussed in the pathogenesis of chronic, particularly hypertrophic, neuropathies (105).

REFERENCES

1. Dyck PJ, Lofgren EP: Nerve biopsy. Choice of nerve, method, symptoms and usefulness. *Med. Clin North Am.* 52; 885, 1968.

2. Ochoa J, Mair WGP: The normal sural nerve in man. I. Ultrastructure and number of fibers and cells. *Acta Neuropathol* 13:197, 1969.

3. Estable-Puig JF, Bauer WC, Blumberg JM: Paraphenylenediamine staining of osmium-fixed plastic, embedded tissue for light and phase microscopy. *J Neuropathol Exp Neurol* 24:531, 1965.

4. Dyck PJ, Lais A: Electron microscopy of teased nerve fibers: Method permitting examination of repeating structures of same fiber. *Brain Res* 23:418, 1970.

5. Olsson Y, Reese TS: Permeability of vasa nervorum and perineurium in mouse sciatic nerve studied by fluorescence and electron microscopy. *J Neuropathol Exp Neurol* 30:105, 1971.

6. Waggener JP, Bunn SM, Beggs J: The diffusion of ferritin within the peripheral nerve sheath: an electron microscopic study. *J Neuropathol Exp Neurol* 24:430, 1965.

7. Olsson Y: Mast cells in the nervous system. *Int Rev Cytol* 24:27, 1968.

8. Sunderland S: The connective tissue of peripheral nerves. *Brain* 88:841, 1965.

9. Aker FD: A study of hematic barriers in peripheral nerves of albino rabbits. *Anat Rec* 174:21, 1972.

10. Ashbury AK: Renaut bodies. A forgotten endoneural structure. *J. Neuropathol Exp Neurol* 32:334, 1973.

11. Thomas PK: The connective tissue of peripheral nerve: an electron microscopic study. *J. Anat* 97:35, 1963.

12. Weller RO, Herzog I: Schwann cell lysosomes in hypertrophic neuropathy and in normal human nerves. *Brain* 93:347, 1970.

13. Napolitano LM, Scalleen, TJ: Observations on the fine structure of peripheral nerve myelin. *Anat Rec* 163:1, 1969.

14. Livingston RB, Pfenninger K, Moor H, Akert K: Specialized paranodal and interparanodal glial-axonal junctions in the peripheral and central nervous system: A freeze etching study. *Brain Res* 58:1, 1973.

15. Harkin, JC: A series of desmosomal attachments in the Schwann sheath of myelinated mammalian nerves. *Z Zellforsch Mikrosk Anat* 64:189, 1964.

16. Hall SM, Williams PL: Studies on the "incisures" of Schmidt and Lanterman. *J Cell Sci* 6:767, 1970.

17. Ochs S: Axoplasmic transport—a basis for neural pathology, in Dyck PJ, Thomas PK, Lambert EH (eds): *Peripheral Neuropathy*. Philadelphia, WB Saunders, 1975, Vol 1, p 213.

18. Richardson KC: The fine structure of the albino rabbit iris with special reference to the identification of adrenergic and cholinergic nerves and nerve endings in its intrinsic muscles. *Am J Anat* 114:173, 1964.

19. Weiss P: Damming of axoplasm in constricted nerve: a sign of perpetual growth of nerve fibers. *Anat Rec* 88:464, 1944.

20. Friede RL: Axon swellings produced in vivo in isolated segments of nerves. *Acta Neuropathol* 3:229, 1964.

21. Webster H de F: Transient focal accumulations of axonal mitochondria during the early stages of Wallerian degeneration. *J Cell biol* 12:361, 1962.

22. Zelena J, Lubinska L, Gutmann E: Accumulation of organelles at the ends of interrupted axons. *Z Zellforch Mikrosk Anat* 91:200, 1968.

23. Holtzman E, Novikoff AB: Lysosomes in the rat sciatic nerve following crush. *J Cell Biol*, 27:651, 1965.

24. Schlaepfer WW: Vincristine-induced axonal alterations in rat peripheral nerve. *J Neuropathol Exp, Neurol* 30:488, 1971.

25. Bray GM, Peyronnard JM, Aguayo AJ: Reactions of unmyelinated nerve fibers to injury—an ultrastructural study. *Brain Res* 42:297, 1972.

26. Waller A: Experiments on the section of the glosso-pharyngeal and hypo-glossal nerves of the frog, and observations on the alterations produced thereby in the structure of their primitive fibers. *Philos Trans R Soc Lond B* 140:423, 1850.

27. Webster H deF: The relationship between Schmidt-Lanterman incisures and myelin segmentation during Wallerian degeneration. *Ann NY Acad Sci* 122:29, 1965.

28. Nathaniel EJH, Pease DC: Degenerative changes in rat dorsal roots following Wallerian degeneration. *J. Ultrastruct Res* 9:511, 1963.

29. Ashbury AK: The histogenesis of phagocytes during Wallerian degeneration, in *Proceedings VI International Congress Neuropathology.* Paris, Mason, 1970, p 666.

30. Thomas PK: Changes in endoneurial sheaths of peripheral myelinated nerve fibers during Wallerian degeneration. *J Anat* 98:175, 1964.

31. Blümcke S, Niedorf HR: Electron microscope studies of Schwann cells during the Wallerian degeneration with special reference to cytoplasmic filaments. *Acta Neuropathol* 6:46, 1966.

32. Ochoa J, Mair WGP: The normal sural nerve in man. II. Changes in the axons and Schwann cells due to aging. *Acta Neuropathol* 13:217, 1969.

33. Cajal SR: *Degeneration and Regeneration of the Nervous Ssytem.* London, Oxford University Press, 1928.

34. Blümcke S, Niedorf NR: Elektronenoptische Untersuchungen an Wachstumsendkolben regenerierender peripherer Nervenfasern. *Virchous Arch* 340:93, 1965.

35. Lampert, P, Cressman M: Axonal regeneration in the dorsal columns of the spinal cord of adult rats. *Lab Invest* 13:825, 1964.

36. Schröder JM: Altered ratio between axon diameter and myelin sheath thickness in regenerated nerve fibers. *Brain Res* 45:49, 1972.

37. Spencer, P. S., Schaumberg H: Central and peripheral distal axonopathy—the pathology of dying back neuropathy; in Zimmerman HM (ed.): *Progress in Neuropathol* New York, Grune & Stratton, 1976, vol 3, p 253.

38. Greenbaum D, Richardson PC, Salmon MV, Urich H: Pathological observations on six cases of diabetic neuropathy. *Brain* 87:201, 1964.

39. Dyck PJ, Johnson WF, Lambert EH, O'Brien PC: Segmental demyelination secondary to axonal degeneration in uremic neuropathy. *Mayo Clin Proc* 46:400, 1971.

40. Dyck PJ, Lambert EH: Polyneuropathy associated with hypothyroidism. *J Neuropathol Exp Neurol* 29:631, 1970.

41. Cavanagh JB: On the pattern of changes in peripheral nerves produced by isoniazid intoxication in rats. *J Neurol Neurosurg Psychiatr* 30:26, 1967.

42. Prineas J: The pathogenesis of dying back polyneuropathies. Part II. An ultrastructural study of experimental acrylamide intoxication in the cat. *J Neuropathol Exp Neurol* 28:598, 1969

43. McLeod JG, Penny R: Vincristine neuropathy: An electrophysiological and histological study. *J Neurol Neurosurg Psychiatr* 32:297, 1969.

44. Bischoff A: The ultrastructure of tri-ortho-cresyl-phosphate poisoning. I. Studies on myelin and axonal alterations in the sciatic nerve *Acta Neuropathol* 9:158, 1967.

45. Prineas J: The pathogenesis of dying back polyneuropathies. Part I. An ultrastructural study of experimental tri-ortho-cresyl-phosphate intoxication in the cat. *J Neuropathol Exp Neurol* 28:571, 1969.

46. Collins, GH, Webster H deF, Victor M: The ultrastructure of myelin and axonal alterations in sciatic nerves of thiamine deficient and chronically starved rats. *Acta Neuropathol* 3:511, 1964.

47. Prineas J: Peripheral nerve changes in thiamine deficient rats. *Arch Neurol* 25:541, 1970.

48. Lampert P, Blumberg JM, Pentchew A: An electron microscopic study of dystrophic axons in the gracile and cuneate nuclei of Vitamin E deficient rats. *J Neuropathol Exp Neurol* 23:60, 1964.

49. Carpenter S., Karpati G., Andermann F, Godl R: Giant axonal neuropathy. *Arch Neurol* 31:312, 1974.

50. Berard-Badier M, Gambarelli D, Pinsard N, Hassoun J, Toga M: Infantile neuroaxonal dystrophy or Seitelberger's disease. II. Peripheral nerve involvement: electron microscopic study in one case. *Acta Neuropathol* Suppl. 5:30, 1971.

51. Schochet SS, Lampert PW, Earle KM: Neuronal changes induced by intrathecal vincristine sulfate. *J Neuropathol Exp Neurol* 27:645, 1968.

52. Lampert P: A comparative electron microscopic study of reactive, degenerating, regenerating and dystrophic axons. *J. Neuropathol Exp Neurol* 26:345, 1967.

53. Powell H, Knox D, Lee S, Charters AC, Orloff M, Garrett R, Lampert P: Alloxan diabetic neuropathy. Electron microscopic studies. *Neurology,* 27:60, 1977.

54. Schochet SS, McCormick, WF, Kovarsky J: Light and electron microscopy of skeletal muscle in type IV glycogenosis. *Acta Neuropathol* 19:137, 1971.

55. Sakai M, Austin J, Whitmer F, Trueb L: Studies in myoclonus epilepsy (Lafora body form). II. Polyglucosans in systemic deposits of myoclonus epilepsy and in corpora amylacea. *Neurology* 20:160, 1970.

56. Schochet SS: Mitochondrial changes in axonal dystrophy produced by Vitamin E deficiency. *Acta Neuropathol* suppl 5:54, 1971.

57. Dyck PJ, Lais AC: Evidence for segmental demyelination secondary to axonal degeneration in Friedreich's ataxia, in Kakulas BK (ed): *Clinical Studies in Myology.* Int. Cong. Series. Amsterdam, Excerpta Medica, 1973, p. 253.

58. Towfighi J, Gonatas NK, McCree L: Hexachlorophene neuropathy in rats *Lab Invest* 29:428, 1973.

59. Suzuki K, DePaul LD: Myelin degeneration in sciatic nerve of rats treated with hypocholesteremic drug AY 9944. *Lab Invest* 26:534, 1972.

60. Lampert PW: Mechanism of demyelination in experimental allergic neuritis. *Lab Invest* 20:127, 1969.

61. Webster H deF, Ulsamer AG, O'connell MF: Hexachlorophene-induced myelin lesions in the developing nervous system of Xenopus tadpoles: morphological and biochemical observations. *J Neuropathol Exp Neurol* 33:144, 1974.

62. Towfighi J, Gonatas NK, McCree L: Hexachlorophene-induced changes in central and peripheral myelinated axons of developing and adult rats. *Lab Invest* 31:712, 1974.

63. Berthold CH, Skoglund S: Postnatal development of feline paranodal myelin sheath segments. *Acta Soc Med Ups* 73:127, 1968.

64. Schröder JM, Krücke W: Zur Feinstruktur der experimentell- allergischen Neuritis beim Kaninchen. *Acta Neuropathol* 14:261, 1970.

65. Ashbury AK, Arnason BG, Adams RD: The inflammatory lesion in idiopathic polyneuritis. *Medicine* (Baltimore) 48:173, 1969.

66. Prineas JW: Acute idiopathic polyneuritis. An electron microscopic study. *Lab Invest* 26:133, 1972.

67. Wisniewski H, Terry RD, Whitaker JN, Cook SD, Dowling PC: Barré syndrome. A primary demyelinating disease. *Arch Neurol* 21:269, 1969.

68. Arnason BGW: Inflammatory polyradiculoneuropathies in Dyck PJ, Thomas PK, Lambert EH (eds): *Peripheral Neuropathy.* Philadelphia, WB Saunders, 1975, vol 2, p 1110.

69. Lampert PW, Schochet SS: Demyelination and remyelination in lead neuropathy. Electron microscopic studies. *J Neuropathol Exp Neurol* 27:527, 1968.

70. Lubinska, L: Region of transition between preserved and regenerating parts of myelinated nerve fibers. *J Comp Neurol* 113:315, 1959.

71. Ochoa J, Fowler TJ, Gilliatt RW: Anatomical changes in peripheral nerves compressed by a pneumatic tourniquet. *J Anat* 113:433, 1972.

72. Allt G, Cavanagh JB: Ultrastructural changes in the region of the node of Ranvier in the rat caused by diphtheria toxin. *Brain* 92:459, 1969.

73. Allt G: Repair of segmental demyelination in peripheral nerves. *Brain* 92:639, 1969.

74. Krücke W: Die Erkrankungen der peripheren Nerven; in Kaufmann E, Staemler M (eds): *Lehrbuch der speziellen pathologischen Anatomie.* Berlin, Walter Gruyter, 1961, vol 3, p 750.

75. Hopkins AP: The effect of acrylamide on the peripheral nervous system of the baboon. *J Neurol Neurosurg Psychiat* 33:805, 1970.

76. Webster H deF: The geometry of peripheral myelin sheaths during their formation and growth in rat sciatic nerves. *J Cell Biol* 48:348, 1971.

77. Friede RL: Control of myelin formation by axon caliber (with a model of the control mechanism). *J Comp Neurol* 144:233, 1972.

78. Raine CS, Wisniewski H, Prineas J: An ultrastructural study of experimental demyelination and remyelination. II. Chronic experimental allergic encephalomyelitis in the peripheral nervous system *Lab Invest* 21:316, 1969.

79. Lubinska L: Demyelination and remyelination in the proximal parts of regenerated nerve fibers. *J Comp Neurol* 117:279, 1961.

80. Fullerton PM, Gilliat RW, Lascelles RG, Morgan-Hughes JA: The relation between fiber diameter and internodal length in chronic neuropathy. *J. Physiol* (London) 178:26, 1965.

81. Dinn JJ: Transnodal remyelination. *J Pathol* 102:51, 1970.

82. Lampert PW, Garrett RS: Mechanism of demyelination in tellurium neuropathy. Electron microscopic observations. *Lab Invest* 25:380, 1971.

83. Bradley WG, Ashbury AK: Duration of synthesis phase in neurilemma cells in mouse sciatic nerve during degeneration. *Exp Neurol* 26:275, 1970.

84. Tomonaga M, Sluga E: Zur Ultrastruktur der π Granula. *Acta Neuropath* 15:56, 1970.

85. Bischoff A: Diabetische Neuropathie: Pathologische Anatomie, Pathophysiologie und Pathogenese auf Grund elektronen mikroskopischer Untersuchungen. *Dtsch Med Wochenschn* 93:237, 1968.

86. Fardeau M, Engel WK: Ultrastructural study of a peripheral nerve biopsy in Refsum's disease. *J Neuropathol Exp Neurol* 28:278, 1969.

87. Lyon G, Evrard P: Sur la présence d'inclusions cristallines dans les cellules de Schwann dans diverses neuropathies périphériques. *CR Acad Sci (Paris) Ser D,* 271:1000, 1970.

88. Job CK: *Mycobacterium leprae* in nerve lesions in lepromatous leprosy. An electron microscopic study. *Arch Pathol* 89:195, 1961.

89. Bischoff A: Neuropathy in leukodystrophies, in Dyck PJ, Thomas PK, Lambert EM (eds), *Peripheral Neuropathy.* 1975, vol 2, p. 891, Philadelphia, Saunders, WB.

90. Webster H deF: Schwann cell alterations in metachromatic leukodystrophy: Preliminary phase and electron microscopic observations. *J. Neuropathol Exp Neurol* 21:534, 1962.

91. Bischoff A, Ulrich J: Peripheral neuropathy in globoid cell leucodystrophy (Krabbe's disease) Ultrastructural and histochemical findings. *Brain* 92:861, 1969.

92. Carpenter S, Karpati G, Andermann F: Specific involvement of muscle, nerve and skin in late infantile and juvenile amaurotic idiocy. *Neurology* 22:170, 1972.

93. Powers JM, Schaumberg HH: Adrenoleukodystrophy: Similar ultrastructural changes in adrenal cortical cells and Schwann cells. *Arch Neurol* 30.406, 1974.

94. Gambetti P, DiMauro S, Baher L: Nervous system in Pompe's disease. Ultrastructure and biochemistry. *J Neuropathol Exp Neurol* 30:412, 1971.

95. Thomas PK, Lascelles RG: Hypertrophic neuropathy. *Q J Med* 36:223, 1967.

96. Weller RO: An electron microscopic study of hypertrophic neuropathy of Dejerine Sottas. *J Neurol Neurosurg Psychiat* 30:111, 1967.

97. Webster, H deF, Schröder JM, Ashbury AK, Adams RD: The role of Schwann cells in the formation of "onion bulbs" found in chronic neuropathies. *J Neuropathol Exp Neurol* 26:276, 1967.

98. Ballin RHM, Thomas PK: Hypertrophic changes in diabetic neuropathy. *Acta Neuropathol* 11:93, 1968.

99. Dyck PJ: Experimental hypertrophic neuropathy. Pathogenesis of onion bulb formations by repeated tourniquet application. *Arch Neurol* 21:73, 1969.

100. Krücke W: Die mukoide Degeneration der peripheren Nerven. *Virch Arch* 304:442, 1939.

101. Dyck JP, Lambert EH: Dissociated sensation in amyloidosis. *Arch Neurol* 20:490, 1969.

102. Bigner DD, Olson WH, McFarlin DE: Peripheral polyneuropathy, high and low molecular weight IgM and amyloidosis. *Arch Neurol* 24:365, 1971.

103. Ohnishi A, Dyck PJ: Loss of small peripheral sensory neurons in Fabry's disease. *Arch Neurol* 31:120, 1974.

104. Lapresle J: Etude anatomique des neuropathies périphériques du diabète sucré *J Ann Diabétol Hôtel Dieu* 9:101, 1968.

105. Krücke W: Oedem and seröse Eutzündung in peripheren Nerven. *Virchows Arch* 308:1, 1941.

8

The Diagnosis of Central Nervous System Disorders by Transmission Electron Microscopy

Julio H. Garcia, M.D.
Professor and Director, Anatomic Laboratories
Head, Division of Neuropathology
Department of Pathology
University of Maryland School of Medicine
Baltimore, Maryland

Hernando Mena, M.D.
Instructor, Department of Pathology
University of Maryland School of Medicine
Baltimore, Maryland

Transmission electron microscopy (TEM) is valuable in the diagnosis of nervous system diseases, as well as in the evaluation of illnesses affecting other organs or tissues. On account of its improved *resolution,* the electron microscope permits a more objective and precise evaluation of certain pathological features in biological samples than any other diagnostic tool.

In the case of the central nervous system (CNS), TEM makes it possible to visualize, simultaneously, all cellular components of the brain or spinal cord. In order to identify neuronal soma, axis cylinders, astrocytes, and myelin sheaths in paraffin-embedded material, one needs many separate histological preparations, each stained by a different method. Although the majority of these histological staining methods are helpful, aesthetic, and pleasing, the procedures involved in many of them are not based on well-established chemical principles; therefore, staining techniques are difficult to apply in a controlled, dependable, and reproducible manner. Moreover, assuming an ideal completion of the staining methods, there are still many nervous tissue components that cannot be visualized by light microscopy, i.e., dendrites and many cell organelles.

The judicious, well-planned, and combined use of light and electron micros-copic methods can make certain a large number of diagnoses that would other-wise remain either uncertain or incomplete.

Examination of samples of human CNS tissues (usually cerebral or cerebellar cortices) by TEM should be preceded by careful selection of appropriate methods for tissue *fixation* and *processing.* In addition to acquiring familiarity with the general principles of techniques and methods for fixation and processing (1), pathologists should become aware of the sources of undesirable effects or "ar-tifacts" upon which one may stumble in the course of analyzing biological sam-ples whose structural features are being evaluated for diagnostic purposes.

Fragments of brain and/or tumors that are excised surgically are abnormal to begin with, and these samples must be subjected to a) *ischemic injury,* an indis-pensable step before excision, which is usually performed with the aid of elec-trocautery; and b) *crushing injury,* either manual or instrumental.

Once the surgical removal of the tissue is completed, the effects of lack of perfusion or ischemia become increasingly evident with advancing time (2); therefore, the sooner the fixation the lesser the effects of ischemia. Ischemic changes in brain biopsy specimens probably are exaggerated by exposing the tissues to the atmosphere (3). It is commonly believed that adequate fixation of CNS tissues must be conducted either in vivo, which is not feasible in humans, or *immediately* after removal of the biopsy specimen. Strict concern for the immedi-acy of the fixation may lead to attempts at mincing the biological samples as soon as they are received from the surgeon's hands. Such practice is to be discour-aged, since it induces pronounced structural alterations secondary to crushing effects (4). It is recommended, instead, that the entire sample (exclusive of necrotic and hemorrhagic components) be reserved, for subsequent mincing, in a vial filled with the appropriate fixative, as indicated below.

Delayed fixation (up to 17 minutes), following ischemic injury, of brain tissues maintained at 37°C results in but slight alteration of ultrastructure; the earliest ischemic changes become apparent mainly in the nuclei, in the form of chroma-tin clumping (Figs. 1–3); myelin sheaths and several other cell membranes are relatively stable for periods of up to 60 minutes (5, 6, 7). Surgically removed fragments of brain are more apt to show signs of *incomplete* ischemia (Fig. 3), or the type of ischemia that follows occlusion of individual arterial branches (8). We suggest that in human surgical pathology, in order to avoid undesirable "ar-tifacts," what is really important is to prevent drying effects and to avoid exces-sive handling or crushing, both before and during fixation.

Peters (9) has discussed several ways to achieve a *good* fixation of the mamma-lian brain; some of the best nervous tissue fixative solutions seem to be those that combine the qualities of two aldehydes, i.e., formaldehyde and glutaraldehyde, as described originally by Karnovsky (10). This combination, with minor varia-tions, has been successfully used in the preservation of samples that subsequently were embedded in either paraffin or epoxy resins (11, 12). Following overnight fixation in such solution, mincing the sample, under the dissecting microscope, is completed in such a manner that tissue dice measuring 3 mm³ are obtained from the surface of the larger specimen, keeping away from areas where crushing is evident, and always preventing drying effects. Rinsing in a buffer solution is followed by postfixation in 1.0% osmium tetroxide and processing in the manner described in detail elsewhere (5, 1, 13).

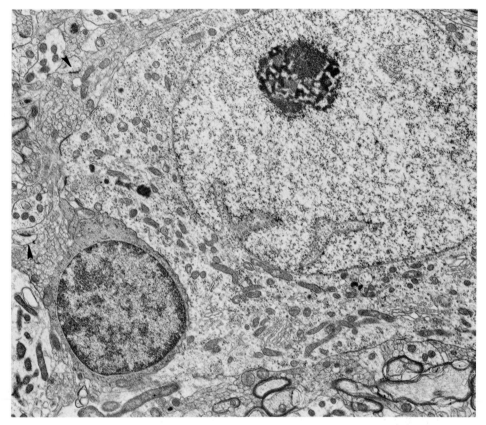

Figure 1. Normal caudate nucleus from adult subhuman primate; in vivo fixation by cardiovascular perfusion. Absence of ischemic effects can be surmised from the even distribution of chromatin granules. Identification of neurons is made by their synaptic contacts (arrowheads). Smaller, darker "satellite" is an oligodendrocyte. (×2,300)

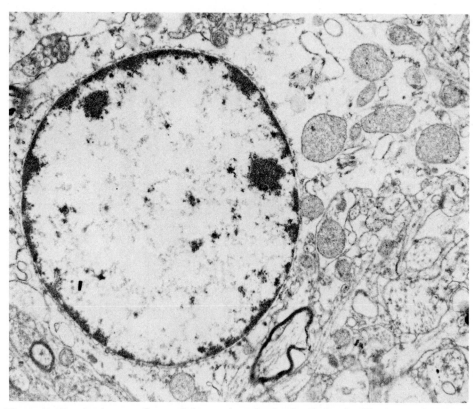

Figure 2. Cerebral cortex from adult cat whose brain circulation had been terminated 60 minutes before fixation. Note nuclear chromatin clumping and clearing of cell sap; mitochondria are not swollen (×5,000)

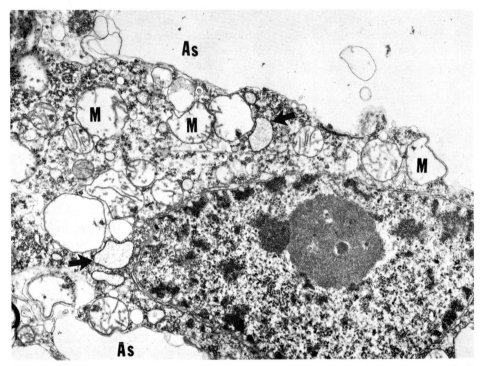

Figure 3. Cerebral cortex from rhesus monkey fixed by perfusion 7 hours after embolic occlusion of the right middle cerebral artery. Incomplete ischemia is evident in the massive dilation of mitochondria (M) and rough ER (arrows). Note massive swelling of the astrocyte (As) that surrounds the neuron. (×6,000)

It is beyond the scope of this chapter to provide a detailed description of the ultrastructure of normal human brain and spinal cord. An excellent and detailed collection of illustrations obtained from normal laboratory animals and a discussion of the current terminology applied to CNS ultrastructure are available in a recently published textbook (14). Our primary objective is to present illustrative material on the *interpretation* that certain ultrastructural abnormalities may be given in diagnostic human pathology. The sequence in the procedure that we recommend is as follows:

1. Collection of the appropriate clinical and paraclinical information, such as age of the patient, nature of the comlaint, location of the lesion, pattern of vascularization (in cases of neoplasia), and other features. An important item is knowledge of the diagnosis or diagnoses considered probable as a result of clinical evaluation and laboratory examinations.
2. Evaluation, by light microscopy, of the paraffin-epoxy-embedded material to determine which features may require further clarification; for example, intranuclear inclusions, secretory granules, presence or absence of cilia, nature of "stored" cytoplasmic lipids, etc.
3. Selection of appropriate areas for ultrathin sectioning
4. Evaluation of the sample at the electron microscope, preferably with the purpose of searching for specific profiles, such as virions, secretory granules, tight intercellular junctions, and others

We present below instances in which the improved resolution of TEM contributed significantly to the ultimate diagnosis of the following types of neurological diseases: a) neoplasms, b) metabolic abnormalities, c) infections, and d)neuropsychiatric disorders.

NEOPLASIA

A very pragmatic approach to the diagnosis of brain and spinal cord tumors may be one that attempts simply a separation of neoplasms into four major categories:

1. Ectoderm- and endoderm-derived neoplasms, or those derived from either lining or secretory epithelium; within the cranium and spinal canal, these fall into the categories of pituitary adenomas, cysts, and metastatic carcinomas.
2. Neuroectodermally derived growths, or neoplasms originating from neurons, glia, and their precursors, including those that originate in leptomeninx, which is considered by many as being derived from neuroectoderm (15).
3. Mesenchymally derived neoplasms, including those that originate from blood vessels, blood cells, cartilage, and bone.
4. Neoplasms derived from germinal, undifferentiated cells, such as medulloblastoma and dysgerminoma.

The origin of many neoplastic cells may be surmised from an analysis of the relationship that cells maintain among them, e.g., formation of rosettes; by the products that neoplastic cells may secrete, e.g., mucin, collagen; by the pigment they contain, e.g., melanin; and by certain intrinsic features, such as nuclear size and stainability. All these characteristics are apparent at light microscopy; however, whenever the size of individual neoplastic cells is very small, the degree of differentiation is very poor, or the sample itself is miniscule, light microscopic methods alone may be inadequate for the appropriate and incontrovertible identification of features that indicate the origin of certain neoplastic cells (Fig. 4).

As in the diagnosis of extracranial neoplasms (16), electron microscopists base the identification of neoplastic cells or their precursors on the presence or absence of features such as: a) intracytoplasmic *secretory granules,* some of which are characteristic of either neurons (17) or adenohypophysis cells (18, 19); b) intracytoplasmic *filaments* that are suggestive of an astroglial derivation; c) certain types of *cellular junctions* that are characteristic of both lining epithelium and arachnoid cells; d) *basal laminae,* which are seen in association with either Schwann cells or pericyte/endothelial cells; e) certain *organelles* found only in endothelial cells of arteries (20); and f) *synaptic contacts* that are specific for neuronally derived cells. Observation of these features contributes to the identification of cells and helps to achieve a precise diagnosis of certain neoplasms; on account of their size, such cell components can be resolved only by TEM.

The following are examples of situations in which TEM evaluation clarified the diagnosis and made possible a more rational treatment of the patient.

Case 1

A 15-year-old girl developed right hemiparesis 2 days before admission to the hospital. A large hypovascular tumor supplied by branches of the internal carotid artery was demonstrated, angiographically, in the left parietal area. A partial removal of this tumor was accomplished after incising the cortex of the parietal lobe; no dural attachment was discerned. Light microscopy of samples stained with hematoxylin-eosin and by several other methods provided no evidence on which a conclusive diagnosis could be based (Figs 5 and 6). Several, disparate diagnostic possibilities were entertained, including medulloblastoma, glioblastoma with sarcomatous changes, malignant meningioma, and malignant undifferentiated neoplasm.

Low-magnification electron micrographs of this neoplasm suggested the formation of imperfect or primitive capillaries (Figs. 7–11), an impression corroborated by the identification, in neoplastic cells, of many features previously described in endothelial cells, particularly those that have been described in human neoplasms of vascular origin (21, 22, 23). The majority of cells in this neoplasm had features of vascular endothelium, atypical mitotic figures were frequent, and cellular differentiation was poor; on these bases, a diagnosis of *hemangiosarcoma* (intracerebral) was tendered; the patient died approximately 10 weeks after the initial craniotomy as a result of both rapid regrowth of the original mass and massive brain swelling; no evidence of neoplastic growth outside the brain was uncovered (24).

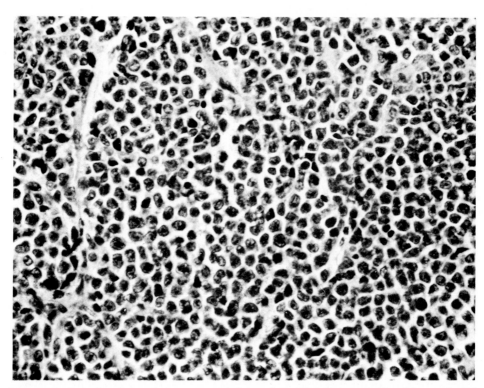

Figure 4. Sample of neoplasm removed from the right thalamus of a 20-month-old infant. Small, undifferentiated cells without any particular arrangement or clues for an accurate diagnosis. Special staining methods did not aid further. (Hematoxylin-eosin, ×25)

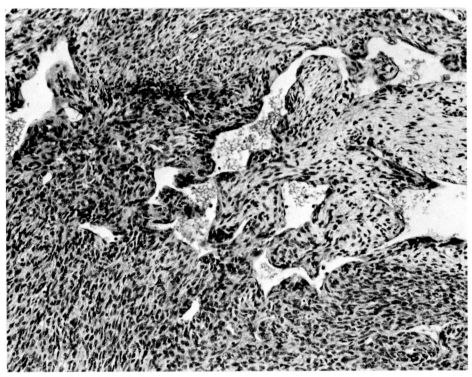

Figure 5. Sample of neoplasm removed from the left parietal lobe of a 15-year-old girl. The cells are small, fusiform shaped, and contain scanty cytoplasm. Vascular channels of different sizes are lined by flat endothelial cells. Atypical mitotic figures were frequent, but there was no coagulative necrosis. (Hematoxylin-eosin ×24.75)

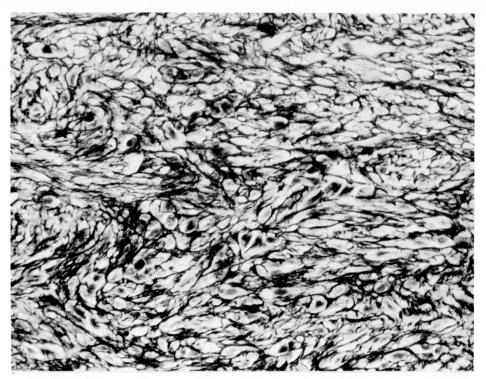

Figure 6. Sample of the neoplasm shown in Figure 5 revealing neoplastic cells interlaced by reticulin fibers. (Hortega's technique, ×61.25)

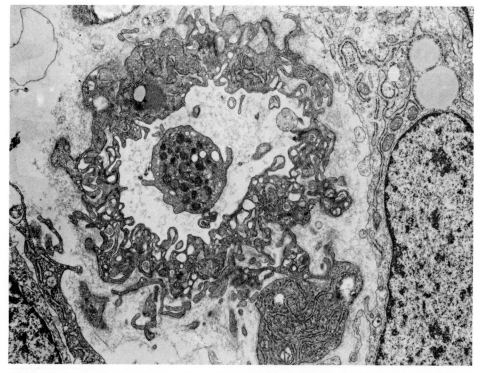

Figure 7. The majority of cells in this neoplasm (same neoplasm as in Figures 5 and 6) are arranged in an organoid pattern that suggests here an imperfect capillary. (×5,375)

360

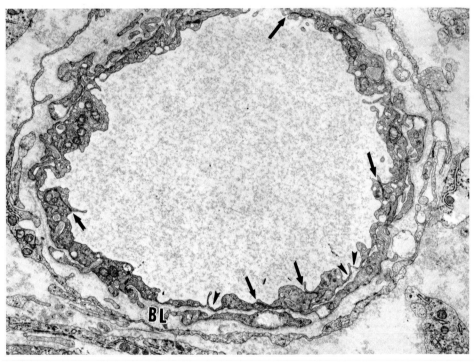

Figure 8. Alternating tight junctions (arrows) and fenestrations (arrowheads) between the neoplastic cells were common. Collagen fibers and the irregular basal lamina (BL) are also visible. Same neoplasm as in Figures 5–7. (×5,500)

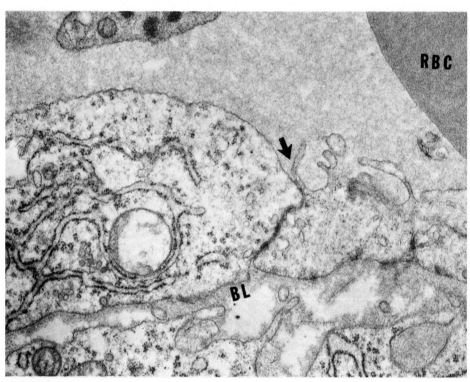

Figure 9. Detail of adjoining neoplastic cells. Note abundant rough endoplasmic reticulum cisternae, free ribosomes, and filaments. Tight junctions (arrow) and basal lamina (BL) are also shown; a red blood cell (BRC) partially occupies the lumen. Same neoplasm as in Figures 5–8. (×12,000)

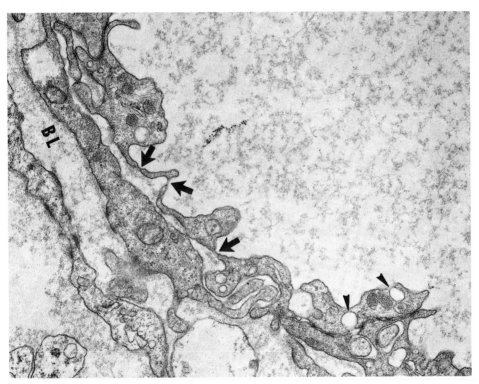

Figure 10. Fenestrations (arrows) and pinocytotic vesicles (arrowheads) are prominent features of these neoplastic cells. Note the imperfect basal lamina (BL) and abundant cytoplasmic filaments. Same neoplasm as in Figures 5–9. (×5,400)

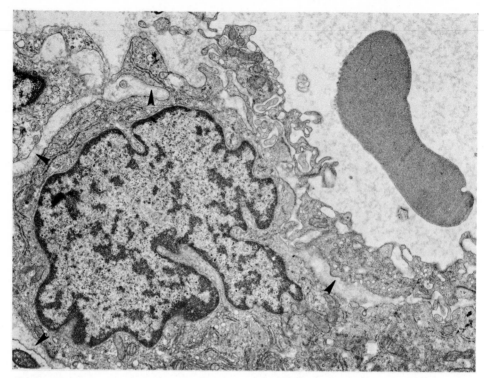

Figure 11. Many interstitial cells such as the one shown here contained lobulated nucleus and abundant rough ER and Golgi cisternae; note the circumferential basal lamina (arrowheads). An erythrocyte is seen in the lumen of this "primitive capillary." Same neoplasm as in Figures 5–10. (×5,400)

Case 2

A 31-year-old man with an 18-month history of generalized seizure disorder was hospitalized because seizures could not be managed medically any longer; neurological examination failed to disclose any deficit. A right carotid angiogram demonstrated a well-vascularized, discretely circumscribed "tumor" that measured about 5.0 cm in largest diameter; it was removed from within the right temporoparietal area. A preoperative diagnosis of meningioma had been suggested by the angiographic pattern, although the mass was not attached to the dura and it lay a few centimeters beneath the cortex of the parietal lobe. Light microscopic features of the neoplasm are illustrated in Figures 12 and 13.

Possible diagnoses included glioma with invasion of the leptomeninges, neuroblastoma, and primary brain sarcoma (not further specified). Reluctance to accept a diagnosis of primary brain sarcoma was based partly on the marked infrequency of this entity, as described in the medical literature (25, 26), and partly on our inability to find adequate illustrative material demonstrating the features of primary brain sarcomas.

The difficulty, in this case, was solved by the fact that many of the ultrastructural characteristics attributed to either normal (27) or neoplastic fibroblasts (28) were easily demonstrated in many of the neoplastic cells (Figs. 14–17), and, thus,

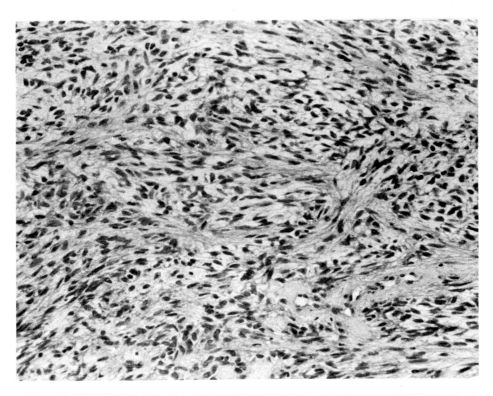

Figure 12. Sample of intracerebral neoplasm removed from the right parietal lobe of a 31-year-old man. The spindle-shaped cells are arranged in interlacing bundles. Note the absence of hemorrhage or coagulative necrosis, which applied also to all specimens examined. (Hematoxylin-eosin ×40)

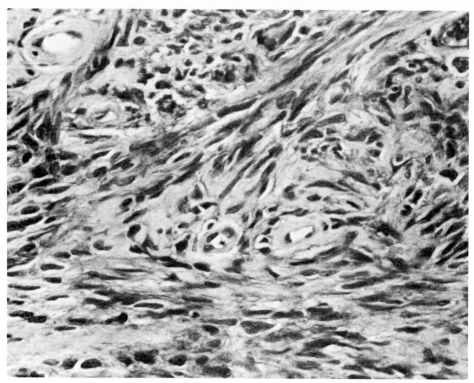

Figure 13. Abundant collagen fibers are seen between neoplastic fibroblasts; blood vessels are unremarkable, and mitotic figures were infrequent. (Masson's trichrome, ×63.75)

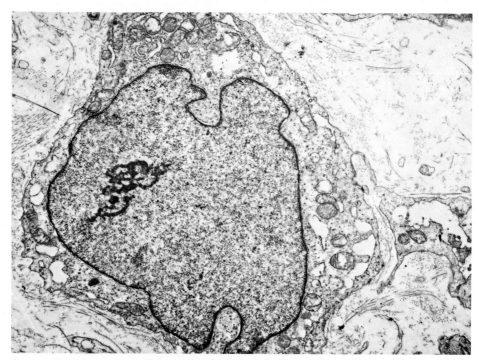

Figure 14. The nucleus of this cell is folded, and the presence of a prominent nucleolus suggests that the cell is engaged actively in protein synthesis. Numerous collagen fibers are seen in the extracellular space. (×2,750)

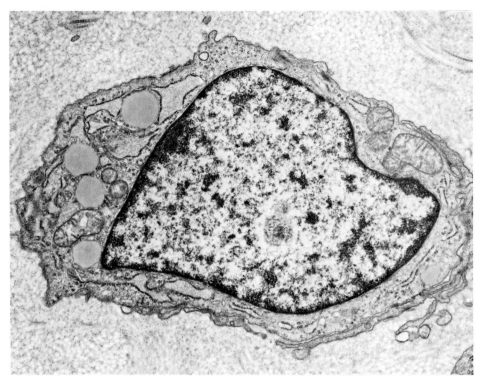

Figure 15. The cytoplasm of this neoplastic cell contains few mitochondria, lipid droplets, and abundant filaments. Note the prominence and abundance of the rough endoplasmic reticulum cisternae and the absence of basal lamina. (×4,900)

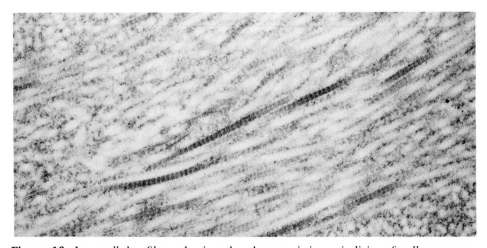

Figure 16. Intercellular fibers, having the characteristic periodicity of collagen, were abundant in this intracerebral neoplasm. (×15,000)

367

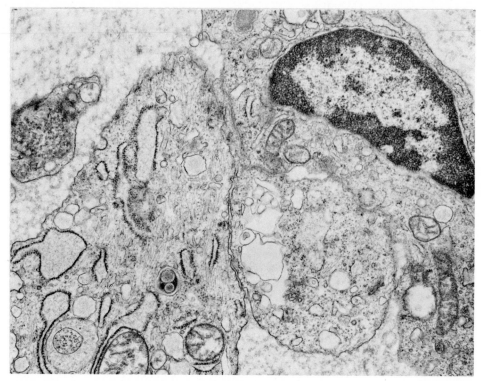

Figure 17. Absence of cell junctional devices between adjoining cells was another constant feature that some (D.D. Copeland, personal communication, 1977) consider typical of fibroblasts, both normal and neoplastic. (×8,000)

a diagnosis of intracerebral *fibrosarcoma* (probably primary) was advanced. The cells of origin, in this case, are probably mesenchymal elements from the subarachnoid sleeves around the larger blood vessels. Approximately 18 months after surgery, the patient is self-supporting and employed in the same occupation he held before the craniotomy, and he is considerably relieved from the seizure disorder (24). Additional, confirmed instances of primary brain sarcomas must be gathered before we can learn to predict accurately the natural history of these neoplasms.

Further data on the usefulness of TEM in the diagnosis of human CNS neoplasms are available in an atlas of brain tumors (29).

METABOLIC DISORDERS

Certain alterations in the metabolism of gangliosides, cerebrosides, animo acids, etc. may constitute the background for various rare neurological disorders that usually become apparent either in childhood or young adolescence; this fact is explained by the hereditary nature of these disorders (30). In the past, the diagnostic evaluation of patients aflicted with many "storage" disorders has required the combined application of biochemical, enzymatic, and structural methods to study fragments of brain obtained by biopsy. At present, thanks to the collective experience gathered by several electron microscopists and the parallel progress made in the biochemical laboratory, a simplified way of making precise diagnoses of several metabolic disorders of the CNS may simply rely on the use of structural techniques. Although in most instances the light microscopist may be able to see that there exists a storage disorder, not further specified, a rather precise identification of many stored substances is accomplished with TEM

Probably the two largest groups of CNS metabolic disorders in whose diagnosis pathologists become involved are a) gangliosidoses and b) ceroid lipofuscinoses. The following are contrasting examples of these diseases.

Case 1

The patient, normal at birth, was noted at the age of 17 weeks to be suffering from psychomotor arrest, cherry-red spot in the ocular fundi and hyperacusis. A brain biopsy sample was obtained from the right parietal lobe; at this time, light microscopy revealed numerous neuronal cells whose soma was distended with a finely granular lipid-like material that appeared faintly positive after PAS reaction and that displaced most of the cytoplasmic organelles, peripherally (Fig. 18).

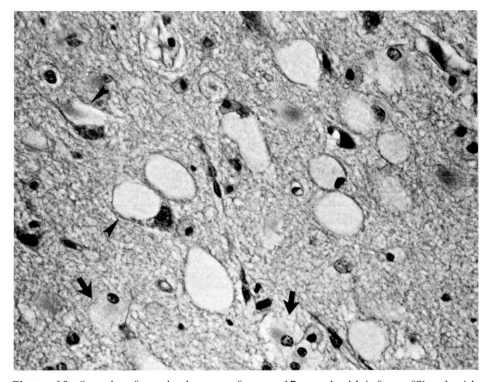

Figure 18. Sample of cerebral cortex from a 17-month-old infant afflicted with "psychomotor retardation." The neuronal (arrowheads) and astroglial (arrow) perikaryon is distended by a finely granular, lipid-like material which displaces the nucleus eccentrically. (Hematoxylin-eosin ×77.5) (*Courtesy of M. Adachi, M.D., Brooklyn, NY*)

These features served to confirm the clinical impression of metabolic (storage) encephalopathy but provided no specific clue on the nature of the stored metabolite.

TEM demonstrated, in the neuronal perikaryon, numerous membranous cytoplasmic bodies (MCB) 0.5 μm to 2.0 μm in diameter, composed of concentrically arranged membranes, each measuring about 25 Å (Figs. 19 and 20). The combination of the above clinical features and the presence of MCBs (31) in this location, i.e., the cerebral cortex, made certain the diagnosis of *GM-2 gangliosidosis*, a condition also called infantile amaurotic idiocy or Tay-Sachs disease (32). As in similar cases of gangliosidosis GM-2, the patient became both blind and progressively isolated from the environment; death occurred at the age of 4½ years. Chemically, MCBs are made of gangliosides, cholesterol, and phospholipids, and their genesis has been traced to the cisternae of the rough endoplasmic reticulum (32). Gangliosidosis GM-2 is the result of a marked deficiency in hexosaminidase-A, and this may be detected in amniotic fibroblasts very early in gestation (30). A relatively late phase of megalencephaly in GM-2 gangliosidosis has been attributed to marked glial proliferation in the brain (33).

Case 2

A previously healthy girl developed, at the age of 3 years, progressive psychomotor deterioration, generalized seizures, and myoclonic jerks. Several laboratory determinations in blood, urine, and cerebrospinal fluid yielded normal results; the patient became blind at the age of 6 years—when a right parietal brain biopsy specimen was obtained. By the time of her death, 2 years later, a younger sibling had begun to develop a similar neurological deficit. A clinical impression of "juvenile amaurotic idiocy" could be supported by light microscopic evidence of "storage" in the perikaryon of cortical neurons (Fig. 21). However, elucidating the nature of the stored substance required examination by electron microscopic methods (Figs. 21–27), especially in view of the similarity in the structural and tinctorial features shared by *lipofuscin* (aging pigment) and *ceroid*, a poorly characterized substance that is considered the cellular hallmark of *ceroid lipofuscinosis* (34, 35). The most reliable ultrastructural indicator of ceroid lipofuscinosis seems to be "curvilinear bodies" that may be demonstrated in brain (36) and retina (37) as well as in biopsy specimens of peripheral nerve, skeletal muscle (Figs. 28 and 29), and skin (38). It has been emphasized that in many of these biopsy specimens, the abnormal pigment may not be detectable at the light microscopic level, whereas detection with TEM is relatively simple (39).

Ceroid lipofuscinosis is a disorder of juvenile or adult onset. Clinical symptoms include progressive dementia, epilepsy, blindness (sometimes accompanied by retinal pigmentation), and increasing motor disability. Transmission in familial cases is autosomal recessive (40). The pigment can be detected in white blood cells of both patients and heterozygotes (41, 42), and ceroid storage in vascular walls has also been described (43). A classification of the clinical and pathological expressions of ceroid lipofuscinoses has proposed to separate them into four groups (44); and an infantile variant—with a different clinical course—has been reported from Finland (45). Both the nature of the stored "ceroid" and the etiology of the disease remain unknown, although it has been established that the stored material is not a ganglioside (46).

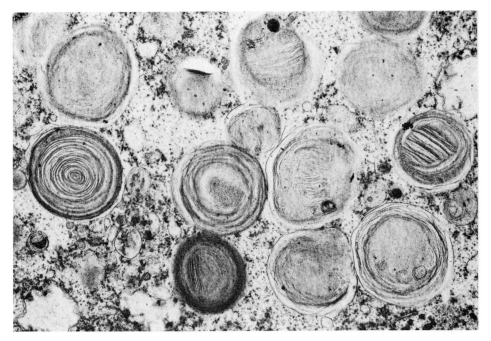

Figure 19. Detailed view of perikaryon from same type of neuron shown in Figure 18 demonstrating the features of membranous cytoplasmic bodies (MCB). (×5,500)

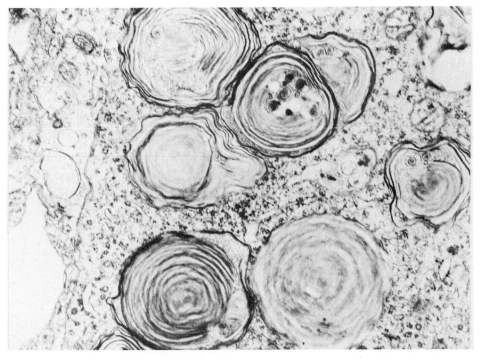

Figure 20. Membranous cytoplasmic bodies such as these may be found in either neuronal perikaryon or axoplasm and, occasionally, in astrocytes. (×7,500) (*Courtesy of U. Sandbank, M.D., Tel Aviv, Israel*)

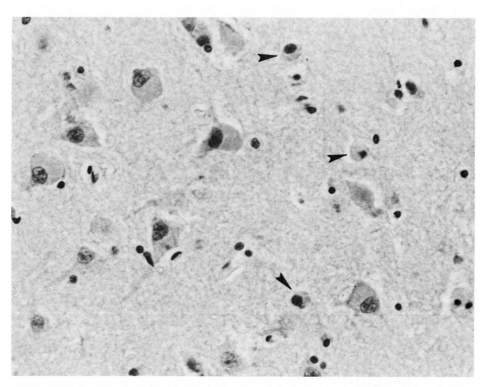

Figure 21. Sample of cerebral cortex from a 6-year-old girl with a diagnosis of "juvenile amaurotic idiocy." Neuronal perikaryon is slightly globular and contains pale, golden-yellow pigment; storage is suggested by the eccentric position and condensed appearance of the nuclei. Pigment is also visible in macrophages (arrowheads). (Hematoxylin-eosin, ×62)

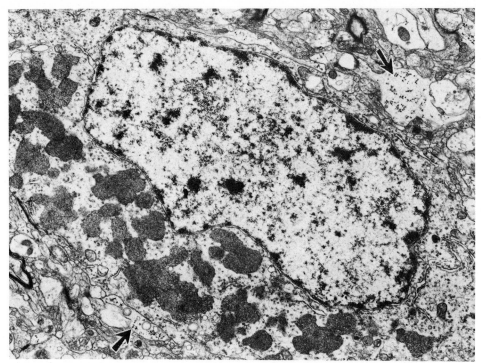

Figure 22. Cortical neuron with perikaryon partly filled with stored material; note ischemic effect in the nucleus. Glycogen granules are visible in astrocytic processes (arrows). (×17,500)

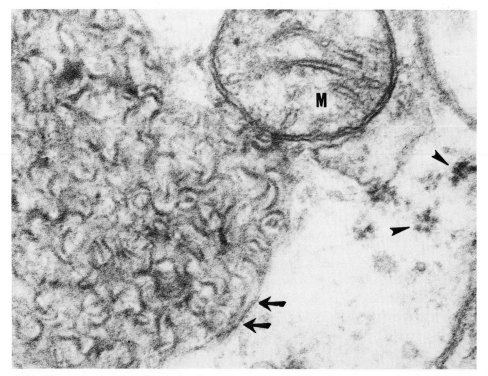

Figure 23. Detail of curvilinear bodies to demonstrate their relationship to a mitochondrion (M) and polyribosomes (arrowheads); note suggestion of unit membrane (arrows). (×100,000)

373

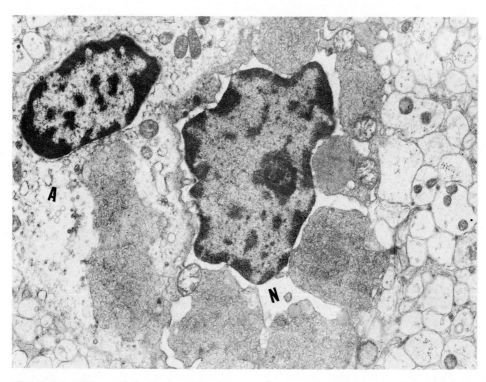

Figure 24. This cortical neuron (N) and the adjacent astrocyte (A) are both filled with "lipofuscin," which here adopts the form of curvilinear profiles. Relatively normal neuropil is visible at the right end. (×12,000)

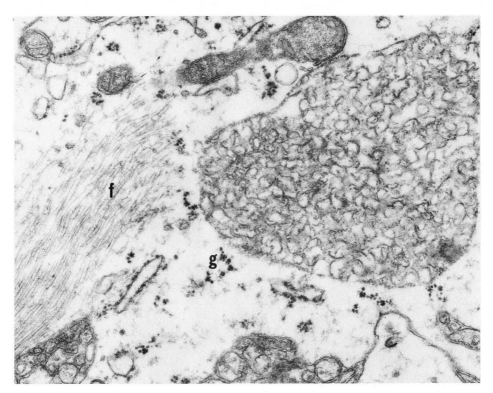

Figure 25. Detail of lipofuscin in astrocytic cytoplasm that also contains abundant filaments (f) and glycogen rosettes (g). (×30,000)

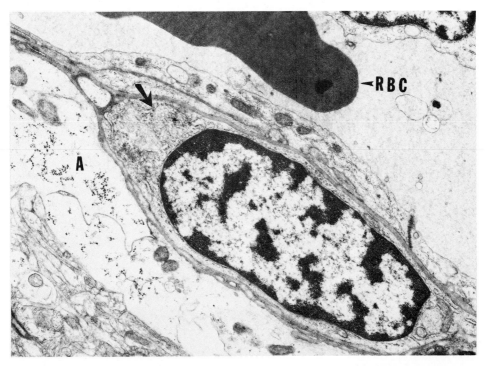

Figure 26. Cerebral capillary with adjoining astrocytic process (A) and lipofuscin in pericyte (arrow); a fragment of a red blood cell (RBC) is visible in the lumen. (×10,000)

375

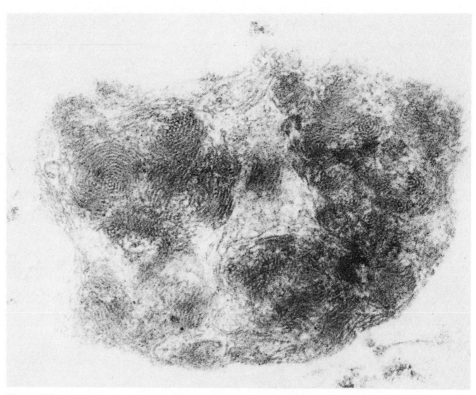

Figure 27. Detail of storage material in cortical neuron from a 5-year-old boy afflicted with progressive dementia, blindness, and seizures; hyperpigmentation in the macular area of ocular fundi was also noted. Here, lipofuscin adopts a "fingerprint"-like appearance. (×87,000) (*Courtesy of H. H. Goebel, M.D.; Goettingen, West Germany*)

28

29

Figures 28 and 29. Samples from a skeletal muscle biopsy. The patient, an 8-year-old girl, began having seizure disorder at the age of 20 months; subsequently, she also became blind and demented. The characteristic subsarcolemmal, paranuclear deposits of lipofuscin (arrows) are shown in Figure 28: a close-up is seen in Figure 29. (Figure 28, ×29,000; Figure 29, ×87,000) (*Courtesy of H.H. Goebel, M.D., Goettingen, West Germany*)

Case 3

At birth this infant boy exhibited numerous somatic deformities including high-arch palate, small mandible, macroglossia, and large bilateral linguinal hernias. His neurological condition was considered retarded, and he suffered from frequent convulsive seizures. At the time of his death, at the age of 14 weeks, light miscroscopy of the brain showed, as the most prominent abnormality, numerous vacuoles or sponginess that was particularly prominent in cerebral subcortical areas, tegmentum of the midbrain and pons, and—to a lesser extent—the striatum (Fig. 30); the cerebellum was involved only focally, but the optic nerves and spinal cord showed abnormalities similar to those seen in the cerebral hemispheres. At this point, it became impossible, by light microscopy, to advance beyond a diagnosis of "spongiform brain degeneration," a designation that encompasses several disorders of heterogeneous etiologies, including cutaneous hexachlorophene intoxication (47, 48, 49).

TEM permitted the demonstration of enlarged watery astrocytes and enlarged intramyelinic spaces, together with structurally abnormal mitochondria and some seemingly specific cytoplasmic bodies in astrocytes (Figs. 31–33). These observations, coupled with the biochemical demonstration of an enzymatic brain defect (i.e., *N*-acetyl-galactosaminyl transferase), made possible a precise diagnosis of *GM-3 gangliosidosis* (50, 51). In contrast with previously described sphingolipidoses in which catabolic enzyme defects are incriminated for cellular disorders of this type (52), GM-3 gangliosidosis appears to be a condition in which the synthesis of gangliosides GM-1 and GM-2 is interfered with due to the lack of a synthesizing enzyme (30).

INFECTIOUS DISEASES

When found in a brain biopsy specimen, monocytic infiltrates, microglial nodules, lymphocytic cuffing, astroglial proliferation, and evidence of neuronophagia are collectively designated "encephalitis." The probability that viruses may be the etiological agent of this condition is strongly supported by the finding of certain intranuclear inclusions designated Cowdry type A (53) (Fig. 34).

The current classification of viruses is based on the definition of a number of features, such as a) site of nucleocapsid assembly (myxoviruses are always intranuclear); b) number of capsomeres (72 for papovaviruses); and c) virion diameter (100 nm for herpesviruses). All can be evaluated accurately in electron micrographs of good quality (54). On the basis of present-day knowledge, only a limited number of viruses need be considered among the diagnostic possibilities in biopsy material from patients in whom intranuclear inclusion bodies (Cowdry type A) are a significant component of a general picture of encephalitis. These viruses are herpes simplex hominis (55), myxoviruses (56), papovaviruses (57, 58), and, possibly, cytomegaloviruses (particularly in newborns).

TEM has proved very valuable in the accurate identification, in tissue samples, of viruses and other microorganisms; even autopsy-derived and paraffin-embedded materials are suitable for electron microscopy conducted with the

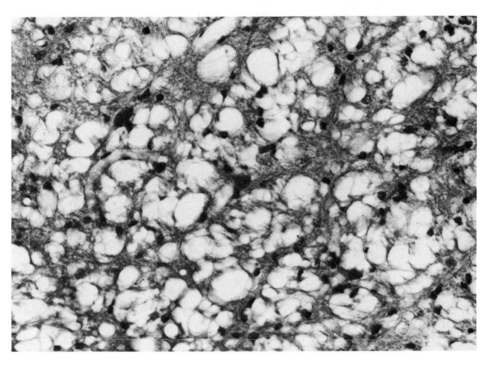

Figure 30. Sample from midbrain (brachium conjunctivum) from a 14-week-old infant who died with a tentative diagnosis of GM-1 gangliosidosis (Tanaka et al., ref. 50). There are abundant spaces of various diameters giving the appearance of status spongiosus. No gliosis is apparent. (Hematoxylin-eosin, ×200) (*With permission from the Editors of The Journal of Neuropathology and Experimental Neurology*)

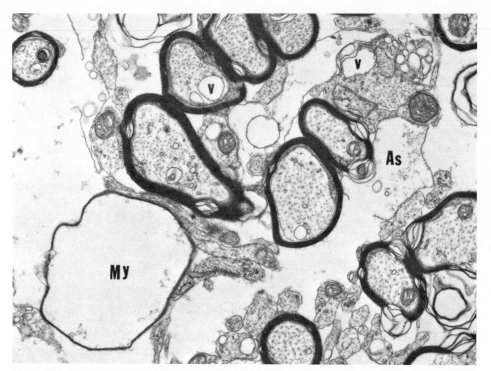

Figure 31. Sample from the internal capsule; same patient as in Figure 30. Sponginess is due to swelling of astrocytic process (As) and enlarged extracellular space; many axons contain vacuoles (V) and some myelinated fibers (My) appear "empty," suggesting disintegration of axis cylinder. (×20,000)

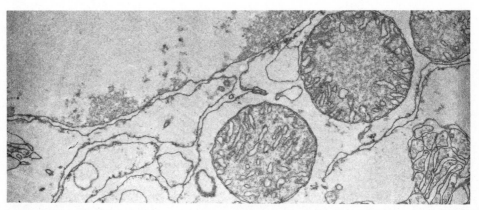

Figure 32. Left temporal cortex of the same patient as in Figures 30 and 31. Next to this astrocytic nucleus there are mitochondria whose inner matrix is filled with a granular-fibrillar material; cristae are short and stubby. (×8,750)

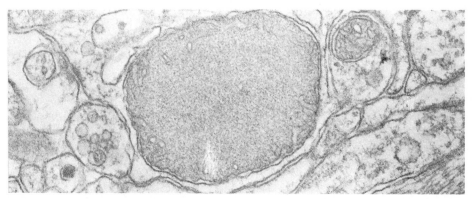

Figure 33. Spinal cord; same patient as in Figures 30–32. This type of astrocytic cytoplasmic body, bounded by a single-unit membrane, suggests an autophagosome; remnants of mitochondria are suggested by the structures resembling cristae. An enzymatic deficiency and GM-3 gangliosidosis were reported in this patient (Max et al., ref. 51). (×22,000)

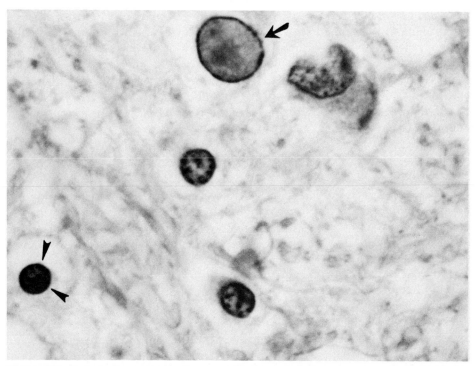

Figure 34. Sample from cerebral cortex of a 54-year-old woman with progressive multifocal leukoencephalopathy (PML). The cell at the top (arrow), tentatively identified as an oligodendrocyte, contains a large, homogeneous intranuclear inclusion body, Cowdry type A, displacing the chromatin to the periphery. Compare these nuclear abnormalitites with those of a normal oligodendrocyte (arrowheads). (Hematoxylin-eosin, ×400)

381

purpose of searching for or evaluating either viral or viral-like particles. Many of these microorganisms are highly resistant to the effects of prolonged ischemia and prolonged formalin storage (59, 60), which are characteristic of most human brain samples obtained at autopsy.

A conclusive diagnosis of viral encephalitis can be based on the actual TEM demonstration of the virions. Although electron microscopic methods allow only the identification of the group of viruses, the combined appraisal of the TEM-derived data and either epidemiological information or evidence in immunofluorescent techniques (61) permits a fairly precise identification of the viral species. Two examples are provided below:

Case 1

A 3-year-old boy developed an abrupt illness consisting of vomiting, decreased level of consciousness, fever (103°F), and partial seizures, which subsequently became generalized. Shortly thereafter, he lapsed into coma and died 3 weeks after the beginning of symptoms. Brain samples were collected at autopsy, which was completed 6 hours after death, from the right temporal lobe, close to the hippocampal gyrus, Light microscopy confirmed the clinical impression of encephalitis, and intranuclear inclusion bodies (Cowdry type A) were demonstrated in many glial cells and neurons (Fig. 35).

Transmission electron microscopy demonstrated many of the cytopathic effects commonly associated with the presence of viruses in cells, such as redistribution of nuclear chromatin and reduplication of the nuclear envelope (55); the ultrastructural features of the virions were those characteristic of herpes viruses (Figs. 36 and 37). Such virions' features (62) narrow down the identification to only to viral species of which the only human pathogen is *Herpesvirus hominis;* hence, the etiological diagnosis of this form of encephalitis was no longer in doubt.

Case 2

A 54-year-old woman afflicted for several years with lymphoblastic lymphoma developed multiple, localized neurological deficits, accompanied by impairment of consciousness and seizures, approximately 4 months before death. Naked-eye inspection of cerebral white matter showed numerous haphazardly distributed, sometimes confluent gray patches where myelin sheaths were seemingly disintegrating or lost. Microscopically, in addition to patchy demyelination, samples of brain tissue showed gliosis (Alzheimer I) and numerous intranuclear abnormalities, including the presence of clear-cut nuclear inclusions in oligodendrocytes or other cells (Figs. 34 and 38). At the electron microscope, these inclusions appeared to be made up almost exclusively of 30 to 35nm virions (Figs. 39 and 40). It is interesting that fixation of these tissues were not completed until about 6 hours after death.

Zu Rhein and Chou (59) called attention to the fact that viral particles preserve their ultrastructural features even after long-term formalin storage and that the features of virions such as those demonstrated in Figures 39 and 40 are consis-

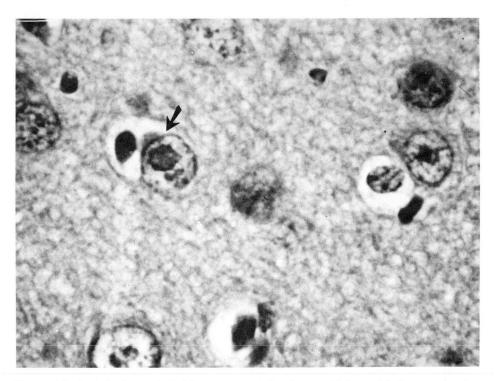

Figure 35. Sample from right hippocampus of a 3-year-old boy with herpes simplex encephalitis. A homogeneous, eosinophilic intranuclear inclusion body is present in one one of the neurons (arrow). (Hematoxylin-eosin ×62)

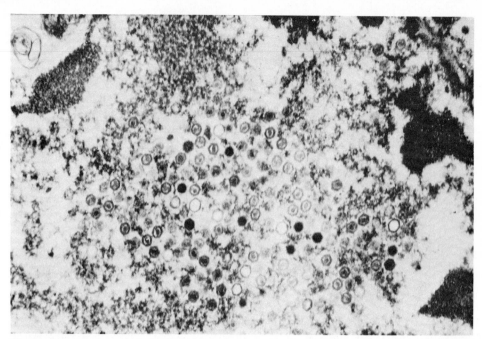

Figure 36. Same patient as in Figure 35. The intranuclear viral particles consist of a) capsid with visible core and halo, b) capsids with homogeneous electron-dense contents, and c) empty capsids. (×15,750)

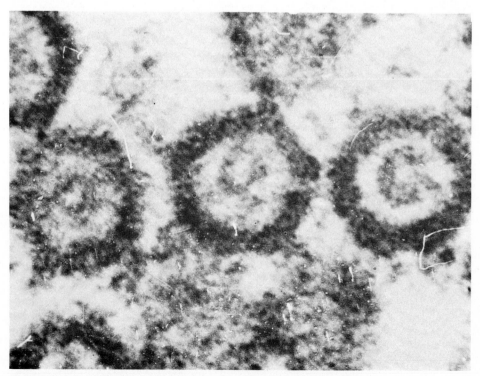

Figure 37. Same patient as in Figures 35 and 36. Viral particles composed of capsids containing a core separated from the capsid by a clear halo. (×31,500)

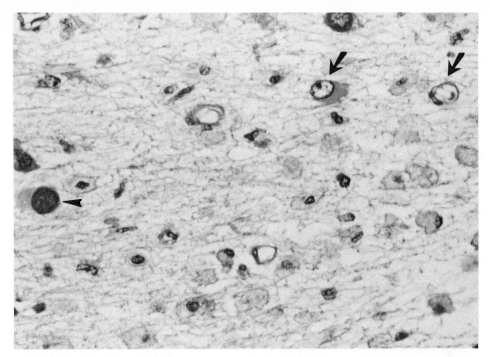

Figure 38. Progressive multifocal leukoencephalopathy; same patient as in Figure 34. In this area there is loss of myelin sheaths and astrocytic proliferation (arrows). The nucleus of an oligodendrocyte (arrowhead) is enlarged and intensively basophilic. (Hematoxylin-eosin, ×62.5)

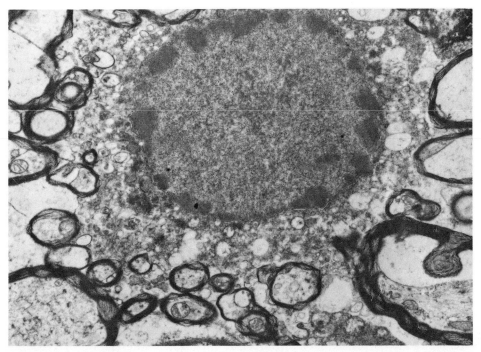

Figure 39. Same patient as in Figures 34 and 38. The nucleus of this glial cell is filled with virions. The chromatin is displaced against the nuclear membrane. (×12,375)

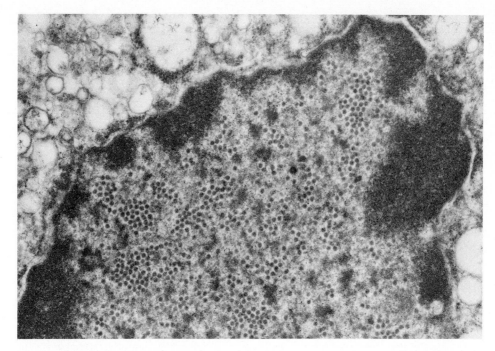

Figure 40. Higher magnification of papova virions from same patient as in Figures 34, 38, and 39. Each measures approximately 30 to 35 nm in diameter. (×31,000)

tent with those characteristic of the family of papilloma, polyoma, vacuolating agents, as defined by Melnick (63). These viruses are thought to be pathogenic to patients with altered immunological status (64).

Some years earlier, Astrom et al (65) had described the clinical and anatomic pathological features of a newly recognized disorder for which they proposed the name of progressive multifocal leukoencephalopathy (PML); clinical features of the disorder include the following in most patients: insidious onset, relentless progression, asymmetrical neurological deficits pointing to hemispheric involvement, impaired mentation, and, frequently, visual abnormalities such as hemianopsia and blindness (64).

Padgett et al. (66) isolated a papovavirus from a patient with Hodgkin's lymphoma and the clinical features of PML. They suggested the name of *JC virus* for this newly isolated agent after the patient's initials. Agents identical or closely related to simian virus-40, a member of the papova family, were isolated from two patients afflicted both with immunological disorders and PML (67), thus confirming the original suspicious that PML was a demyelinating disorder due to an atypical form of viral infection (64).

At present, the diagnosis of PML may be safely made on the basis of this triad of light microscopic features: a) patchy demyelination, b) gliosis, including Alzheimer I astrocytes, and c) intranuclear inclusion bodies (Cowdry type A); characteristic abnormalities in chromatin staining described by Richardson (64) are shown in Figure 38.

NEUROPSYCHIATRIC DISORDERS

A third variety of infectious, or rather "transmissible" disease, of the nervous system in which TEM may aid in the ultimate diagnosis is the group of *spongiform encephalopathies* whose etiology is currently attributed to "slow" viruses. The spongiform designation refers to the morphological alterations seen by light microscopy in the brain cortex; the etiological agent(s) is said to be "slow" because of the protracted incubation period (1 to 2 years) that follows the inoculation of brain extracts into experimental animals (68). At least two separate human neuropsychiatric diseases, Creutzfeldt-Jakob disease and kuru, share the structural and, possibly, epidemiological features of spongiform encephalopathies as described in detail by Gajdusek (69).

Many patients afflicted with spongiform encephalopathies are brought to the attention of the pathologist when he receives a brain biopsy specimen from an adult (usually over the age of 30) with a clinical diagnosis of organic dementia of unknown origin (as illustrated by the next case).

Case 1

A 72-year-old man, thought to be in excellent physical and mental health, showed rather abrupt and pronounced, although intermittent, behavioral aberrations; after a few weeks he became permanently demented and lapsed into coma. All routine laboratory tests of cerebrospinal fluid and serum yielded normal results. A presumptive diagnosis of Creutzfeldt-Jacob disease was based on the analysis of the clinical investigations and the nature of the alterations evident in the computerized tomogram scan, namely, diffuse, symmetrical, rapidly developing, and pronounced atrophy of cerebral cortex in a patient with progressive dementia, extrapyramidal signs, and decreased level of consciousness (70). Brain biopsy material was obtained about 7 weeks after the beginning of symptoms; light microscopy disclosed rather minimal alterations such as "clearing" of neuronal perikaryon and vacuolation of the neuropil (Fig. 41); the typical spongiform changes were not seen in any of over 20 histology samples examined, although they could be demonstrated subsequently in the material collected at autopsy (Fig. 42).

TEM of the biopsy material demonstrated clearly increased electron lucency of neuronal perikaryon, "herniation" of plasma membranes into adjoining cells, and intracytoplasmic vacuoles (Figs. 43–46) all of which are considered characteristic features of this type of encephalopathy (68). The diagnosis of Creutzfeldt-Jacob syndrome in this patient was conclusively proved at the time of the postmortem examination, when the abnormalities in the cerebral cortex were fully developed and no longer offered any difficulty in the interpretation. The patient died approximately 5 months after the beginning of symptoms.

An important variation of the rapidly progressive course has been seen in a woman whose symptoms began at age 29, although she did not require hospitalization until 2 years later, when the classical findings of Creutzfeldt-Jacob syndrome were demonstrated on brain biopsy. The patient died at the age of 33.

Heat and formaldehyde do not inactivate the agent of spongiform encephalopathy; therefore, a number of special precautions are recommended for places where tissues from demented persons are handled; such procedures in-

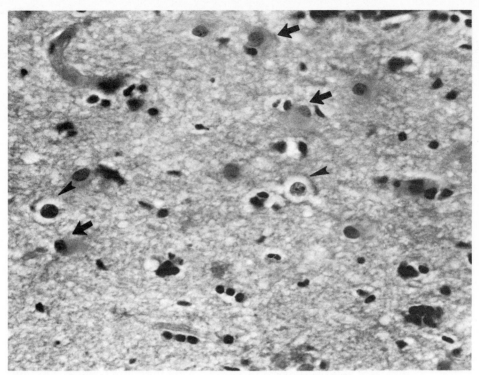

Figure 41. Cerebral cortex from a 72-year-old man with transmissible dementia (Creutzfeldt-Jacob disease). There is vacuolization of the neuropil and gliosis with presence of hypertrophic astrocytes (arrows). Few neurons (arrowheads) shown "clearing" of the perikaryon. (Hematoxylin-eosin, ×61.25)

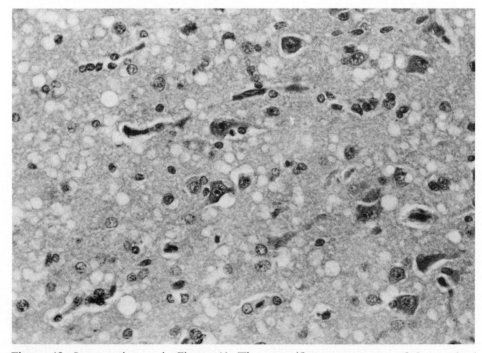

Figure 42. Same patient as in Figure 41. The spongiform appearance of the cerebral cortex has become quite evident in the few weeks elapsed since the time of the biopsy. (Hematoxylin-eosin ×65)

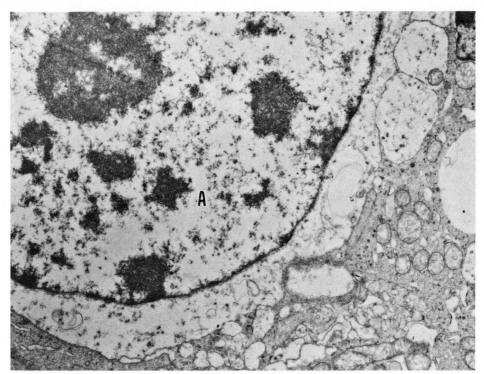

Figure 43. Same patient as in Figures 41 and 42, showing cytoplasmic clearing of astrocyte (A). This, by itself, could be due to the effects of incomplete ischemia (Garcia et al., ref. 6). (×9,300)

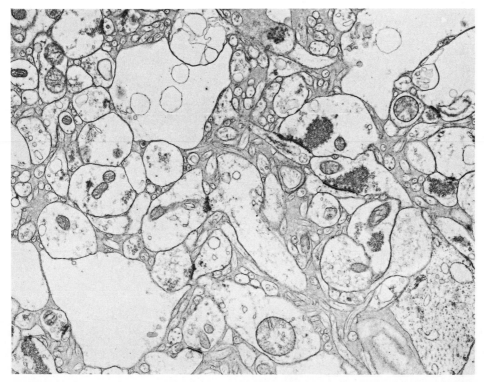

Figure 44. Same patient as in Figures 41–43. The status spongiosus seen under the light microscope corresponds to clusters of swollen dendritic and astrocytic processes. (×3,400)

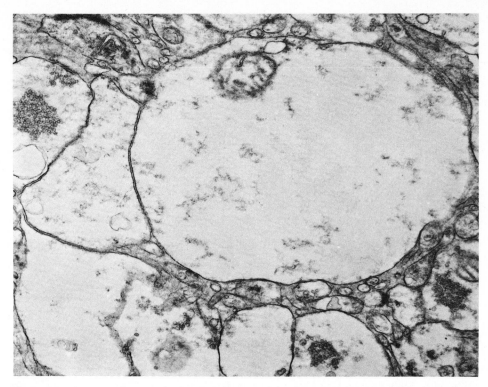

Figure 45. Same patient as above, showing swollen astrocytic processes. (×5,800)

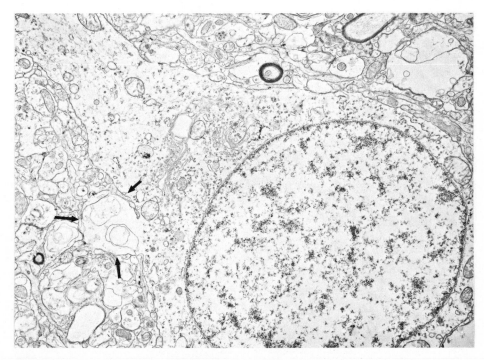

Figure 46. Same patient as above, showing herniation of the neuronal cytoplasm (arrow) into the adjacent neuropil. (×3,125)

clude the use of special disinfectants and the autoclaving of all cutting tools. Human-to-human transmission seemingly requires either cannibalism, organ transplantation, or subepithelial innoculation (71).

Further evidence of the contributions that TEM may make to the understanding of human disease is apparent in studies of brains from patients with demyelinating disorders, through which it has been suggested that the process of demyelination in multiple sclerosis may be the result of the interaction between a myelin sheath (that contains progressively decreasing numbers of lamellae) and a microglial cell (72).

SUMMARY

Illustrated examples of instances in which the improved *resolution* of TEM helped to establish a diagnosis in patients with primary neurological symptoms have been presented. Definition of certain cellular features, invisible by light microscopic methods, not only is significant for the exact diagnosis of some neoplasms but also can be of considerable practical value in the diagnosis of otherwise obscure metabolic disorders that commonly become manifest in childhood. TEM permits not only the characterization of poorly definable light microscopic abnormalities, such as "inclusion bodies" and storage substances, but also, even in autopsy material, the identification of virions or other infectious agents and of cytopathic effects that are currently attributed to "slow" viruses.

Before attempting electron microscopic evaluation of brain material, it is important to have adequate information on the nature of the neurological disorder under investigation and a clear idea of the features that require TEM confirmation, i.e., intracytoplasmic inclusions, intercellular junctions, cytoplasmic granules, etc. The benefits of TEM may not be fully reaped in diagnostic pathology if the technique is used for scanning purposes and in the absence of knowledge of the clinical predicament under investigation. The maximal usefulness of TEM is obtained when it is combined with observation of light microscopic preparations and when it is undertaken for the specific purpose of determining the presence or absence of a cellular organelle, such as secretory granules. Optimal results may be expected only when rigorous attention is paid to the use of appropriate methods for collecting, fixing, and processing the samples.

ACKNOWLEDGMENT

The authors acknowledge with gratitude the excellent secretarial collaboration of Liz Tinnell.

REFERENCES

1. McDowell EM: Fixation and processing, in Trump BF, Jones RT (eds): *Diagnostic Electron Microscopy.* New York, John Wiley, 1978, Vol 1, p 113.
2. Trump BF, Goldblatt PJ, Stowell RS: An electron microscopic study of early cytoplasmic alterations in hepatic parenchymal cells of mouse liver during necrosis *in vitro* (autolysis). *Lab Invest* 11:986, 1962.

3. Hudgins WR, Garcia JH: The effect of electrocautery, atmospheric exposure and surgical retraction on the permeability of the blood brain barrier. *Stroke* 1:375, 1970.

4. Garcia JH, Kamijyo Y, Kalimo H, Tanaka J, Viloria JE, Trump BF: "Dark" neurons and red neurons: Ultrastructural evaluation. *J Neuropathol Exp Neurol* 34:84, 1975.

5. Kalimo H, Garcia JH, Kamijyo Y: Cellular and subcellular alterations of human CNS: Studies utilizing *in situ* perfusion-fixation in immediate autopsies. *Arch Pathol* 97:352, 1974.

6. Garcia JH, Kamijyo Y, Kalimo H, et al: Cerebral ischemia. The early structural changes and correlation of these with known metabolic and dynamic abnormalities, in Whisnant JP, Sandok B (eds): *Cerebral Vascular Diseases,* Ninth Princeton Conference. New York, Grune & Stratton, 1975, p 313.

7. Kalimo H, Garcia JH, Kamijyo Y, et al: The ultrastructure of brain death: II. Electron microscopy of feline cortex after complete ischemia. *Virchows Arch B* 25:207, 1977.

8. Garcia JH, Kalimo H, Kamijyo Y, et al: Cellular events during partial cerebral ischemia. I. Electron microscopy of feline cerebral cortex after middle cerebral artery occulusion. *Virchows Arch B* 25:191, 1977.

9. Peters A: The fixation of central nervous tissue and the analysis of electron micrographs of the neuropil, with special reference to the cerebral cortex, in Nauta WJG, Ebbesson W (eds): *Contemporary Research Methods in Neuroanatomy.* New York, Springer-Verlag, 1970, p 56.

10. Karnovsky MJ: A formaldehyde-glutaraldehyde fixative of high osmolality for use in electron microscopy. *J Cell Biol* 27:137A, 1965.

11. McDowell EM, Trump BF: Histological fixatives suitable for diagnostic light and electron microscopy. *Arch Pathol Lab Med* 100:405, 1976.

12. Paljarvi L, Kalimo H, Garcia JH: The efficiency of aldehyde perfusion-fixation in the osmotic stabilization of brain tissue: Electron microscopy. *Histochem J,* to be published.

13. Lossinsky AS, Garcia JH: Vascular perfusion fixation of the central nervous system for light and electron microscopy, in preparation.

14. Peters A, Palay SL, Webster HF: *The Fine Structure of the Nervous System: The Neurons and Supporting Cells.* Philadelphia, WB Saunders, 1976.

15. Gil DR, Ratto GD: Contribution to the study of the origin of leptomeninges in the human embryo. *Acta Anat* 85:620, 1973.

16. Gyorkey F, Min K-W, Krisko I, et al: The usefulness of electron microscopy in the diagnosis of human tumors. *Hum Pathol* 6:421, 1975.

17. Azzarelli B, Richards DE, Anton AH, et al: Central neuroblastoma. Electron microscopic observations and catecholamine determinations. *J Neuropathol Exp Neurol* 36:384, 1977.

18. Garcia JH, Kalimo H, Givens J: Human adenohypophysis in Nelson syndrome: Ultrastructural and clinical study. *Arch Pathol Lab Med* 100:253, 1976.

19. Horvath E, Kovacs K: Ultrastructural classification of pituitary adenomas. *Can J Neurol Sci* 3:9, 1976.

20. Weibel ER, Palade GE: New cytoplasmic components in arterial endothelia. *J Cell Biol* 23:101, 1964.

21. Kawamura J, Garcia JH, Kamijyo Y: Cerebellar hemangioblastoma: Histogenesis of stroma cells. *Cancer* 31:1528, 1973.

22. Kawamura J, Kamijyo Y, Sunaga T, et al: Tubular bodies in the vascular endothelium of a cerebellar neoplasm. *Lab Invest* 30:358, 1974.

23. Hirano A, Matsui T: Vascular structures in brain tumors. *Hum Pathol* 6:611, 1975.

24. Mena H, Garcia JH: Primary brain sarcomas: light and electron microscopic features. *Cancer,* in press.

25. Russell DS, Rubinstein LJ: *Pathology of Tumours of the Nervous System.* Baltimore, Williams & Wilkins, 1977, p 92.

26. Rubinstein LJ: Tumors of the central nervous system, in *Atlas of Tumor Pathology,* Second Series. Washington DC, Armed Forces Institute of Pathology, 1972, p 190.

27. Movat HZ, Fernando NVP: The fine structure of connective tissue. I. The fibroblast. *Exp Mol Pathol* 1:509, 1962.

28. Brucher JM, Hizaw K, Wechsler W: Fibrome et fibrosarcome intracerebraux: Etude histologique et ultrastructurale. *Pathol Eur* 6:227, 1971.

29. Poon TP, Hirano A, Zimmerman HM: *Electron Microscopic Atlas of Brain Tumors.* New York, Grune & Stratton, 1971.

30. Brady RO: Biochemical genetics in neurology. *Arch Neurol* 33:145, 1976.

31. Terry RD, Weiss M: Studies in Tay-Sachs disease. II. Ultrastructre of the cerebrum. *J Neuropathol Exp Neurol* 28:18, 1963.

32. Volk B, Adachi M, Schneck L: The gangliosidoses. *Hum Pathol* 6:555, 1975.

33. Aronson SM, Lewitan A, Rabiner AM, et al: The megalencephalic phase of infantile amaurotic familial idiocy. *Arch Neurol Psychiatr* 79:151, 1958.

34. Zeman W, Donahue S, Dyken P, et al: The neuronal ceroid-lipofuscinoses (Batten-Vogt syndrome), in Vinken JP, Bruyn GW (eds): *Handbook of Clinical Neurology.* Amsterdam, North Holland Publishing Co, 1970, vol 10, p 558.

35. Zeman W: The neuronal ceroid-lipofuscinoses, in Zimmerman HM (ed): *Progress in Neuropathology.* New York, Grune & Stratton, 1976, vol 3, p 203.

36. Duffy PE, Kornfeld M, Suzuki K: Neurovisceral storage disease with curvilinear bodies. *J Neuropathol Exp Neurol* 27:351, 1968.

37. Goebel HH, Fix JD, Zeman W: The fine structure of the retina in neuronal ceroid-lipofuscinosis. *Am J Ophthalmol* 77:25, 1974.

38. Carpenter S, Karpati G, Andermann F: Specific involvement of muscle, nerve, and skin in late infantile and juvenile amaurotic idiocy. *Neurology* 22:170, 1972.

39. de Baecque CM, Pollack MA, Suzuki K: Late infantile neuronal storage disease with curvilinear bodies. *Arch Pathol Lab Med* 100:139, 1976.

40. Crome L, Stern J: Inborn lysosomal enzyme deficiencies, in Blackwood W, Corsellis JAN (eds): *Greenfield's Neuropathology.* London, Edward Arnold, 1976, p 531.

41. Witzleben CL, Smith K, Nelson JS, et al: Ultrastructural studies in late-onset amaurotic idiocy: Lymphocyte inclusions as a diagnostic marker. *J. Pediatr* 79:285, 1971.

42. Merritt AD, Smith SA, Strouth JC, et al: Detection of heterozygotes in Batten's disease. *Ann NY Acad Sci* 155:860, 1968.

43. Kristensson K, Rainer S, Sourander P: Visceral involvement in juvenile amaurotic idiocy. *Acta Neuropathol* 4:421, 1965.

44. Dekaban AS, Herman MM: Childhood, juvenile and adult cerebral lipidoses. *Arch Pathol* 97:65, 1974.

45. Santavuori P, Haltia M, Rapola J, et al: Infantile type of so-called neuronal ceroid-lipofuscinosis. A clinical study of 15 patients. *J Neurol Sci* 18:157, 1973.

46. Suzuki, K, Johnson AB, Marquet E, et al: A case of juvenile lipidosis: electron microscopic, histochemical and biochemical studies. *Acta Neuropathol* 11:122, 1968.

47. Curley A, Hawk RE, Kimbrough RD, et al: Dermal absorption of hexochlorophene in infants. *Lancet* 2:296, 1971.

48. Lampert PW, Schochet SS: Electron microscopic observation on experimental spongy degeneration of the cerebellar white matter. *J Neuropathol Exp Neurol* 27:210, 1968.

49. Adachi M, Schneck L, Cara J, et al: Spongy degeneration of the central nervous system (Van Bogaert and Bertrand Type: Canavan's disease). *Hum Pathol* 4:331, 1973.

50. Tanaka J, Garcia JH, Max SR, et al: Cerebral sponginess and GM3 gangliosidosis: Ultrastructure and probable pathogenesis. *J Neuropathol Exp Neurol* 34:249, 1975.

51. Max SR, Maclaren NK, Brady RO, et al: GM-3 (hematoside) sphingolipodystrophy. *N Engl J Med* 291:929, 1974.

52. O'Brien JS: Ganglioside Storage disease. *Adv Hum Genet* 3:39, 1972.

53. Cowdry EV: The problem of intranuclear inclusions in virus diseases. *Arch Pathol* 18:527, 1934.

54. Dalton AJ, Haguenan F: *Ultrastructure of Animal Viruses and Bacteriophages.* New York, Academic Press, 1973.

55. Viloria JE, Garcia JH: Human herpes simplex encephalitis: Cellular events. *Beitr Pathol Bd* 157:14, 1976.

56. Tellez-Nagel I, Harter DH: Subacute sclerosing leukoencephalitis: Ultrastructure of intranuclear and intracytoplasmic inclusions. *Science* 154:899, 1966.

57. Zu Rhein GM: Association of papova-virions with a human demyelinating disease (progressive multifocal leukoencephalopathy). *Prog Med Virol* 11:185, 1969.

58. Weiner LP, Narayan O, Penney JB Jr, et al: Papovavirus of JC type in progressive multifocal leukoencephalopathy. *Arch Neurol* 29:1, 1973.

59. Zu Rhein GM, Chou SM: Particles resembling papova virus in human cerebral demyelinating disease. *Science* 148:1477, 1965.

60. Hirano A: Electron microscopy in neuropathology, in Zimmerman HM (ed): *Progress in Neuropathology.* New york, Grune & Stratton, 1971, vol 1, p 1.

61. Weller TH, Coons AH: Fluorescent antibody studies with agents of varicella and herpes zoster propagated *in vitro. Proc Soc Exp Biol Med* 86:789, 1954.

62. Kaplan AS: Recent studies of herpes viruses. *Am J Clin Pathol* 57:783, 1972.

63. Melnick JL: Papova virus group. *Science* 135:1128, 1962.

64. Richardson EP, Jr: Progressive multifocal leukoencephalopathy. *N Engl J Med* 265:815, 1961.

65. Astrom KE, Mancall EL, Richardson EP Jr: Progressive multifocal leukoencephalopathy. *Brain* 81:93, 1958.

66. Padgett BL, Zu Rhein GM, Walker DL, et al: Cultivation of papova-like virus from human brain with progressive multifocal leukoencephalopathy. *Lancet* 1:1257, 1971.

67. Weiner LP, Herndon RM, Narayan O, et al: Isolation of virus related to SV40 from patients with progressive multifocal leukoencephalopathy. *N Engl J Med* 286:3850, 1972.

68. Lampert PW, Gajdusek DC, Gibbs CJ Jr: Subacute spongiform virus encephalopathies. Scrapie, kuru and Creutzfeldt-Jakob disease: A review. *Am J Pathol* 68:626, 1972.

69. Gajdusek DC: Unconventional viruses and the origin and disappearance of kuru. *Science* 197:943, 1977.

70. Rao CVG, Brennan TG, Garcia JH: Computed tomography in the diagnosis of Creutzfeldt-Jacob disease. *J Comp Assis Tom* 1:211, 1977.

71. Gajdusek DC, Gibbs CJ Jr, Asher DM, et al: Precautions in medical care of, and in handling materials from, patients with transmissible virus dementia (Creutzfeldt-Jakob disease). *N Engl J Med* 297:1253, 1977.

72. Prineas J: Pathology of the early lesion in multiple sclerosis. *Hum Pathol* 6:531, 1975.

Index